Natural Fibre Polylactic Acid Composites

This text provides readers with a comprehensive understanding of the properties, processing techniques, and applications of natural fibre-reinforced PLA composites, enabling them to develop sustainable and high-performance materials for a range of industries. It encompasses a wide range of topics within the field, spanning fundamentals, manufacturing processes and techniques, and applications.

- Covers types, characteristics, and sources of natural fibres.
- Delves into the unique properties of PLA as a matrix material and examines natural fibre-reinforced PLA biocomposites, hybrid biocomposites, and nanofiber-reinforced bionanocomposites.
- Explores various processes and techniques for fabricating natural fibre composites, emphasizing the influence of fibre–matrix interactions and surface modifications on the resulting properties.
- Discusses techniques for manufacturing components from these composites, including injection molding, extrusion, and 3D printing.
- Offers lifecycle assessments of natural-fibre reinforced PLA composites, evaluating their environmental impact and sustainability.
- Showcases a broad range of applications in industries such as automotive, construction, packaging, aviation, and consumer goods.

With its comprehensive coverage, scientific approach, and technical depth, this book serves as an invaluable resource for researchers, engineers, and practitioners seeking to advance their knowledge and expertise in the field of natural fibre-reinforced PLA composite materials.

S. M. Sapuan is Professor of Composite Materials at Universiti Putra Malaysia.

Vasi Uddin Siddiqui is currently working as Postdoctoral Researcher at the Department of Mechanical and Manufacturing Engineering, Universiti Putra Malaysia.

R. A. Ilyas is Deputy Director at the Centre for Advanced Composite Materials (CACM), Universiti Teknologi Malaysia.

Natural Fibre Polylactic Acid Composites

Properties and Applications

S. M. Sapuan, Vasi Uddin Siddiqui, and
R. A. Ilyas

CRC Press
Taylor & Francis Group
Boca Raton London New York

CRC Press is an imprint of the
Taylor & Francis Group, an **informa** business

Designed cover image: Shutterstock

First edition published 2025
by CRC Press
2385 NW Executive Center Drive, Suite 320, Boca Raton FL 33431

and by CRC Press
4 Park Square, Milton Park, Abingdon, Oxon, OX14 4RN

CRC Press is an imprint of Taylor & Francis Group, LLC

© 2025 S. M. Sapuan, Vasi Uddin Siddiqui, and R. A. Ilyas

Reasonable efforts have been made to publish reliable data and information, but the authors and publisher cannot assume responsibility for the validity of all materials or the consequences of their use. The authors and publishers have attempted to trace the copyright holders of all material reproduced in this publication and apologize to copyright holders if permission to publish in this form has not been obtained. If any copyright material has not been acknowledged please write and let us know so we may rectify in any future reprint.

Except as permitted under U.S. Copyright Law, no part of this book may be reprinted, reproduced, transmitted, or utilized in any form by any electronic, mechanical, or other means, now known or hereafter invented, including photocopying, microfilming, and recording, or in any information storage or retrieval system, without written permission from the publishers.

For permission to photocopy or use material electronically from this work, access www.copyright.com or contact the Copyright Clearance Center, Inc. (CCC), 222 Rosewood Drive, Danvers, MA 01923, 978-750-8400. For works that are not available on CCC please contact mpkbookspermissions@tandf.co.uk

Trademark notice: Product or corporate names may be trademarks or registered trademarks and are used only for identification and explanation without intent to infringe.

ISBN: 978-1-032-67887-0 (hbk)
ISBN: 978-1-032-67888-7 (pbk)
ISBN: 978-1-032-67889-4 (ebk)

DOI: 10.1201/9781032678894

Typeset in Times
by Apex CoVantage, LLC

Contents

Acknowledgements

The authors, S.M. Sapuan and Vasi Uddin Siddiqui would like to express their gratitude and sincere appreciation to Universiti Putra Malaysia (UPM) for offering the essential resources, facilities, and useful assistance that enabled them to complete this work. The author, RA Ilyas would like to express gratitude for the financial support received from the Universiti Teknologi Malaysia for the project "The impact of Malaysian bamboos' chemical and fibre characteristics on their pulp and paper properties", grant number PY/2022/02318 – Q.J130000.3851.21H99. The co-authors would also like to acknowledge the substantial contributions of UPM students: Umaibala A/L Rajendiran, Divesh A/L Makendren, Ahmad Fathi Bin Mohd Hasbullah, Hazwan Teh Bin Muhammad Shariffudin the, Tharushan A/L Arumugam, Denesh A/L Ravi, Mahesh A/L Ilangovan, Vickneswaran A/L Kuppusamy. Their committed efforts were crucial in accomplishing the successful completion of this work. The co-authors express their gratitude to everybody who provided support, guidance, and assistance.

1 Natural Fibers

1.1 INTRODUCTION

Natural fibers are derived from natural resources like plants, animals, and minerals. Animal fibers contain protein as the main component, but the fibers can also be obtained from wool, feathers, silks, etc. Mineral-based fibers can be found in asbestos and basalt, but the former was prohibited due to health concerns and is no longer in use. Natural fibers from plants are the major fibers used in engineering and industrial applications due to high strength, abundance, and lower cost.

Scientists are increasing the use of natural fibers in the current generation due to global technological improvement. Sustainable solutions and innovation in a variety of industries are shaped by natural fibers and their composites counterparts, which are now indispensable parts of the constantly evolving area of materials science. Natural fibers, which are produced from renewable resources including plants, animals, and minerals, provide a potent replacement for traditional synthetic materials as awareness of environmentally responsible methods rises. They mainly contain cellulose. Today, different types of plant-based fibers are used including kenaf, oil palm, pineapple, coconut (shell and coir), bamboo, jute, sisal, sugarcane bagasse, roselle, water hyacinth, sugar palm, etc.

Natural fibers continuously elongate into separate pieces, much like filaments. They can be twisted into a variety of forms, including threads, ropes, and filaments (Parbin et al. 2019). Researchers worldwide are concentrating on creating new materials that would enhance the environmental quality of products because the world needs more environmentally friendly materials. This chapter deals with the different aspects of natural fibers with some examples and their applications.

1.1.1 Categorization of Natural Fibers

Figure 1.1 shows a variety of natural fibers. As shown, other mineral fiber sources include asbestos, fibrous brucites, wollastonite, and basalt; other animal fiber sources are sheep, alpaca, goat, silkworms, and horses for wool, silk, and hairs (alpaca, goat and horse hairs). Many, many plants are sources of natural fibers including flax, hemp, kenaf, jute, ramie, nettle, cotton, etc. Figure 1.2 shows some additional sources and breakdowns of plant fibers. Woods are an important source of natural fibers, classified as hardwood and softwood (Parbin et al. 2019).

1.1.2 History of Natural Fibers

Kvavadze et al. (2009) reported the discovery of ancient wild flax as a result of excavations of caves in Georgia. The natural fibers were used as cords for hafting

DOI: 10.1201/9781032678894-1

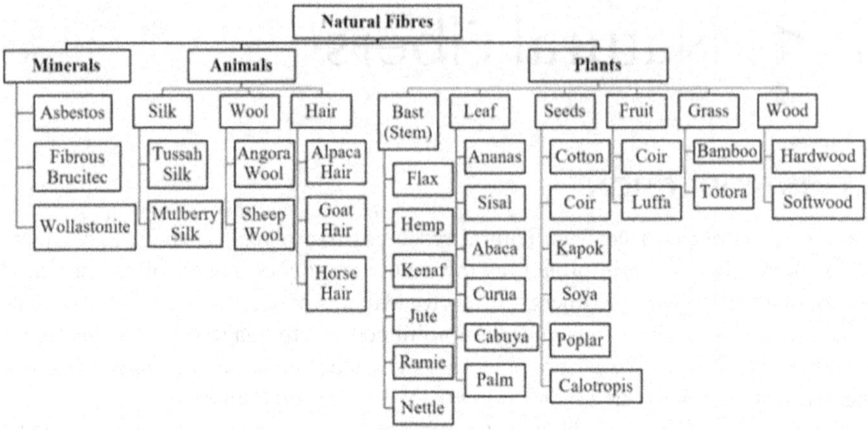

FIGURE 1.1 The categorization of natural fibers.

Source: Parbin et al. (2019).

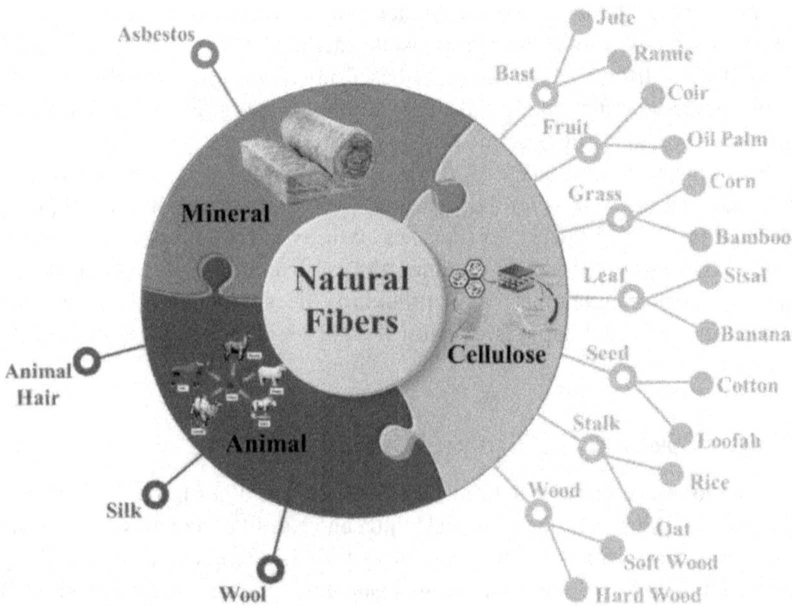

FIGURE 1.2 Classification of plant fibers according to origin.

Source: Elfaleh et al. (2023).

stone tools, weaving baskets, or sewing garments. The fibers were reported to be there for 30,000 years. Flax has been used for thousands of years for linen. Indeed, the linen mills in Belfast, Northern Ireland, UK, came to be called Linenopolis (Figure 1.3).

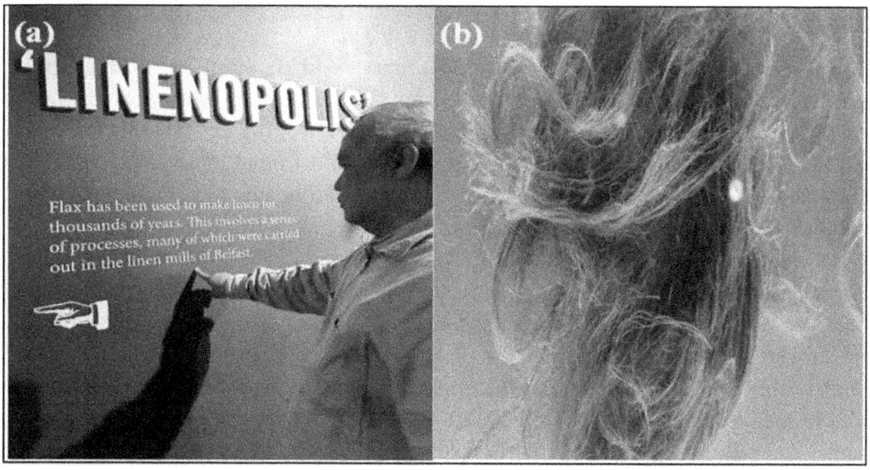

FIGURE 1.3 (a) 'Linenopolis'; (b) flax fibers.

Source: Photos taken at the Titanic Museum, Belfast Northern Ireland, UK by the first author.

1.2 FIBER EXTRACTION

Natural fibers can be extracted from animal, plant, and mineral sources. However, plant-based fibers are regarded as a profitable source of natural fiber due to their high strength, ease of availability, abundance, and lower cost. Extracting natural fibers from their plant sources is a critical step in manufacturing composites materials. Various techniques are employed, each with its own advantages and limitations.

1.2.1 Decortication

Decortication is a common method of removing fibers from plants with fibrous stems, such as flax, hemp, and jute. It involves removing the outer layer (bark) of the stem to expose the underlying fibers. Decortication can be performed mechanically or chemically depending on the plant and processing conditions. The conventional mechanical extraction procedures typically include removing the plant stem, commonly referred to as the Bacnis and Leonit processes.

Modern mechanical extraction employs decortication, in which the plant stems are compressed between two cylindrical rollers to acquire the fibers, following the removal of the pulp. Decorticators significantly enhance fiber production, resulting in a 20- to 25-fold increase compared with the traditional procedure. Biological extraction of fibers from plant stems uses microbes and enzymes to achieve efficient results. However, mechanical extraction is unable to remove the natural binding material (pectin) from the spaces between the fibers in a fiber bundle. Chemical extraction, on the other hand, can remove the pectin but causes significant environmental pollution. In contrast, biological extraction yields more fiber while minimizing negative effects on the environment (Ntenga et al. 2022).

1.2.2 Retting

Retting is a microbiological process that separates fibers from stems. It involves exposing the plant material to microorganisms such as bacteria and fungi that break down the pectic substance binding the fibers together. Retting can be conducted outdoors, in water, or with specialized chemicals, using different methods such as dew and water retting. There are also newer methods including chemical, enzymatic, and steam explosion retting.

Following the harvest, the stems are either stored in the field (dew retting) or submerged in water (water retting) for a period of two to three weeks. During this time, the pectic chemicals that attach the fiber to other plant tissues are softened and broken down by enzymatic activity carried out by microorganisms. Dew and water retting are the conventional techniques for extracting the lengthy bast fibers. Both procedures require 14 to 28 days for the decomposition of pectic components, hemicellulose, and a portion of lignin (Beus et al. 2019). Although the fibers obtained through water retting can be of superior quality, this technique has the drawbacks of prolonged duration and the potential for water pollution (Tahir et al. 2011).

1.2.3 Blow Molding

This technique is used for extracting bamboo fibers. Bamboo fibers are thermally softened using a screw extruder and then forced through a cooling chamber to create continuous fibers. This method produces bamboo fibers with excellent mechanical strength and high quality (Nurul Fazita et al. 2016).

1.2.4 Enzymatic Method

Enzymes can be used to degrade nonfibrous constituents of plants, extracting natural fibers like sisal and kenaf. This technique produces high-quality fibers with fewer impurities (Fu et al. 2012).

1.2.5 Choosing the Right Extraction Technique

The correct extraction technique depends on the type of fiber, its intended use, and the specific constraints of the production process. The quality of the fiber is greatly impacted by the extraction process, so it is important to select it carefully (Parbin et al. 2019). Natural fibers can be processed separately into yarns, fabrics, and mats.

1.3 PROPERTIES OF NATURAL FIBERS

Natural fibers decompose naturally, minimizing the environmental consequences associated with their disposal. Renewable resources are derived from organic sources that can be regenerated. Natural fibers also absorb moisture and then release it, rendering them suitable for wearing in hot and humid environments. They possess high strength and durability, which makes them well suited for a wide range of applications, and they insulate, aiding in maintaining body warmth during cold weather. Table 1.1 shows some properties common industrial natural fibers.

TABLE 1.1
Physical and Mechanical Properties of Natural Fibers

Type of Fiber	Density (g/cm^3)	Strain at Break (%)	Tensile Strength (MPa)	Young's Modulus (MPa)
Flax	1.4–1.5	1.2–3.2	345–1500	27.6–80
Hemp	1.48	1.6	550–900	70
Jute	1.3–1.46	1.5–1.8	393–800	10–30
Kenaf	1.2	2.7–6.9	295–930	22–60
Ramie	1.5	2–3.8	220–938	44–128
Banana	1.35	5–6	529–914	27–32
Nettle	1.51	1.7	650	38
Sisal	1.33–1.5	2–14	400–700	9–38
Coir	1.2	15–30	175–220	4–6
Pineapple	1.5	1–3	170–1627	60–82
Abaca	1.5	2.9	430–813	33.1–33.6
Bamboo	0.6–1.1	1.3–8	140–441	11–36
Cotton	1.21	3–10	287–597	5.5–12.6

Source: Koohestani et al. (2019).

1.4 OVERVIEW OF SOME NATURAL FIBERS

1.4.1 Kenaf (*Hibiscus Cannabinus L.*)

Kenaf is a perennial plant native to West Africa reaching a height of 1.5–3.5 meters with a woody central stem. The stem has a diameter of 1–2 cm and is often branched. The fruit is a 2 cm diameter capsule containing several seeds. The stem consists of a bast fiber component making up 26–35% of its dry weight. The mean fiber length of 2.5 mm is well suited for various pulp and paper applications. Kenaf bast fiber is used in cordage, composites materials, and coarse fabric (Pari et al. 2015). Known for its strong and durable mechanical properties, kenaf has gained increased interest in recent years due to its ability to assimilate nitrogen and phosphorus from the soil and its remarkable carbon sequestration capabilities (Aji et al. 2009).

In recent years, there has been growing interest in using kenaf for paper pulp, construction materials, biocomposites, bedding material, and oil absorbents; Figure 1.4 shows a more consumer-based application of kenaf. Additionally, kenaf has gained recognition as a medicinal crop due to the health benefits of its seed oil, including regulation of blood pressure and cholesterol levels (Monti & Alexopoulou 2013).

The FAO categorizes kenaf among the category "allied fibers." According to FAO data from 2016, India and China are the leading producers of kenaf, with India producing 110,000 tons and China producing 60,000 tons. These two countries accounted for three quarters of the world's kenaf output in 2015–2016. Despite reports of regular imports of kenaf from Bangladesh, it is not officially recognized as a kenaf producer (Carus et al. 2015, pp. 54–55). Kenaf is also cultivated in Indonesia.

FIGURE 1.4 (a) The kenaf plant; (b) a kenaf fiber tote bag.

1.4.2 HEMP

1.4.2.1 Latin Name: *Cannabis sativa L*

Hemp is a tall, annual herbaceous plant with a taproot system. It can reach heights of up to 4 meters and is a promising crop for sustainable fiber production (Amaducci & Gusovius 2010). Known for its weed suppression, disease resistance, and low pesticide requirements, hemp is a versatile crop that thrives in various climates. It requires minimal fertilizer and irrigation, making it an environmentally friendly choice. Hemp cultivation can even improve soil health by controlling weeds and certain soil-borne diseases (González-García et al. 2007).

While hemp originated in the temperate regions of central Asia, its cultivation has spread worldwide. The primary hemp-producing regions today are China, Canada, and Europe. In 2018, the global hemp cultivation area reached approximately 150,000 hectares, with Canada leading in production at 56,000 hectares, followed by China (47,000 hectares) and Europe (43,000 hectares). Within Europe, France, Estonia, Romania, and Italy are the major hemp-producing countries. Europe produces 30,000 to 40,000 metric tons of hemp fibers. The demand for hemp seeds and cannabinoids, rather than fibers, is the primary driver of European hemp production.

Hemp fiber is used in various industries, including textiles, paper, insulation, and biocomposites. In Europe, the focus is primarily on the production of technical short fibers through the complete fiber line, with long-fiber processing for textiles no longer practiced. Hemp fibers are primarily used in pulp and paper production, as well as insulation materials and compression molding parts for the automotive industry. Hemp shives are commonly used as animal bedding, especially for horses. Hemp seeds are primarily used in animal feed (Carus et al. 2015).

1.4.3 Jute

1.4.3.1 Latin Name: *Corchorus capsularis L./Corchorus olitorius L*

Jute, a renewable natural fiber, is widely cultivated in India, primarily Tossa jute (*Corchorus olitorius L.*) and white jute (*Corchorus capsularis L.*). The two species share a similar overall appearance, with elongated stems and a uniform circumference of around 3 cm. They also both branch at the upper portion, but they differ in their fruit characteristics. White jute has a coarse, creased, spherical seed capsule measuring around 0.75 cm in diameter, resembling a small cucumber, while Tossa jute has an elongated pod. Additionally, white jute is typically shorter than Tossa jute with a length of approximately 5 cm.

According to Rahman (2010), white jute is cultivated in low-lying areas, whereas Tossa jute is cultivated in higher areas. Jute thrives in tropical and subtropical regions with ample rainfall and alluvial soils. It is primarily cultivated in the Bengal Basin of India and Bangladesh. India is the world's leading jute producer, followed by Bangladesh. In 2016, global jute production reached 3.3 million tons, with India and Bangladesh accounting for the majority (Sobhan et al. 2010).

Jute is the primary natural fiber among the bast fibers and the second most prevalent natural fiber in the global market, following cotton. The global production of jute in 2016 amounted to 3.3 million tons. India is the leading producer with a production of 1.9 million tons, while Bangladesh closely follows with a production of 1.3 million tons. In 2016, China (mainland) ranked third in importance, with a total of 40,000 tons. The total production area is approximately 1.5 million hectares. The production area in India is approximately 765,000 hectares, whereas in Bangladesh, it is over 678,000 hectares, according to FAOSTAT 2018 (Ullah et al. 2021).

Jute has a diverse array of applications, mostly as packing materials, including hessian, sacking, ropes, twines, carpet backing fabric, and similar products. In addition, jute is utilized for "diversified jute products" as a solution to counter the decreasing demand for traditional jute products. These goods are typically designed for innovative, unconventional, and nontraditional use of jute, for example floor coverings, home textiles, technical textiles, geotextiles, automobile interior parts, particle board, shopping bags, handicrafts, and garments (Rahman 2010).

1.4.4 Roselle

Roselle is scientifically known as *Hibiscus sabdariffa L.*, and it belongs to the Malvacea family. It is a member of the hibiscus family, a tropical plant commonly found in various regions. It is cultivated for both infusion and bast fiber production. There are two primary varieties of roselle: *H. sabdariffa var. Altissimac Wester* and *H. sabdariffa var. sabdariffa*. The stems yield strong fibers that are used in making ropes, sacks, coarse fabrics and paper (Nadlene et al. 2016).

1.4.4.1 *H. Sabdariffa Var. Altissimac Wester*

This variety is primarily cultivated for its fiber, which resembles jute. It grows as a tall, unbranched annual, reaching heights of up to 4.8 meters. The stems are typically green or red, while the leaves are mostly green with sometimes crimson veins.

The blossoms are yellow, but the calyces can be red or green. The calyces of this variety are not fleshy, have thorns, and are not suitable for consumption.

1.4.4.2 *H. Sabdariffa var. Sabdariffa*

This variety is known for its compact, dense growth patterns. It is classified into four distinct races: bhagalpuriensi, intermedius, albus, and ruber (Howard & Howard 1911). All of these races are cultivated from seeds. While bhagalpuriensi has green calyces with red streaks that are not edible, intermedius and albus have yellow-green calyces that are suitable for consumption. These two varieties also produce fiber.

In recent years, roselle fibers have attracted the attention of researchers exploring their potential as reinforcement materials. These fibers exhibit properties comparable with those of well-established natural fibers like jute (Singh et al. 2018). The outermost layer, the bark, protects the plant from temperature fluctuations and excessive moisture loss, contributing to stem hardening. Beneath the bark, the phloem contains the fibers, which provide structural support to the conducting cells within the stem. The xylem, or woody core, is located in the inner region of the plant (Kalia et al. 2011). Figure 1.5 illustrates a roselle's life cycle from plant to fiber.

1.4.5 Flax

Flax, *Linum usitatissimum L.*, is an annual plant utilized for the fiber from its stems and seeds (Elfaleh et al. 2023). The plant has the capacity to reach heights between 0.6 and 1.2 meters, while its stem can have diameters ranging from 1 to 3 millimeters. Flax fiber is highly valued for its exceptional quality and productivity in comparison with other fibers now on the market [86]. It is mostly cultivated in temperate climates, such as the Mediterranean (Europe and Egypt) and the Indian subcontinent, making it one of the oldest fibers used by humans. Flax fiber exhibits numerous benefits such as its considerable length, 33 mm average; flax also shows favorable mechanical characteristics, low density, exceptional toughness, and high strength. Indeed, flax fiber is twice as strong as cotton and five times as strong as wool, and its strength increases by 20% when it's wet (Tahir et al. 2011).

Flax is primarily produced in regions with suitable climates for its cultivation. The EU, Belarus, the Russian Federation, and China are the major flax producers globally. Within the European Union, France, the United Kingdom, the Netherlands, and Belgium are the leading flax-producing countries. According to FAOSTAT 2018, France produced the most flax straw in 2016, followed by Belgium, the UK, and the Netherlands (Ullah et al. 2021).

The successful cultivation of flax requires careful consideration of several factors. The specific variety and its resistance to diseases are crucial determinants of flax output stability. To prevent soil depletion and the buildup of harmful organisms, it is recommended to rotate flax crops with other plants every six years. Proper nutrient management, particularly nitrogen, is essential for healthy growth and optimal yield.

Flax fiber is primarily used in the production of textiles such as linen. The traditional processing of flax involves field-retting, which is suitable for regions with high humidity levels. China is a major importer and processor of long-fiber flax from Europe, transforming it into yarn, textiles, and clothing.

FIGURE 1.5 Extracting roselle fiber: (a) roselle plant, (b) stalks in bundle form, (c) water retting, (d) removing the fibers from the stalks, and (e) final roselle fibers.

Source: Ilyas et al. (2021).

The by-product of flax processing, known as tow or short fiber, has various industrial applications. It is used in biocomposites for automotive components and insulation materials. In times of high demand for linen fashion, short fibers may be mechanically processed to resemble cotton and blended with other fibers like cotton or viscose/lyocell.

1.4.6 Sugar Palm

The sugar palm (*Arenga pinnata*) is a widely recognized and versatile tropical tree belonging to the Palmae family, which has about 181 genera with around

FIGURE 1.6 Natural fibers as reinforcements in polymer composites for construction.

Source: Photo taken by first author.

2600 known species (Ishak et al. 2013a). It offers a variety of valuable products, including edible fruits, fibers, and sago flour. The fibers from the fruit can be used for weaving, making ropes and brooms, constructing roads, and producing roof materials. The stem core can be processed into sago flour, while the root can be brewed into tea or used as an insect repellent. The stem can also be used for various purposes, including making posts for pepper, boards, tool handles, water pipes, and musical instruments (Ishak et al. 2013b; Adawiyah et al. 2013; Ilyas et al. 2018a).

Sugar palm fiber (SPF) is a natural fiber extracted from various parts of the sugar palm tree, including the bunch, fronds, and trunk (Ishak et al. 2013b). Traditionally, SPF has been used in rural communities for making ropes, brooms, and roofing materials (Figure 1.6). In recent years, it has gained attention as a sustainable energy source for bioethanol from the and a reinforcement material in composites materials, and in underground and underwater cables, road construction, and material engineering (Ishak et al. 2013b). *A. pinnata* is cultivated in tropical regions.

Recently, sugar palm has been utilized as a sustainable energy source by converting sugar palm sap into bioethanol through fermentation. In addition to traditional uses, SPF is currently being widely accepted as a strengthening material in composites materials (Figure 1.7; Ilyas et al. 2018b). SPF, a comparatively recent natural fiber in comparison to other natural fibers, is present in many domains based on the categorization of natural fibers.

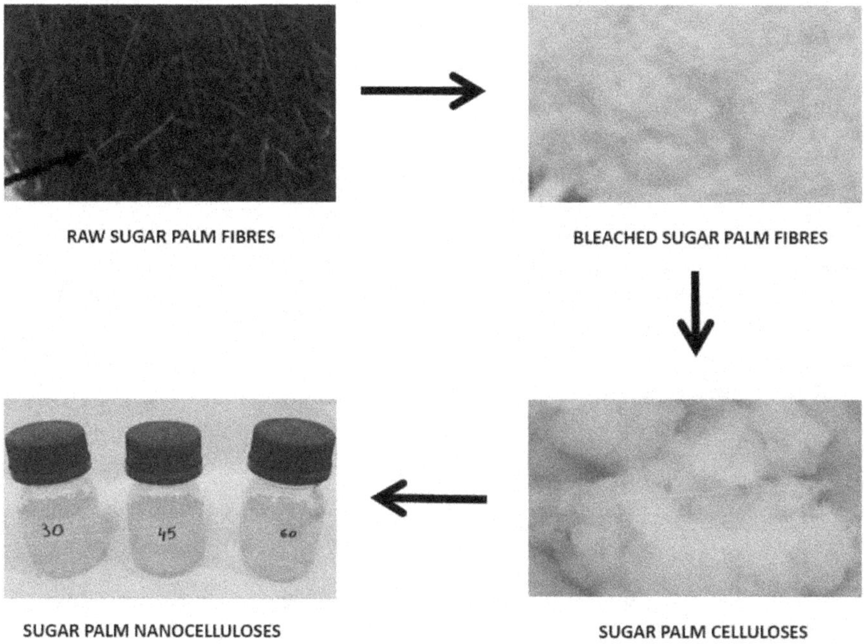

FIGURE 1.7 Extracting sugar palm nanocellulose from raw sugar palm.

Source: Photos taken by first author.

1.5 ADVANTAGES OF NATURAL FIBERS

Natural fibers are ecofriendly and fully biodegradable. They are also less dense than synthetic fibers, and their cultivation causes less damage to the environment than does producing synthetic fibers. Natural fibers are abundantly available, so producing them costs less than producing synthetic fibers. Their production requires little energy and has low CO_2 emissions.

1.5.1 Carbon Footprint of Natural Fibers

The agricultural practices for growing natural fibers generally require few chemical inputs and energy, contributing to a lower carbon footprint. The concept of "carbon footprint" refers to human impacts on climate change and the environment emissions of greenhouse gases such as carbon dioxide (CO_2), methane (CH_4), nitrous oxide (N_2O), and chlorofluorocarbons (CFCs) (Beus et al. 2019). The methodology corresponds to the standards ISO 14040 and ISO 14044 (PAS 2050 2011). When compared to the manufacture of synthetic fibers, the cultivation of these fibers often results in a lower carbon footprint and requires less energy input. This naturally sustainable quality is consistent with international initiatives to promote ecologically friendly production methods. Table 1.2 shows that natural fibers have a low carbon

TABLE 1.2
Carbon Footprints of Some Natural Fibers Based on Cradle-to-Gate Assessment

Fiber	Note	Carbon Footprint (CO_2-eq/ton)	
		Mass allocation	Economic allocation
Flax	-	349	902
Hemp	Using mineral fertilizers	406	846
	Using organic fertilizers	364	759
	Using mineral + organic fertilizers	366	530
Jute	-	479	976
Kenaf	Traditional fiber production	418	975
	Based on DRÅXLMAIER Group	385	770

Source: Parbin et al. (2019).

footprint and are environmentally sustainable because they do not harm the environment (Parbin et al. 2019).

1.6 NATURAL FIBERS IN THE AUTOMOTIVE INDUSTRY

Natural fibers can be used as reinforcements in various polymer matrices such as thermoplastics, thermosets, and elastomers in composites with good physical, mechanical, thermal, and acoustic properties.

1.6.1 Flax

Flax has significant traction in the automotive industry due to its strength, durability, and sustainability. In 2012, flax accounted for 50% of the market share for natural fibers used in composites in the European Automotive Industry. The continued production of technical short fibers as a byproduct of textile production ensures a steady supply of flax for industrial applications (Zhu et al. 2013).

1.6.2 Hemp

In 2012, hemp accounted for 12% of the market share for natural fibers used in composites in the European Automotive Industry. In 2005, hemp fibers accounted for 9.5% of the market share in the European Automotive Industry for composites. From 2005 to 2012, there was growth in the market share of hemp, though its future growth was proposed to depend on a number of factors (Carus et al. 2015).

1.6.3 Jute

Jute has the potential to play a significant role as a natural fiber in the automotive industry. The jute industry is currently experiencing huge volumes and efficient logistics. However, the reputation of jute has been significantly harmed due to the fogging caused by the use of batching oil in the textile process chain, which was intended to facilitate processing. Today, it is rather simple to obtain substantial quantities of jute free residual batching oil. Furthermore, the production capabilities frequently exceed the demand for these fibers, particularly due to the declining usage in conventional applications. Nevertheless, the environmentally and socially controversial practice of water retting, along with the absence of a contemporary processing technology, continues to pose problems (Carus et al. 2015).

1.7 CONCLUSIONS

From this chapter, it can be concluded that natural fibers are promising alternatives to replace synthetic fibers. However, they have some limitations, although these can be rectified in the processing. Natural fibers can be selected specifically for the benefits of their characteristics in particular applications.

REFERENCES

Adawiyah, D. R., Sasaki, T., Kohyama, K. (2013). Characterization of Arenga Starch in Comparison with Sago Starch. *Carbohydrate Polymers* 92:2306–2313. https://doi.org/10.1016/J.CARBPOL.2012.12.014

Aji, I. S., Sapuan, S. M., Zainudin, E. S., Abdan, K. (2009). Kenaf Fibres as Reinforcement for Polymeric Composites: A Review. *International Journal of Mechanical and Materials Engineering* 4(3):239–248.

Amaducci, S., Gusovius, H. J. (2010). Hemp–Cultivation, Extraction and Processing. In: Prof. Dr.-Ing Jürg Müssig (ed.) *Industrial Applications of Natural Fibres: Structure, Properties and Technical Applications* 109–134. DOI:10.1002/9780470660324. John Wiley & Sons, Ltd.

Beus, N. de, Barth, M., Carus, M. (2019). *Carbon Footprint and Sustainability of Different Natural Fibres for Biocomposites and Insulation Material: Study Providing Data for the Automotive and Insulation Industry*. Nova-Institut GmbH, Hürth, Germany 1–57.

Carus, M., Eder, A., Dammer, L., et al (2015). *Wood-Plastic Composites (WPC) and Natural Fibre Composites (NFC)*. Nova-Institute, Hürth, Germany 16.

Elfaleh, I., Abbassi, F., Habibi, M., et al (2023). A Comprehensive Review of Natural Fibers and Their Composites: An Eco-Friendly Alternative to Conventional Materials. *Results in Engineering* 19:101271. https://doi.org/10.1016/j.rineng.2023.101271

Fu, J., Li, X., Gao, W., et al (2012). Bio-Processing of Bamboo Fibres for Textile Applications: A Mini Review. *Biocatal Biotransformation* 30:141–153. https://doi.org/10.3109/10242422.2012.650450

González-García, S., Hospido, A., Moreira, M. T., Feijoo, G. (2007). Life Cycle Environmental Analysis of Hemp Production for Non-Wood Pulp. In: *3rd International Conference on Life Cycle Management* 1. https://www.lcm2007.ethz.ch/109.pdf

Howard, A., Howard, G. L. C. (1911). Studies in India Fibre Plants, No. 2. On Some New Varieties of *Hibiscus cannaius L.* and *Hibiscus sabdariffa L.* Memoirs of the Department of Agriculture in India. *Botanical Series* 4(2):36.

Ilyas, R. A., Sapuan, S. M., Ishak, M. R. (2018a). Isolation and Characterization of Nanocrystalline Cellulose from Sugar Palm Fibres (Arenga Pinnata). *Carbohydrate Polymers* 181:1038–1051. https://doi.org/10.1016/J.CARBPOL.2017.11.045

Ilyas, R. A., Sapuan, S. M., Ishak, M. R., Zainudin, E. S. (2018b). Development and Characterization of Sugar Palm Nanocrystalline Cellulose Reinforced Sugar Palm Starch Bionanocomposites. *Carbohydrate Polymers* 202:186–202. https://doi.org/10.1016/J.CARBPOL.2018.09.002

Ilyas, R. A., Sapuan, S. M., Kirubaanand, W., et al (2021). Roselle: Production, Product Development, and Composites. *Roselle: Production, Processing, Products and Biocomposites* 1–23. https://doi.org/10.1016/B978-0-323-85213-5.00009-3

Ishak, M. R., Sapuan, S. M., Leman, Z., et al (2013a). Sugar Palm (Arenga pinnata): Its Fibres, Polymers and Composites. *Carbohydrate Polymers* 91:699–710. https://doi.org/10.1016/J.CARBPOL.2012.07.073

Ishak, M. R., Sapuan, S. M., Leman, Z., et al (2013b). Sugar Palm (Arenga pinnata): Its Fibres, Polymers and Composites. *Carbohydrate Polymers* 91:699–710. https://doi.org/10.1016/J.CARBPOL.2012.07.073

Kalia, S., Dufresne, A., Cherian, B. M., Kaith, B., Avérous, L., Njuguna, J., Nassiopoulos, E. (2011). Cellulose-Based Bio-and Nanocomposites: A Review. *International Journal of Polymer Science* 2011:35. https://doi.org/10.1155/2011/837875

Koohestani, B., Darban, A. K., Mokhtari, P., et al (2019). Comparison of Different Natural Fiber Treatments: A Literature Review. *International Journal of Environmental Science and Technology* 16:629–642. https://doi.org/10.1007/s13762-018-1890-9

Kvavadze, E., Bar-Yosef, O., Belfer-Cohen, A., et al (2009). 30,000-Year-Old Wild Flax Fibers. *Science* 325:1359. https://doi.org/10.1126/science.1175404

Monti, A., Alexopoulou, E., eds (2013). *Kenaf: A Multi-Purpose Crop for Several Industrial Applications: New Insights from the Biokenaf Project.* Springer Science & Business Media.

Nadlene, R., Sapuan, S. M., Jawaid, M., Ishak, M. R., Yusriah, L. (2016). A Review on Roselle Fiber and Its Composites. *Journal of Natural Fibers* 13(1):10–41.

Ntenga, R., Saidjo, S., Wakata, A., et al (2022). Extraction, Applications and Characterization of Plant Fibers. In: Jeon H-Y (ed) *Natural Fiber*. IntechOpen 1–32.

Nurul Fazita, M. R., Jayaraman, K., Bhattacharyya, D., et al (2016). Green Composites Made of Bamboo Fabric and Poly (Lactic) Acid for Packaging Applications – A Review. *Materials* 9:435. https://doi.org/10.3390/ma9060435

Parbin, S., Waghmare, N. K., Singh, S. K., Khan, S. (2019). Mechanical Properties of Natural Fiber Reinforced Epoxy Composites: A Review. *Procedia Computer Science* 152:375–379. https://doi.org/10.1016/j.procs.2019.05.003

Pari, L., Baraniecki, P., Kaniewski, R., Scarfone, A. (2015). Harvesting Strategies of Bast Fiber Crops in Europe and in China. *Industrial Crops and Products* 68:90–96.

Rahman, M. S. (2010). Jute-a Versatile Natural Fibre. Cultivation, Extraction and Processing. In: Prof. Dr.-Ing Jürg Müssig (ed.) *Industrial Applications of Natural Fibres: Structure, Properties and Technical Applications*. Print ISBN:9780470695081 | Online ISBN:9780470660324 | DOI:10.1002/9780470660324. John Wiley & Sons, Ltd.

Singh, H., Singh, J. I. P., Singh, S., Dhawan, V., Tiwari, S. K. (2018). A Brief Review of Jute Fibre and Its Composites. *Materials Today: Proceedings* 5(14):28427–28437.

Sobhan, M. A., Sur, D., Amin, M. N., Ray, P., Mukherjee, A., Samanta, A. K., Guha Roy, T. K., Abdullah, A. B. M., Bhuyan, F. H., Roul, C., Roy, S. (2010). *Jute Basics*. IJSG Publication.

Tahir, M. M., Khurshid, M., Khan, M. Z., et al (2011). Lignite-Derived Humic Acid Effect on Growth of Wheat Plants in Different Soils. *Pedosphere* 21:124–131. https://doi.org/10.1016/S1002-0160(10)60087-2

Ullah, I., Mao, H., Rasool, G., et al (2021). Effect of Deficit Irrigation and Reduced N Fertilization on Plant Growth, Root Morphology and water Use Efficiency of Tomato Grown in Soilless Culture. *Agronomy* 11:228.

Zhu, J., Zhu, H., Njuguna, J., Abhyankar, H. (2013). Recent Development of Flax Fibres and Their Reinforced Composites Based on Different Polymeric Matrices. *Materials* 6: 5171–5198. https://doi.org/10.3390/ma6115171

2 Natural Fiber Composites

2.1 INTRODUCTION

Composites materials have replaced many conventional materials such as metals, ceramics, woods, and polymers, due to the advantages that they offer, such as specific strength and stiffness. The use of composites materials in our lives has become very common, including in our furniture, transportation, education, sport and leisure, and buildings and infrastructures. Composites materials are engineered materials made of fiber reinforcements and matrices bonded together to produce materials that have superior properties to those of the constituent materials alone. Composites are unique in that they combine material properties in a manner not existing in nature. Such combinations most often yield lightweight materials with high stiffness and tailored properties for specific applications.

The plastic going into the oceans is on course to rise from 11 million tons now to 29 million by 2040, according to a study published in June by Pew Trusts, an independent public interest group. Cumulatively, this would reach 600 million tons, the weight of 3 million blue whales. One response to the problem is using biocomposite materials to produce sustainable, green products. Renewable biocomposite products are high quality (mechanically and thermally) and ecofriendly, and biopolymers and natural fiber-reinforced biopolymer composites made from natural resources that degrade faster in the environment have attracted the attention of many researchers.

Biocomposites or natural fiber composites are materials composed of two or more distinct constituent materials where one is naturally derived (plant/animal) that are combined to fabricate new materials with superior properties to those of synthetic materials. Lignocellulosic composites are ecofriendly and economically feasible and minimize negative environmental impacts. Natural fiber composites are highly effective substitutes for conventional composites and traditional engineering materials.

There are two types of natural fiber composites: synthetic polymer and biopolymer; we only address the latter in this book. Research on biocomposites has focused on characterizing them and has intensified in recent years. Researchers have also conducted a variety of life cycle assessments (LCAs) of different natural fiber composites materials as well as investigating materials selection and product design and development, although this research is still limited.

Growing environmental awareness has directed the attention of most researchers to ecofriendly, renewable, recyclable, sustainable, and biodegradable materials. Because we are Malaysian authors, we emphasize that we focused on mainly tropical natural fibers because we are in the tropics. Malaysia is blessed with an abundance of natural resources suitable for natural fibers such as oil palm, banana,

DOI: 10.1201/9781032678894-2

coconut, sugarcane, cassava, roselle, betel nut, pineapple, durian skin, cocoa pod, kenaf, rice husk and straw, water hyacinth, seaweed, sugar palm, and bamboo. These fibers are suitable to be used as reinforcements and fillers in polymer composites.

2.2 BIOPOLYMERS USED IN NATURAL FIBER COMPOSITES

Biocomposites are polymeric composites reinforced with natural fibers (biomass) in various orientations, sizes, forms, and weight/volume percentages. These sustainable materials offer advantages including reduced environmental impact, cost-effectiveness, and improved properties.

Manufacturing biocomposite materials involves a variety of polymers as the matrix including thermoplastics, polymers that can be repeatedly softened and reshaped when heated, making them versatile for processing. Examples include polypropylene, polyethylene, and PLA. Thermosets are polymers that cure irreversibly when heated, forming a rigid structure; examples include epoxy, polyester, and polyurethane resins. Soy-based polymers offer renewable and biodegradable alternatives to traditional plastics, and starches and natural carbohydrates found in plants are used as matrix material in biocomposites. Rubber is also a natural polymer obtained from the latex of rubber trees that is effective matrix material.

Biopolymers are naturally occurring polymers synthesized by living organisms. They are biodegradable and offer a sustainable alternative to traditional synthetic polymers. Figure 2.1 classifies biopolymers by source as natural, meaning extracted directly from natural sources, such as cellulose, starch, chitosan, and proteins, or synthetic, meaning chemically synthesized from renewable resources, such as polyhydroxybutyrate and polycaprolactone.

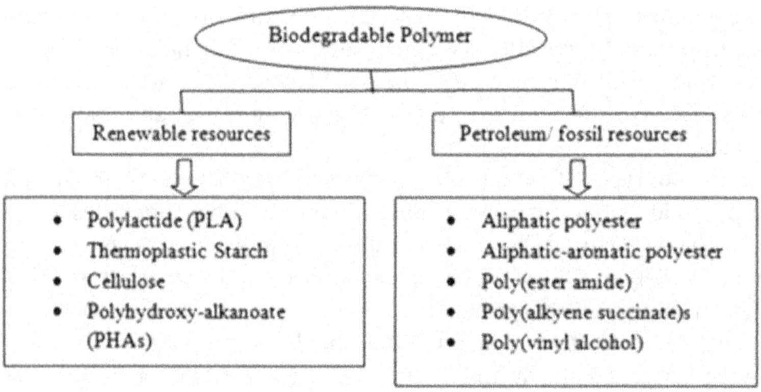

FIGURE 2.1 Biopolymer classification by source.

Source: Salit, M.S. (2014).

2.3 ADVANTAGES OF NATURAL FIBER COMPOSITES

2.3.1 Sustainability and Environmental Benefits

Sustainability is the ongoing ability of people to live in peace and harmony on Earth. Natural fiber composites are an attractive option in the industrial landscape because of their substantial environmental and sustainability benefits including that they are renewable. This naturally sustainable quality is consistent with international initiatives to promote ecologically friendly production methods.

Lopez et al. conducted LCAs of natural fiber composites and their synthetic equivalents and confirmed the less negative environmental impacts from producing natural fiber composites, including that no processes produce greenhouse gases. The study sheds light on the many advantages of selecting natural fibers in terms of sustainability and highlights the significance of life cycle assessments in assessing the environmental performance of materials (Lopez et al., 2020).

2.3.2 Renewability

In the context of fabrication, the renewability of natural fibers addresses concerns related to the depletion of finite resources. Unlike synthetic fibers that rely on nonrenewable petrochemical sources, natural fibers can be sustainably harvested, reducing the environmental impact associated with resource extraction. This renewable characteristic is particularly crucial in industries that are heavily reliant on fiber-reinforced composites, such as automotive and construction, where the demand for sustainable materials is growing. Indeed, the automotive industry is expanding its use of natural fiber-based composites in linings (roof, rear wall, side panel lining), and they are being used as well in furniture, construction, packaging, and shipping (Sanjay et al., 2018).

Additionally, cultivating natural fibers often involves less environmental burden than producing synthetic fibers. The agricultural practices for growing natural fibers require fewer chemical inputs and energy, contributing to a smaller carbon footprint. This makes natural fiber composites an attractive choice in fabrication processes that prioritize reduced environmental impact. Natural fibers are extracted from different plants and animals (chicken feather, hair, etc.) (Aziz & Ansell, 2004).

2.3.3 Biodegradability

The biodegradability of natural fiber composites stands out as a significant advantage, offering an environmentally friendly solution in fabrication. As these composites often consist of plant-based fibers such as jute, flax, or hemp, they possess inherent biodegradable properties, contributing to a reduced environmental impact at the end of their lifecycle.

The biodegradability of a material refers to its intrinsic capacity to be degraded by microbial attack, resulting in a gradual simplification of its structure. The organic nature of natural fibers enables them to break down naturally, providing

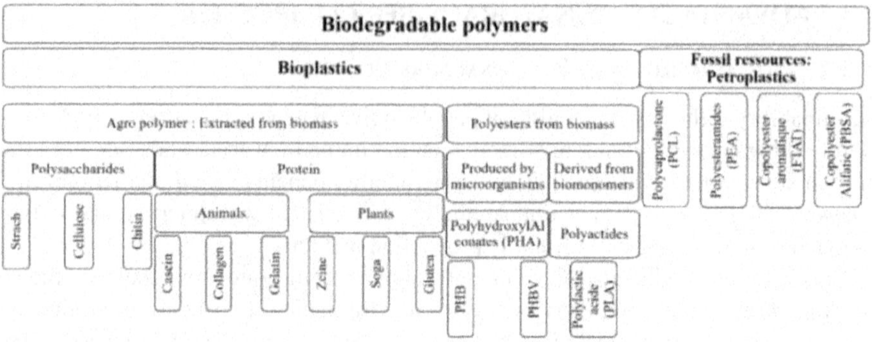

FIGURE 2.2 Biodegradable polymers.

Source: Corrales-Ureña et al. (2018).

an ecofriendly alternative to nonbiodegradable synthetic materials (Corrales-Ureña et al., 2018). This characteristic becomes particularly relevant in industries where reducing the environmental impact of materials after their useful life is a crucial consideration. Biodegradable polymers can be produced from renewable or fossil resources. Hence, based on Figure 2.2 it can be seen that natural fibers that come from plants and animals are biodegradable polymers.

Due to their organic origins, plant fibers possess high biodegradability that helps maintain a healthy balance in the ecosystem. Moreover, plant composites products are easily recycled because of the extremely low abrasivity of the plant fibers. Because they can fully dissolve at the end of their life cycle, natural fiber-reinforced plastics are the most environmentally benign materials when combined with biodegradable polymers as the matrix material (Mugdha et al., 2022). Glass is substituted with natural fiber composites in non-structural applications. One such instance is the current replacement of glass-based vehicle components with natural fiber-reinforced composites.

2.4 NATURAL FIBER COMPOSITES: FORMATION AND COMPOSITION

2.4.1 Matrix Materials in Natural Fiber Composites

Natural fiber composites are prepared on matrices classified as either thermoplastics or thermosets based on the type of bonding in the material (Table 2.1).

Polymer matrix composites are made of different organic polymers with different reinforcing agents that result in short or continuous fibers with properties like improved stiffness, high strength, and improved fracture resistance. Polymer matrix composites also support heavier mechanical loads. The function of the matrix is to adhere the fibers to more efficiently transfer loads (Cao & Wu, 2008). Table 2.2 presents examples of polymer matrix composites including their natural fiber components.

TABLE 2.1
Major Polymers Used as Matrix for Composites

Polymers	
Thermoplastics	**Thermosets**
Nylon	Phenolic
Cellulose acetate	Epoxy
Polystyrene	Polyester
Polyethylene	Polyurethane
Polycarbonate	
Polyvinyl chloride	

Source: Gowda et al. (2018).

TABLE 2.2
Polymers Reinforced with Natural Fibers

Thermoplastics		Thermosets	
Polymer	**Fibers**	**Resin**	**Fiber**
PP	Cura, coconut husks, hemp, jute, sisal, sugarcane bagasse	Polyester	Bamboo, banana, coconut, flax, pineapple, hemp
PE	Banana, rice husk, sugarcane bagasse	Polyurethane	Coconut, banana, *Curcuma*, sisal
HDPE	Banana, *Curcuma*, sisal, wood	Epoxy	Cotton, flax, hemp, jute, sisal, pineapple
PS	Coconut husk, sisal, sugarcane bagasse	Phenolic	Flax, sisal, jute, banana
		Vinylester	Pineapple, sisal, jute, coconut, hemp

Source: Gowda et al. (2018).

2.4.2 Natural Fibers as Reinforcements

Biological fillers in polymeric composites improve filler/matrix interactions and prevent leaching. Natural fiber fillers were already documented as various reinforcements in ancient Egypt. However, their rediscovery can be traced to the early 20th century. In recent years, discussions on balancing carbon footprints have increased interest in natural fibers/additives derived from agricultural products, especially annual plants (Islam et al., 2015).

Natural fibers can be extracted from various sources such as plants, animals, and minerals and processed separately into yarns, fabrics, and mats. Natural fibers can be used as reinforcements in various matrices such as thermoplastics, thermosets, and elastomers into composites with mechanical, thermal, and acoustic properties. Moreover, natural fibers improve interfacial adhesion between fiber and matrix (Sanjay et al., 2019).

2.4.3 Composites Fabrication Methods

Natural fiber composites can be made using a variety of techniques, including hand lay-up, injection, compression, and resin transfer molding. Fabrication, fiber and matrix mixing, and moisture removal from natural fiber are the three main processing methods; researchers have identified 80 °C as the ideal temperature for drying out natural fibers. The two most widely used methods for combining fiber and matrix are internal mixing machines and extrusion. Natural fiber-reinforced polymer composites are typically made through compression and injection molding. Several advantages of injection molding over other processes include tighter tolerances, a shorter production cycle, and the capacity to create complex parts in large quantities. Apart from its quick cycle time and high reproducibility, compression molding also has low cost and waste (Kerni et al., 2020; Faruk et al., 2012).

Table 2.3 lists the different methods for manufacturing and mixing composites according to the type and length of fiber. According to the table, automatic mixing and compression molding are more appropriate for producing composites with short natural fibers. In contrast, the manual lay-up technique and mixing are better suited for producing composites with long natural fibers (Kerni et al., 2020).

TABLE 2.3
Natural Fiber Manufacturing and Mixing Processes

Natural Fiber	Matrix	Mixing Process	Manufacturing Process
Hemp	Polyester	Manual mixing	Vacuum bagging
Kenaf	Unsaturated polyester	Kenaf was put into a mold into which resin was injected with a pressure of 1.3 bar	Resin transfer molding
Pineapple	High-impact polystyrene	Internal mixing	Compression molding
Banana	Epoxy	Epoxy resin was put into the mold and fiber was manually laid	Hand lay-up
Kenaf	Soy based resin	Extrusion	Compression molding
Bamboo	Starch	Manual mixing	Dried in an oven
Kenaf-PALF	HDPE	Internal mixing	Compression molding
Basalt	Polyester	Manual mixing	Compression molding
Agave	Epoxy	Stirring	Compression molding

Source: Kerni et al. (2020).

2.5 PREPARATION AND PROCESSING

2.5.1 Selection of Natural Fibers

Natural fibers are sourced from abundant and easily accessible materials, and their cost is notably lower than that of synthetic and traditional fibers. Green natural fibers, being fully ecofriendly and biodegradable, demonstrate a reasonable level of reliability when juxtaposed with synthetic fibers. These ecofriendly natural fibers present substantial advantages, including widespread availability, low density, and notably favorable mechanical properties. Currently, natural fibers have gained increased attention within research fields, especially concerning their utilization in reinforcement materials, owing to their appealing characteristics (Thyavihalli Girijappa et al., 2019).

Kenaf fibers, a type of bast fiber, hold significant importance and are primarily utilized in the production of paper and ropes (Girijappa et al., 2019). These fibers are extracted from various parts of the plant, including the flowers, outer fiber, and inner core. The outer fiber, known as bast, constitutes approximately 40% of the stalks' dry weight, while the inner core comprises about 60%. Notably, kenaf is environmentally friendly and entirely biodegradable. Historically, it was employed in textiles, cords, ropes, storage bags, and even for boat construction by the Egyptians (Suharty et al., 2016). Presently, kenaf is utilized as composites in collaboration with diverse materials, finding applications in automotive, construction, packaging, furniture, textiles, mats, paper pulp, and more (Thyavihalli Girijappa et al., 2019).

Hemp, a plant species primarily cultivated in Europe and Asia, typically reaches heights ranging from 1.2 to 4.5 meters and attains a diameter of about 2 centimeters (Bhoopathi et al., 2014). From a structural perspective, the inner core is surrounded by an outer layer called the bast fiber, which is joined to the inner layer by a binding substance that resembles pectin or glue. These fibers are used to make fabrics, animal bedding, garden mulch, ropes and different building supplies. Recent advancements have also led to the utilization of hemp in the creation of diverse composites (Li et al., 2007). Harvesting hemp plants involves separating the woody core from the bast fibers through a series of mechanical procedures. The woody core undergoes cleaning to achieve the desired core content and sometimes gets trimmed to specific sizes. Simultaneously, the separated bast fibers undergo further treatment to create yarn or bundles (Clarke, 2010).

Growing to a maximum height of 15 to 20 centimeters in about four months, jute is a major natural fiber that is grown throughout Asia, including China, India, Bangladesh, and Myanmar (Khan et al., 2006). After four months of growing, the fibers are extracted after harvest. Retting is an essential step in the extraction of fibers and can be done chemically $\left(N_2H_8C_2O_4\right)$ and $\left(Na_2SO_3\right)$ or biologically (Rahman, 2010). Harvested stalks are bundled together and immersed in water for around 20 days in biological retting. The pectin between the woody core and the bast breaks down during this soaking, making it easier to separate the fibers, which are then dried for use (Banik et al., 2003).

2.5.2 Processing Techniques

Moisture, fiber type, fiber volume fraction, and composites temperature are the primary factors affecting the processing of natural fiber composites. The four main processes used in the production of natural fiber composites are injection, extrusion, compression, and resin transfer molding.

The use of compression molding for thermoplastic composites dates to the 1990s, when demand for lightweight, high-performing materials increased (Gholampour & Ozbakkaloglu, 2020). Materials heated to a certain temperature are added to the molding cavity to begin this process. The mold's core side then applies extreme pressure to the cavity, compressing and deforming the material. The material is taken from the mold, and the mold is opened when the composites solidifies under high pressure. Key considerations while employing this technique include the amount of material, heating time, mold pressure, and cooling period.

Next, extrusion molding is one of the most widely used methods for creating composites made of natural fibers. This method is preferred because of the excellent strength and stiffness of the final composites as well as how simple it is to manufacture them (Campilho, 2015). This procedure starts with the thermoplastic material being stored as pellets or grains in a hopper. They are then put to melt in a hot barrel. The molten polymers are then used to give the composites the required form. The final stage is to cool the product. Next to extrusion molding, injection molding is the most widely used process for mass making composites. This process begins with heating the polymers until they melt in the hopper in the form of pellets or granules. Subsequently, the melted materials are kept within the mold and injected into a chamber made using a split die mold. The melt solidifies and the mold is opened once it has cooled.

Lastly is resin transfer molding. Rather than feeding the polymer into an open mold, in resin transfer molding, the resin is loaded into a holding chamber after it has been preheated (Dai & Fan, 2014). Figure 2.3 graphically displays the

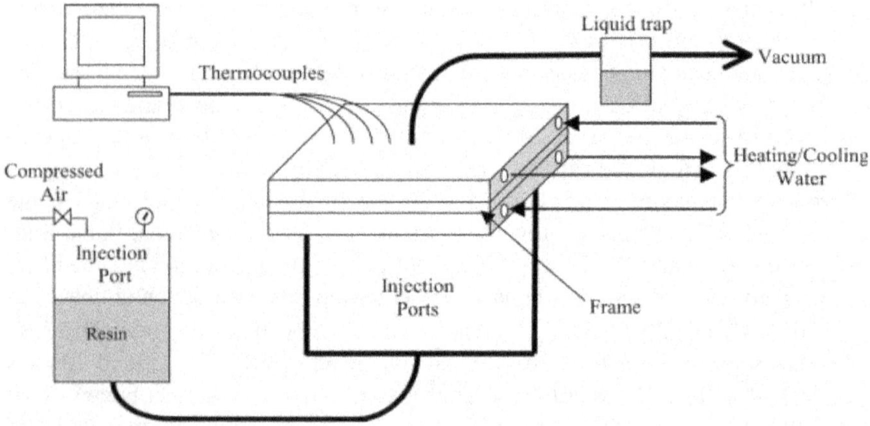

FIGURE 2.3 Resin transfer molding.

Source: Gholampour and Ozbakkaloglu (2020).

process (Gholampour & Ozbakkaloglu, 2020). Large-component production at medium volume is best suited for this technology. After the textile layers are positioned in the solid mold, the preforms are impregnated with resin through injection. Vacuum is frequently utilized to help draw the resin into the cavity and prevent air bubbles.

2.5.3 Surface Modification and Treatment

Important steps in improving the qualities and suitability of natural fibers for use in a range of industrial applications include surface modification and treatment. Plant-based natural fibers provide several intrinsic benefits, including renewable nature, biodegradability, and sustainability, but frequently, they are not strong enough, durable enough, or compatible enough with the matrices used in composites materials that absorb a lot of moisture, which causes brittle matrices to shatter. There are numerous ways to alter the surface of natural fibers, mainly to increase their adherence to polymeric matrix and decrease their absorption of water (Cruz & Fangueiro, 2016). For instance, cellulosic fibers at the micro- and nanoscale are a desirable replacement for synthetic fibers in the production of environmentally friendly goods (Cruz & Fangueiro, 2016).

First, the physical method of treating natural fibers is called plasma treatment (Figure 2.4), and it has been effectively applied to alter the surfaces of a range of natural fibers. Plasma treatment greatly improves natural fibers' mechanical qualities greatly. Furthermore, different functional groups can be introduced onto the natural fiber surface by plasma treatment, and these functional groups can establish strong covalent connections with the matrix to create a strong fiber/matrix interface. Additionally, plasma-induced surface etching enhances surface roughness and produces better mechanical interlocking with the matrices (Cruz & Fangueiro, 2016).

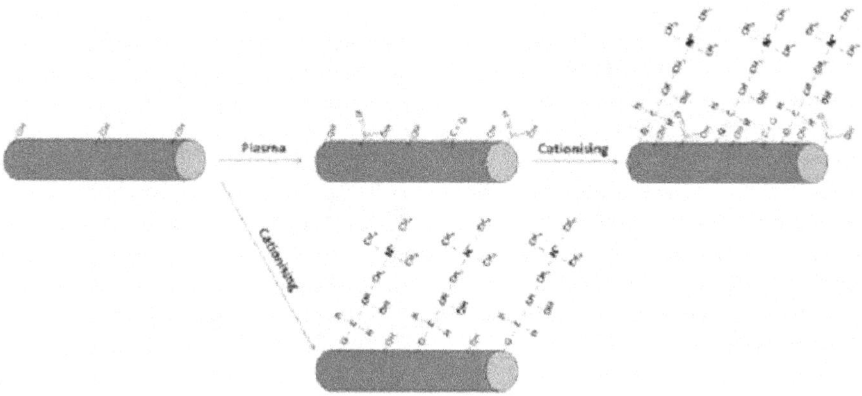

FIGURE 2.4 Plasma treatment and actionization.

Source: Shahidi et al. (2013).

Chemical treatment is the second method of treating natural fibers; fibers have been treated with chemicals like silane, alkali, peroxides, permanganates, and water-repellent agents. Chemical treatments improve natural fibers' mechanical properties, such as the alkali treatment depicted in Figure 2.5, which alter the fibers' crystalline structure and removes weak components like lignin and hemicelluloses (Xu & Varna, 2019).

Certain chemical treatments, including applying water-repellent substances, can also reduce the amount of moisture that natural fibers absorb and the ensuing swelling. Moreover, chemical treatments (such those employing silane coupling agents) can greatly improve the fiber/matrix interfacial contacts by forming strong chemical bonds, which will dramatically improve the mechanical performance of composites.

The engineering community has also investigated biological alternatives to chemical and physical methods. As shown in Figure 2.6, cellulose nanofibrils were recently applied to sisal and hemp fibers, using these fibers as bases during the fermenting of bacterial cellulose. According to the study, adding polymeric matrices such as PLA and cellulose acetate butyrate to the surface of these natural fibers with a bacterial cellulose deposit of between 5% and 6% significantly improved the bonding at the interface. The authors developed new-generation natural fiber composites marked by an enhanced interaction between the fiber and matrix components in this inventive surface modification approach (Cruz & Fangueiro, 2016).

Harnessing the benefits of natural fibers within composites materials and effectively employing them across diverse industrial applications constitute key objectives. Nonetheless, the research within this domain encompasses a wide spectrum, with literature discussing multiple methodologies encompassing physical, chemical, and biological approaches.

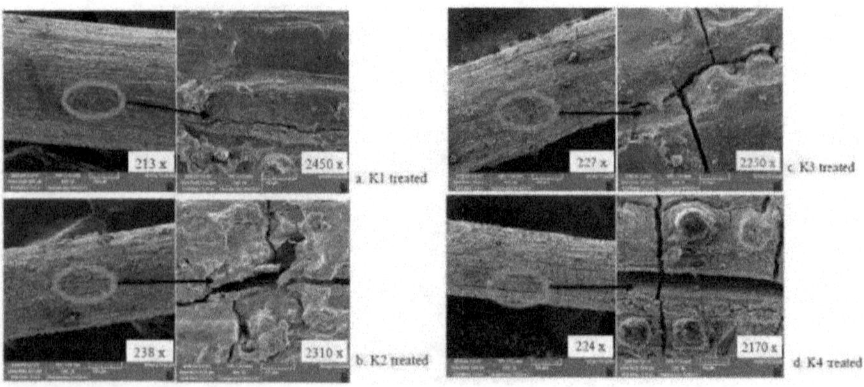

FIGURE 2.5 Surface morphology of coconut fiber with alkali treatment.

Source: Muhammad et al. (2015).

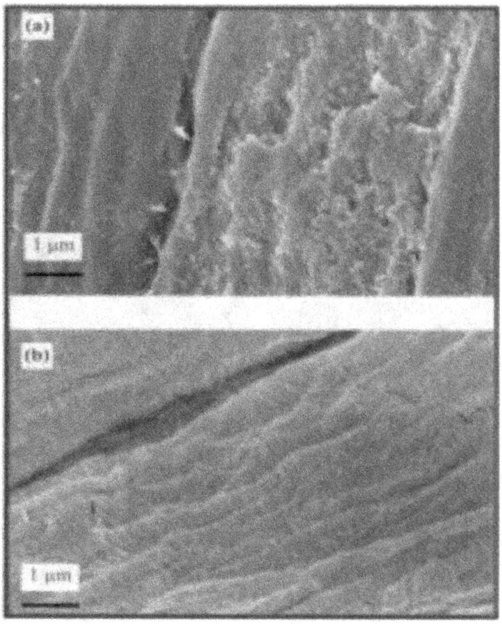

FIGURE 2.6 a) Natural hemp fiber; b) hemp fiber after bacterial cellulose modification.

Source: (Cruz & Fangueiro, 2016).

2.6 CHARACTERIZATION TECHNIQUES

2.6.1 Microscopy and Structural Analysis

The structural analysis of natural fiber composites uses a range of techniques to investigate the microstructure and properties of composites materials. X-ray diffraction (XRD), Fourier transform infrared spectroscopy (FTIR), scanning electron microscopy (SEM), transmission electron microscopy (TEM), and Raman spectroscopy are examples of applied techniques. Applications can look at the mechanical characteristics, morphology, and chemical makeup of natural fibers and composites. For example, SEM surface morphology and fiber–matrix interface can be used to analyze the fibers, and XRD can be used to characterize the fiber crystallization, which can be used to optimize the fibers' performance for various modern applications (Lau et al., 2018). Figure 2.7 illustrates the surface morphology of *Furcraea foetida* fibers under various magnifications using SEM. To act as a positive force resistant against fiber pullout, it was necessary to investigate the surface morphology of the fibers.

2.6.2 Mechanical and Thermal Evaluation

Accurately characterizing the structure-to-structure relationships through thermal analysis is essential when creating new composites. Many parameters, such as cure

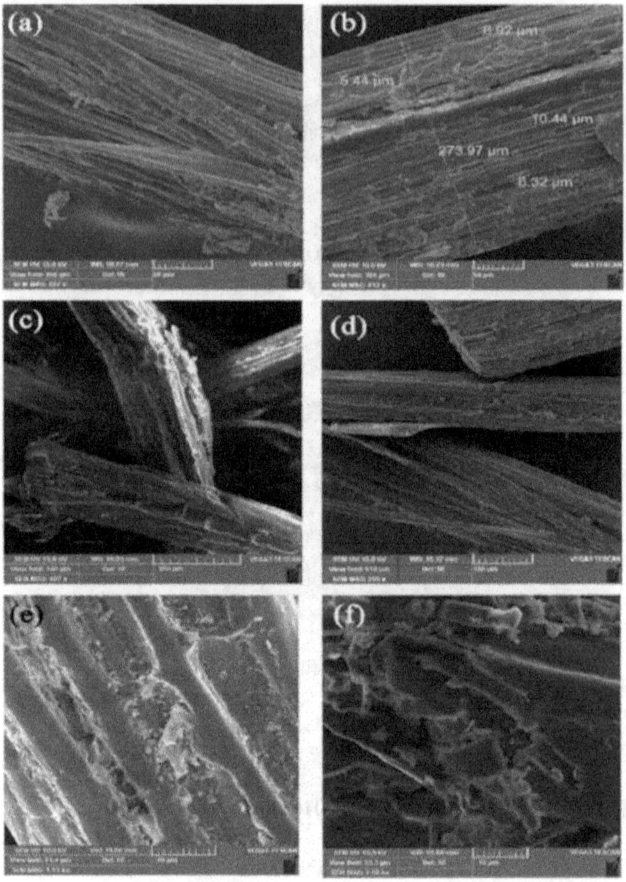

FIGURE 2.7 (a)–(f): Scanning electron microscopy images of cross-sections of *Furcraea foetida* fibers.

Source: Manimaran et al. (2018).

events, post-cure events, thermal transitions (e.g., glass transition, crystallization, melting, and degradation), and possible interactions between different formulation components are typically determined by thermal analysis. Dynamic mechanical analysis, or dynamic mechanical thermal analysis (DMTA), is a technique that allows for monitoring the temperature, oscillation frequency, and time while observing a sample's elastic and viscous responses to an oscillating load. One of the main advantages of DMTA is its capacity to distinguish between the elastic and viscous response when identifying the properties that cannot be achieved with any other method. The literature clearly highlights the enormous potential of this analytical technique in understanding and evaluating the mechanical and thermal properties of this kind of material (Costa et al., 2016).

Mechanical and thermal properties vary among different natural fibers. In this context, I have compared the properties of a few natural fibers from three different research articles. The first paper focuses on the use of doum, a type of natural fiber integrated into a low-density polyethylene matrix: Adding doum fibers significantly increased Young's modulus, bending modulus, and torsional modulus in particular: Young's modulus increased by 145% over neat polymer at 30 wt.% fiber weight; bending modulus gained 135% at 20 wt.%; and torsion modulus increased by 97% at 0.1 Hertz. However, the thermal analysis showed a slight decrease in products from the fine matrix polymer, where the crystallinity and texture of the composites were affected by the incorporation of natural fibers (Srinivasan et al., 2014).

The authors of the second paper examined blends made from banana and flax fibers. The simulation consists of a single banana fiber inserted between two flax fibers by hand layup with 40% volume share using epoxy resin and HY951 hardener. Glass fiber-reinforced polymer (GFRP) was used for lamination on both sides; as mentioned, these composites have higher ultimate tensile strength and flame resistance than monofiber composites. The authors labeled the composites as follows: composite 1, GFRP, banana, and flax; composite 2, flax composite; composite 3, banana fiber composites. The authors reported that composite 1, the hybrid, showed better performance on the double shear test than the other two composites in a linear manner. Additionally, composite 1 had a higher limiting oxygen index than the others, close to 26, and therefore demonstrated higher flame resistance than single-fiber composites 2 and 3 (Arrakhiz et al., 2013).

The third paper was an investigation of the use of hemp obtained from northern Morocco as a reinforcing agent in thermoplastic matrix composites. Polypropylene (PP)/hemp 25 wt.% had a 50% higher Young's modulus than neat PP without a coupling agent, and the PP/hemp 20 wt.% composites showed a 74% gain in Young's modulus with the coupling agent. Thermal analysis of the systems showed the effect of fiber incorporation and compatibilizer. The authors observed shifts to the right at the decomposition temperature peaks of the PP-SEBS-g-MA/25 wt.% hemp fibers and PP/25 wt.% hemp fiber composites by 40 °C and 25 °C, respectively, from the neat PP (Elkhaoulani et al., 2013).

2.6.3 Machinability of Natural Fiber Composites

The complex structure of natural fibers poses serious challenges to machining natural fiber-reinforced polymer (NFRP) composites, necessitating in-depth tribological research. To distinguish the shearing energy from the friction energy, Chegdani et al. investigated how natural fiber orientation affects the machinability of NFRP composites using the Merchant model. They experimentally studied NFRP composites machining by examining the impacts of process variables (feed, cutting speed, and tool geometry) on the composites delamination and the machined surface using standard machining techniques like drilling and milling (Chegdani et al., 2020). Machining is influenced by machining time and the cost of production. Under these constraints, authors proposed using a genetic algorithm to optimize the process parameter and achieved near-range parameters. Reinforcement with filler material led to differences in machining time and unit production cost (Jani et al., 2021).

Machining processes, particularly drilling, present a number of noteworthy issues such as delamination, fiber peel-up, pullout, spalling, hole shrinkage, fuzzing, and thermal degradation. The undesirable damage caused by drilling lowers the composite's strength against fatigue, which can have a severe negative impact on its long-term performance. Therefore, consideration should be given to the choice of machining parameters and cutting tools in addition to the fibers to be used in composites as filler or reinforcement as well as the mechanical and thermal properties of the fibers (Lotfi et al., 2021).

In addition to failure of machined materials, traditional machining can be dangerous to workers because dust is created during the process, and one of the most crucial factors in traditional machining is the tool. The tool materials wear down during cutting due to the abrasive nature of the fiber reinforcement and can even experience plastic deformation during machining. The variables that affect NFRP composites' machining performance are the matrix used, the state of the machine, the fiber reinforcement, the machining process, the fabrication technique, and the fiber and matrix bonding. When determining the optimal machining parameters, modeling can be used to predict composites' machining performance, potentially saving time and money (Vigneshwaran et al., 2021).

2.7 MECHANICAL PROPERTIES OF NATURAL FIBER COMPOSITES

2.7.1 Tensile Strength and Modulus

The tensile strength and modulus of natural fiber composites vary depending on factors such as the type of fiber, processing methods, and matrix materials. Due to its high specific strength, jute has become an increasingly popular natural fiber reinforcement in composites, for instance in strengthening the mechanical characteristics of cement mortar and the tensile qualities of oil palm composites. Furthermore, adding jute shifts the damping factor toward the high-temperature zone and increases the storage modulus. When the composite is made using compression molding, the molding temperature affects its mechanical characteristics. During the composites fabrication, the degradation of jute fiber with rising molding temperature decreased the achieved tensile strength (Sinha et al., 2017).

Composites materials were created by mixing and hot-pressing nonwoven mats made of hemp and PP fibers in different ratios. The tensile characteristics of the resulting composites materials were used to investigate the effects of anisotropy and hemp fiber content. As seen in Figure 2.8, the tensile strength decreased with increasing hemp content (highest loss of 34% at 70% hemp) with fibers oriented perpendicularly. On the other hand, the tensile strength exhibited a distinct pattern with parallel fibers, and a maximum value was discovered with increasing fiber loading. Tensile strength of composites with fibers oriented perpendicularly was 20%–40% lower than that of composites with fibers oriented parallel. The fibers in the composite matrix structure could not support weight because they were oriented perpendicularly to the direction of load, a flaw that could lead to failure (Ku et al., 2011).

2.7.2 Flexural Strength

Bend strength and rupture modulus are other names for flexural strength. The flexural strengths of flax and kenaf are nearly equal, ranging from 167 to 169 MPa, as shown in Figure 2.9. These are bast fibers, and between 40% and 45% of plants have

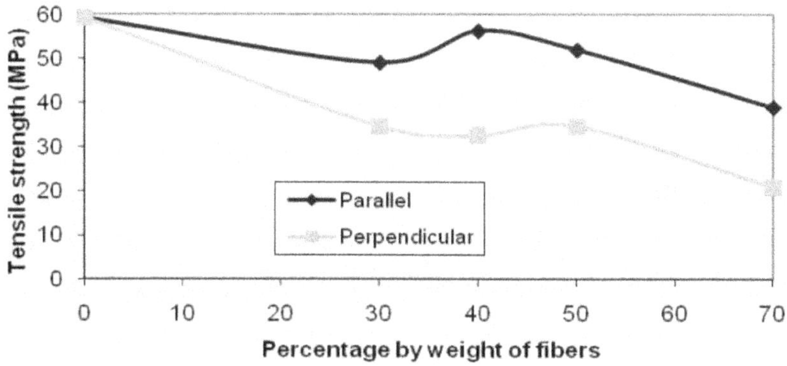

FIGURE 2.8 Tensile strength of polypropylene/hemp with varying percentage by weight of failure.

Source: Ku et al. (2011).

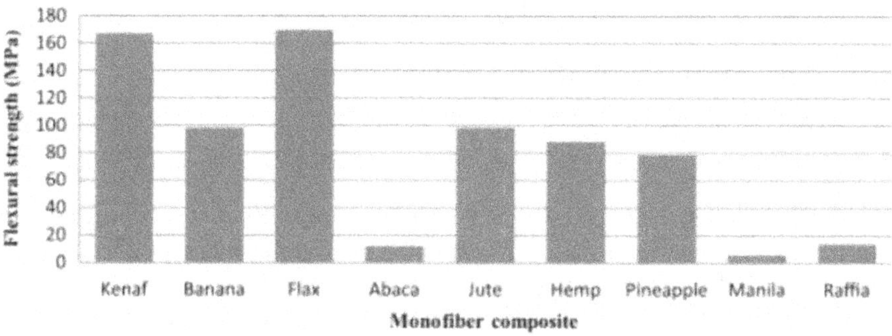

FIGURE 2.9 Flexural properties of monofiber composites.

Source: Wypych (2018).

multicellular layers that are pressed together. They have far greater flexural strength than other fiber-based composites as a result. Flexural strength is very low in manila fiber. These fibers are derived from the leaves of a particular kind of banana tree found in specific regions of the world, known as Manila hemp (Wypych, 2018).

Adding carpet waste fibers in rectangular cuttings markedly reduced the flexural strength of concrete but improved the toughness and energy absorption of concrete under flexural load. The recycled carpet cuttings could also be used in higher amounts than the short-length fibers in concrete (Pakravan et al., 2019).

2.7.3 Impact Resistance

Impact resistance is the primary feature of composites reinforced with natural fibers (Cantwell & Morton, 1991). As previously mentioned, even when a sizable amount of elastomer impact modifier was added to the polymer, it remained rather small. Figure 2.10 shows a plot of the quantity of PET fibers against the fracture strength

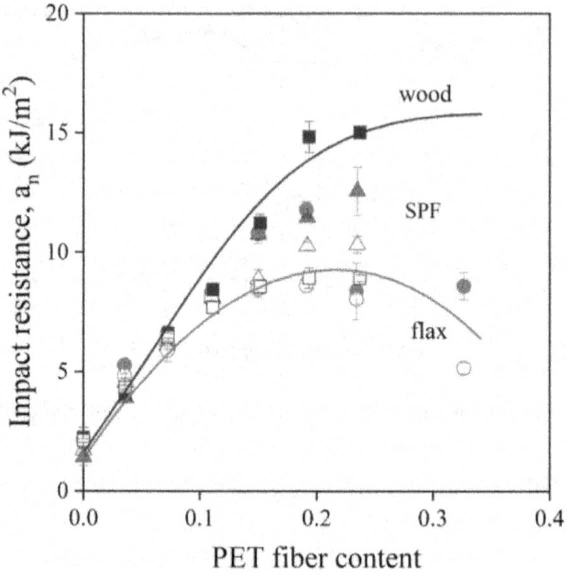

FIGURE 2.10 The impact resistance of flax, SPF, and wood.

Source: Várdai et al. (2021).

of hybrid composites reinforced with natural fibers. Impact resistance increases dramatically as PET fiber content increases; at high PET content, it reaches a commendable level of 15 kJ/m^2, having previously been as low as less than 2 kJ/m^2. At low PET levels, the type of natural fiber has minimal effect on impact resistance; however, as the composite's synthetic fiber content increases, the difference becomes more noticeable (Shahzad, 2009). Wood fibers with good adhesion show the strongest impact resistance, whereas impact resistance is lowest when flax is used without the MAPP coupling agent. Impact strength shows a peak dependency on composition, indicating the influence of numerous processes or variables (Várdai et al., 2021).

2.7.4 Hardness and Durability

Natural fiber composites, whose mechanical qualities depend on the kind of natural fibers utilized, have drawn a lot of interest lately as environmentally friendly substitutes for conventional synthetic composites. Composites reinforced with flax and jute fibers perform admirably in terms of hardness. For example, flax fiber composites frequently exhibit a good balance between stiffness and flexibility, producing a material with a respectable degree of hardness. Jute fibers also help create composites that have desirable hardness, which makes them appropriate for uses where rigidity and strength are necessary (Várdai et al., 2021). Crucially, though, it should be noted that a number of variables, including matrix material, processing methods, and fiber orientation, affect how hard natural fiber composites are, emphasizing the importance of careful design and optimization.

When assessing the mechanical properties of natural fiber composites, durability is an important element; long-term stability and resistance to environmental changes are two examples of the factors that affect durability. Because sisal and hemp are natural fibers with inherent strength and resilience, they are frequently used in composites that display noteworthy durability. Composites made of hemp, in particular, are well-known for being resistant to microbiological and moisture invasions, which increases the material's overall endurance. Sisal gives composites good structural stability and abrasion resistance, which makes them appropriate for applications where extended exposure to harsh environments is expected (Pakravan et al., 2019). However, even with these advantages, natural fiber composites' longevity could be jeopardized if appropriate processing methods and protective coatings are not used, highlighting the significance of thorough material engineering in guaranteeing their long-term performance.

2.8 THERMAL PROPERTIES

2.8.1 Melting and Glass Transition Temperatures

Thermal properties of a material, such as glass transition (Tg), crystallization, and melting temperatures and thermal stability, are significantly influenced by the length of polymeric chains and variations in the size, distribution, and morphology of reinforcing fibers. Thermogravimetric analysis and differential scanning calorimetry are frequently used to measure these thermal properties, which have direct impacts on materials' overall recyclability and mechanical performance (i.e., how well a material may hold up in a particular application over several cycles of reprocessing) (Zhou et al., 2022).

The Tg of neat PLA decreased by 7% following seven cycles of extrusion and injection molding. When the composite was recycled using the same techniques, adding 30wt% flax fiber reduced its by 23% (Bourmaud et al., 2016). Similarly, Chaitanya et al. (2019), Åkesson et al. (2016), and Le Duigou et al. (2008) reported lower Tg in PLA composites reinforced with sisal, cellulose, and flax after multiple cycles of extrusion and injection molding. Similar results were noted after recycling wood fiber/PP composites (Wang et al., 2020). In each case, lower Tg after recycling was attributed to a rise in the mobility of molecular chains attributable to polymer degradation. Polymer degradation can increase the number of chain ends (Gewert et al., 2015). Another possible reason for Tg reduction is that some extractives (e.g., wax) detach from natural fibers and act as a plasticizer.

2.8.2 Thermal Conductivity

The impact of fiber surface condition on natural fiber composite's thermal conductivity. The chemical treatment increases the composite's thermal conduction. It is a fact that natural fiber has lower thermal conductivity than matrix. Experimental and FEM simulation results of the hemp fiber reinforced composites showed that the angle of the fiber arrangement has a significant impact on the composite's thermal conductivity, although the authors neglected to discuss the impact of fiber direction.

In contrast to the rising trend of in-plane thermal conductivity, the composite's effective transverse thermal conductivity fell as the fiber content increased (Liu et al., 2012).

In a different study, researchers investigated banana fiber in insulation and found that the banana fiber experimentally determined thermal conductivity with a 50.4 mm thick specimen for various densities from 20 to 120 kg/m3. They also compared the calculated and experimental thermal conductivity at 25 °C and found that banana fiber showed the lowest thermal conductivity value of 0.04415 W/m K at a density of 70.4 kg/m^3 (Muthukumar et al., 2020).

Several modeling techniques have been used to study the transverse thermal conductivities of natural fiber composites, including fractal models, finite element-based computational models, two-dimensional square-arrayed pipe filament models, and thermal–electrical analogy. Furthermore, researchers have investigated the thermal, vibration-damping, and biodegradation characteristics of natural fiber composites and found that they have significant anisotropic transverse thermal conductivities that allow for effective thermal dissipation along specific directions. To better understand the thermo-mechanical properties of natural fibers, the researchers suggested theoretical and numerical investigations as alternatives to difficult, expensive, and time-consuming experimental measurements (Cao & Wu, 2008).

2.8.3 Heat Stability and Flame Resistance

The fire performance of natural fiber composites still needs to be improved to satisfy the stringent regulations for engineering applications: Vertical burn experiments have identified that most composites burn out after 10 seconds of flame application. The need to achieve desirable fire performance has stimulated numerous investigations in the field of fire retardation in natural fiber-based polymeric composites. Fire retardation is essential for lowering flammability and comply with tight regulations because natural fiber and polymer-based composites burn easily above the temperatures at which their constituents ignite and continue to burn in the presence of oxygen and heat sources (Kim et al., 2018).

To render such composites flame retardant, a traditional method used in textile applications was treating fabrics with fire-retardant solutions using pad/dry-cure. Pad/dry-cure is a textile finishing process widely used to apply flame retardants to fibers and fabrics. In the process, fibers/fabrics are impregnated in flame retardant solution and passed through rollers to squeeze the excess solution (Pornwannachai et al., 2018). The heat stability of natural fibers can be studied, among others, by thermogravimetric analysis. Researchers took measurements on a Setaram instrument in helium flow using a heating ramp of 58 °C/min. To improve thermal stability, they coated or grafted the fibers with monomers. They found that introducing flame retardants to composites extended time to sustainable ignition, limited or eliminated the flame phase of combustion, reduced heat release rate and surface flame spread, and eliminated hazardous condensation (Kozłowski & Władyka-przybylak, 2008).

Although NFRC is cheaper and can replace glass fiber in some applications, improving the flammability will likely increase the total cost of the composites. Gas-phase flame retardants minimize the heat release in the gas phase from combustion by releasing the free radicals that react and inhibit the chain reactions whereby the decomposition products of combustible materials propagate combustion. Selecting suitable flame retardants supports the creation of flame-resistant composites and prolongs their variety of uses. Using relatively low amounts of flame retardants with polymers shows promising results. Among non-halogenated flame-retardant additives, phosphorus and nitrogen-based compounds have proved to be compelling solutions, especially in matrices containing oxygen or nitrogen atoms in their backbone. At the same time, silicon-based additives also provide effective solutions (Prabhakar et al., 2015).

2.9 APPLICATIONS IN VARIOUS INDUSTRIES

2.9.1 Automotive

Automobile parts made of natural composites are already being produced and used (Luthada, 2019), primarily composites made of flax, hemp, and sisal and polyester or polypropylene. Natural composites are used mostly for marketing purposes rather than technical ones, with weight reduction being the main driver.,

Germany is a leader in using composites made of natural fibers. Mercedes, BMW, Audi, and Volkswagen are all using these composites in both exterior and interior applications (Parbin et al., 2019; Sanjay et al., 2018). For instance, the 1999 S-Class Mercedes-Benz inner door panel was created in Germany using 35% Baypreg F semi-rigid (PUR) elastomer from Bayer and 65% a blend of sisal, hemp, and flax. It should be noted that even though luxury automotive manufacturers are on board with natural fibers in their materials, it is not for cost reasons; rather, they appear to be recognizing the need to do more toward environmental responsibility. When Audi debuted the A2 midrange vehicle in 2000, the door trim panels were composed of polyurethane and strengthened with a flax/sisal mixture. Toyota created an ecoplastic from sugar cane that will be used to line the automobiles' interiors. The Araco Corporation in Japan even introduced the Grasshopper in 2003, a fully electric vehicle made entirely from plant-based composites, mainly kenaf in the body (Ramli et al., 2018).

2.9.2 Building and Construction

Natural fiber composites have found compelling applications in the building and construction industry, contributing to sustainable practices and ecofriendly structures. Building materials such as laminate panels and boards are a significant application area for incorporating diverse natural fibers such as bamboo, coir, and jute into composites materials. These lightweight materials are ecofriendly substitutes for conventional composites or synthetic reinforcements. The use of natural fiber composites in construction materials aligns with the industry's increasing focus on minimizing the environmental impact of building processes and materials (Di Schino & Corradi, 2020).

Compressed earth blocks reinforced with banana fibers and jute fiber composites, both used to retrofit structures and both showing remarkable tensile strength, are two more great applications of using natural fiber to reinforce structural elements. Banana fiber, a naturally occurring resource that was allowed to decompose and release carbon dioxide, created strong earth-compressed blocks in response to Egypt's need to develop new building techniques that would outperform more affordable and traditional methods like cement-filled bricks. In addition, this effort promoted the materialization of architectural contextualization by encouraging the use of culturally significant materials in nonvernacular urban architecture.

2.9.3 Packaging

Packaging is an industry with considerably applications of natural fiber composites in materials. Packaging is the process of getting something ready for preservation, transit, storage, and exhibition (Marsh & Bugusu, 2007). The next logical step in providing the packaging industry with reasonably priced, environmentally friendly solutions is biodegradable polymers reinforced with natural fibers. Paper goods and corrugated boards are made using natural fiber composites (McCracken & Sadeghian, 2018) incorporating sisal and coir. These composites provide a sustainable solution with advantageous mechanical qualities by strengthening and stiffening packing materials. As the packaging industry faces increasing pressure to adopt environmentally friendly practices, natural fiber composites present a promising avenue for developing packaging materials that are both functional and ecoconscious, catering to the growing consumer demand for sustainable and recyclable packaging options (Eskandari et al., 2021).

2.9.4 Aviation

Because natural fiber composites offer a special combination of lightweight qualities and environmental sustainability, they are seeing increasing use in the aircraft industry (Thyavihalli Girijappa et al., 2019) in interior airplane parts such as panels, seat backs, and overhead compartments. In aircraft bodies, lighter airframes use less fuel and therefore require less maintenance, making composites materials attractive for use in commercial transport aircraft (Quilter, 2004). The vertical tail fin of the Airbus A310 in 1985 was the company's first significant use of composites material in a commercial airplane. In the latter case, the composite fin's weight and production cost reduced the metal fin's 2,000 components (not including fasteners) to fewer than 100 (Quilter, 2004). These composites offer a lightweight alternative that lowers pollutants and improves fuel economy. The usage of natural fiber composites in the aviation industry is consistent with the sector's goal of creating aircraft parts that adhere to strict safety regulations while reducing the environmental effect of materials used in interior applications.

2.9.5 Consumer Goods

In the consumer goods sector, natural fiber composites are gaining traction as environmentally friendly and sustainable alternatives to traditional production materials for a variety of commodities; one well-known use is in furniture and home decor. Tables, chairs, and other decorative elements are often made of composites materials that use natural fibers such as rattan, bamboo, and jute (Sydow & Bieńczak, 2018). These composites provide ecofriendly choices for consumers seeking practical, aesthetically pleasing products that also align with their sustainability ideals. The overall strength and visual appeal of furniture are enhanced by the combination of natural fibers and polymer matrices, which contributes to the growing demand for environmentally responsible household goods.

Natural fiber composites have also been used in apparel, purses, wallets and eyewear among other accessories and fashion products. Composites that contain fibers like flax, hemp, or pineapple leaf are seeing increasing use in these accessories (Yurduseven, 2023). One especially notable example is Piñatex, which offers the fashion industry a sustainable leather substitute manufactured from pineapple leaf fibers. Incorporating natural fiber composites into consumer goods not only meets the demands of environmentally conscious consumers but also symbolizes a larger shift toward integrating sustainable materials into everyday things in response to the growing demand for ecofriendly products.

2.10 ENVIRONMENTAL IMPLICATIONS

The UN General Assembly proclaimed 2009 the International Year of Natural following intensive planning by the FAO. The main goal was to highlight how natural fibers support both food security and poverty reduction. To direct the International Yarn Fiber Foundation's operations, the International Steering Committee was established 2005 (Adekomaya et al., 2016).

The linear economy leaves many economic opportunities untapped, pollutes and degrades the environment, puts pressure on scarce resources, and strains the economies of individual societies. There is an urgent need to switch from linear to circular economies around the world to address this issue (Gardetti, 2019). In brief, a circular economy is driven by three major activities and approaches (Figure 2.11): reduce, reuse, and recycle, the three-legged stool of traditional waste management (Manickam & Duraisamy, 2019). The economic value of the circular economy is evident: For instance, the world economy could be improved by \$192 billion by 2030 if the fashion industry addressed the problems presented by its current linear economy (Chen et al., 2021).

Biobased materials such as agricultural and forestry waste can be used as reinforcements or fillers in biocomposites. These materials are renewable and biodegradable and are suitable alternative to fossil-based materials; in addition, producing biobased plastics and composites properly utilizes biomass waste. Over the course of the forecast period in one study, 2017 to 2030, researchers reported that the demand for biobased plastics and composites is expected to rise at a compound annual growth rate of 11.2%. The main objective of circular economies is to maintain sustainability by minimizing waste and maintaining resources at their highest value. Although the concept is concerned with reducing the carbon footprint at every stage of a product's life cycle, one way to integrate the concept is to integrate wastes into composites to divert from the disposable mindset that is currently rampant in contemporary

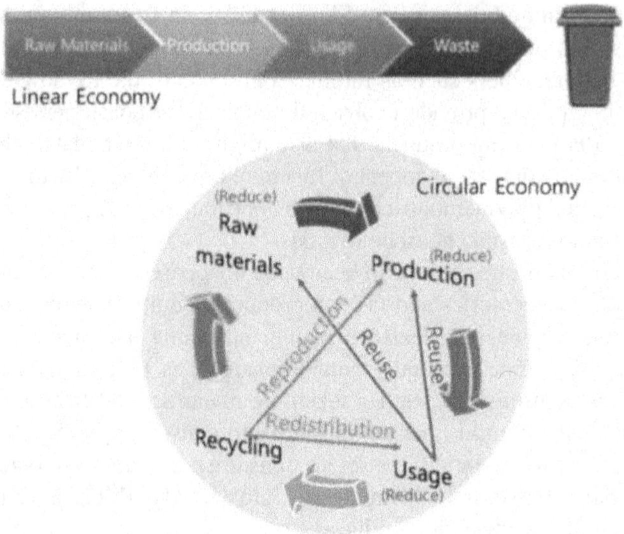

FIGURE 2.11 Key differences between linear and circular economies.

Source: Chen et al. (2021).

society. A circular economy includes converting biocomposite waste into valuable products, energy, or secondary materials to reduce biocomposite waste landfilling (Shanmugam et al., 2021).

For a considerable amount of time, debate in academia and other institutions has focused on sustainability. The 2030 Agenda for Sustainable Development reaffirmed the need for a shift in the sustainable development paradigm. The agenda is a set of guidelines that requires the governments of its member nations to treat people and, more appropriately, the environment with respect. It compiles 17 goals, known as the Sustainable Development Goals (SDGs), for all 169 member states to have fulfilled the goals by 2030. The 2030 Agenda is contained in a Transforming Our World document that emphasizes the three pillars of sustainable development: environmental protection, social inclusion, and economic growth. The circular economy is gaining the interest of academics and organizations because it is highly relevant. This is a multidisciplinary problem that has been approached from various angles in the literature (Gazzola et al., 2020).

Biological treatment and conventional methods are the two recognized approaches for managing biodegradable materials at the end of their useful lives. Anaerobic digestion and aerobic digestion (composting) are two types of biological treatment. Biodegradable materials that undergo aerobic composting produce carbon dioxide, water, methane, and compost that can be applied as fertilizer. Conventional methods like landfilling and incineration can also be used to treat biodegradable waste. If no biological solution is available, burning is possible. Large-scale burning of biodegradable materials releases thermal energy and yields ash, carbon dioxide, and water (Staikos et al., 2008).

2.11 CHALLENGES AND FUTURE DIRECTIONS

2.11.1 Current Challenges with Natural Fiber Composites

The manufacturing of natural fiber composites has a number of difficulties in the modern, globalized world. Even though natural fiber composites are starting to gain traction as more ecofriendly alternatives to synthetic materials, a number of challenges remain before their widespread application (Thyavihalli Girijappa et al., 2019). One of the primary difficulties is the inherent variations in natural fibers, which is caused by factors including climate, soil composition, and plant varieties. This unpredictable nature can lead to variations in fiber characteristics and make it challenging to achieve consistent performance with composites materials.

The main issue with natural fiber composites is that they are highly hygroscopic; they absorb moisture from their surroundings, which alters their dimensional stability and mechanical qualities. Therefore, it is critical for natural fiber composites to incorporate effective moisture management (Mwaikambo & Ansell, 2002). Surface treatments and coatings lessen the effects of moisture absorption and guarantee the long-term performance and durability of these composites.

A larger significant obstacle in work with natural fiber composites is the lack of standardization about testing procedures, which creates difficulties in precisely evaluating their qualities (Adekomaya et al., 2016). The comparability of results is hampered by differences in testing protocols and standards, which makes it challenging to produce generally accepted criteria. Standardizing testing procedures for natural fibers is essential to enabling precise and insightful comparisons (Verma et al., 2014). To overcome this obstacle and encourage uniformity in natural fiber composites investigations, research community collaboration is essential in creating universally applicable standardized testing methodologies.

Weathering and durability are also issues with natural fiber composites that can affect the materials' long-term performance and applicability. Because they come from plants, natural fibers can deteriorate when they come into contact with moisture, sunshine, and changes in temperature. One of the factors influencing the weathering and durability of natural fiber composites is UV radiation from sunshine, which results in discoloration, surface erosion, and poor mechanical properties (Faruk et al., 2012). In summary, the resolution of concerns pertaining to moisture absorption, standardization of testing, and durability in natural fiber composites is crucial for their further advancement and widespread application. Further research is required to address these challenges and make natural fiber composites more trustworthy, predictable, and environmentally sustainable for use in a variety of sectors.

2.11.2 Emerging Trends and Innovations

The emphasis on sustainable materials and environmentally friendly production techniques has led to new trends in natural fiber composites. Using sophisticated processing methods to improve the characteristics of composites made of natural fibers is one notable trend. In particular, nanotechnology has drawn attention for its ability to reinforce natural fibers at the nanoscale; the application of nanotechnology to natural fiber composites aims to enhance their mechanical strength, thermal stability, and barrier qualities. This pattern demonstrates a desire to increase performance and broaden the variety of applications for natural fiber composites (Kaushik & Singhal, 2018).

Additionally, advances in surface modification methods have been a key development in improving natural fiber compatibility with polymer matrices. One innovation that improved adherence with polymer matrices was plasma treatment on the surfaces of natural fibers. Polymer powders treated with plasma have numerous uses in a range of industries. The hydroxyl groups of cellulose fibers interact with the functional groups on the modified surface to enhance the interfacial adhesion and characteristics of the composites. By addressing the difficulty of attaining the best possible interfacial bonding in natural fiber composites, these developments improve mechanical performance (Sari et al., 2019).

As we noted earlier, the automotive industry is also finding widespread uses for natural fiber composites in response to the push for lightweight, biobased materials. Hemp and flax are two common natural fibers used in the production of automobile parts. For instance, Mercedes-Benz was the first to produce door panels for their A-class cars using polymer laminates based on jute (Puttegowda et al., 2021). This tendency is consistent with the larger movement toward environmentally friendly and energy-efficient modes of transportation. In a nutshell, the emerging trends and innovations in natural fiber composites encompass nanotechnology integration, applications in the automotive industry, surface modification, and the use of waste-derived fibers. These trends collectively contribute to the ongoing evolution of natural fiber composites, expanding their potential applications and reinforcing their status as sustainable alternatives in various industries.

2.11.3 Prospects for Sustainable Composites

The potential for sustainable composites made of natural fibers signals a bright future for environmental responsibility and innovation in materials research and engineering today. Growing awareness of the critical role natural fiber composites can play in environmentally acceptable alternatives is a result of the increasing worldwide necessity for sustainable practices.

The potential for natural fiber-based sustainable composites is enhanced by ongoing developments in processing techniques (Thyavihalli Girijappa et al., 2019). The goal of ongoing research is to improve manufacturing processes so that natural fiber composites perform better and are more affordable. New developments in processing techniques – like 3D printing, injection molding, and optimized extrusion – address the scalability and adaptability of these composites (Park et al., 2022). Furthermore, surface modification methods are being developed to solve issues with moisture absorption and interfacial bonding and enhance the compatibility of natural fibers with polymer matrices. These advancements in technology point to a bright future for the broad use of sustainable composites in a range of manufacturing techniques.

Given the growing attention being paid to environmental issues and sustainability worldwide, sustainable composites made of natural fibers have a bright future. Growing demand exists for substitutes that lessen carbon footprints and encourage environmentally responsible behaviors due to increased awareness of the negative ecological effects of conventional materials (Shabir et al., 2023). To encourage the development of sustainable composites, governments and regulatory agencies across the globe are enforcing more stringent environmental rules. Renewable resource-based and biodegradable natural fiber composites fit nicely with these sustainability objectives, making them attractive options for a range of sectors looking for environmentally friendly substitutes.

Cooperation and knowledge sharing between academics, business leaders, and legislators are critical for the future of sustainable composites made of natural fibers. Because this subject is interdisciplinary, it takes a team effort to solve problems, exchange best practices, and spur innovation. The creation and use of sustainable composites are accelerated by collaborative efforts like industry forums, conferences, and research alliances. The outlook for sustainable composites made of natural fibers is improving as stakeholders work together to improve standards, procedures, and materials. This will pave the way for more sustainable materials research and engineering in the future.

2.12 CONCLUSIONS

In conclusion, in this chapter, natural fiber composites are introduced in more detail, including their manufacture and characterization. These environmentally friendly substitutes for traditional materials like wood and metal are made from plant sources including bamboo, bananas, jute, hemp, kenaf, flax, and coir and reduce waste while optimizing the advantages of natural resources. These composites perform better by showing more effective internal bonding and fewer voids when the compression pressure and temperature are increased during fabrication. The need for sustainable resources is rising in the current global environment, and natural cellulosic fibers are becoming a suitable replacement for synthetic fibers in a number of applications.

For those who are new to the field of natural fiber composites, in particular, this chapter offers an extensive overview of the characteristics of natural fibers and the methods used to characterize them. There is a lot of potential for creating new natural fiber-reinforced polymer composites because there are so many natural fibers available in the world.

REFERENCES

Adekomaya, O., Jamiru, T., Sadiku, R., & Huan, Z. (2016). A review on the sustainability of natural fiber in matrix reinforcement – a practical perspective. *Journal of Reinforced Plastics and Composites*, *35*(1), 3–7. https://doi.org/10.1177/0731684415611974.

Åkesson, D., Vrignaud, T., Tissot, C., & Skrifvars, M. (2016). Mechanical recycling of PLA filled with a high level of cellulose fibres. *Journal of Polymers and the Environment*, *24*, 185–195.

Arrakhiz, F. Z., El Achaby, M., Malha, M., Bensalah, M. O., Fassi-fehri, O., Bouhfid, R., Benmoussa, K., & Qaiss, A. (2013). Mechanical and thermal properties of natural fibers reinforced polymer composites: Doum/low density polyethylene. *Journal of Materials & Design*, *43*, 200–205. https://doi.org/10.1016/j.matdes.2012.06.056.

Aziz, S. H., & Ansell, M. P. (2004). The effect of alkalization and fiber alignment on the mechanical and thermal properties of kenaf and hemp bast fibre composites: Part 1 – polyester resin matrix. *Composites Science and Technology*, *64*, 1219–1230. https://doi.org/10.1016/j.compscitech.2003.10.001.

Banik, S., Basak, M. K., Paul, D., Nayak, P., Sardar, D., Sil, S. C., et al. (2003). Ribbon retting of jute – a prospective and eco-friendly method for improvement of fibre quality. *Industrial Crops and Products*. *17*, 183–190. https://doi.org/10.1016/S0926-6690(02)00097-3.

Bhoopathi, R., Ramesh, M., & Deepa, C. (2014). Fabrication and property evaluation of banana-hemp-glass fiber reinforced composites. *Procedia Engineering*, *97*, 2032–2041. https://doi.org/10.1016/j.proeng.2014.12.446.

Bourmaud, A., Åkesson, D., Beaugrand, J., Le Duigou, A., Skrifvars, M., & Baley, C. (2016). Recycling of L-poly-(lactide)-poly-(butylene-succinate)-flax biocomposite. *Polymer Degradation and Stability*, *128*, 77–88.

Campilho, R. D. S. G. (2015). *Natural Fiber Composites*. CRC Press, Boca Raton.

Cantwell, W. J., & Morton, J. (1991). The impact resistance of composite materials – a review. *Composites*, *22*(5), 347–362. https://doi.org/10.1016/0010-4361(91)90549-v.

Cao, Y., & Wu, Y. Q. (2008). Evaluation of statistical strength of bamboo fiber and mechanical properties of fiber reinforced green composites. *Journal of Central South University of Technology (English Edition)*, *15*(1 Suppl.), 564–567. https://doi.org/10.1007/S11771-008-0422-Z/METRICS.

Chaitanya, S., Singh, I., & Song, J. I. (2019). Recyclability analysis of PLA/Sisal fiber biocomposites. *Composites Part B: Engineering*, *173*, 106895.

Chegdani, F., Takabi, B., El Mansori, M., Tai, B. L., & Bukkapatnam, S. T. S. (2020). Effect of flax fiber orientation on machining behavior and surface finish of natural fiber reinforced polymer composites. *Journal of Manufacturing Processes*, *54*, 337–346. https://doi.org/10.1016/j.jmapro.2020.03.025.

Chen, X., Memon, H. A., Wang, Y., Marriam, I., & Tebyetekerwa, M. (2021). Circular economy and sustainability of the clothing and textile industry. *Materials Circular Economy*, *3*(1), 1–9. https://doi.org/10.1007/s42824-021-00026-2.

Clarke, R. C. (2010). Traditional fiber hemp (Cannabis) production, processing, yarn making, and weaving strategies-functional constraints and regional responses. Part 2. *Journal of Natural Fibers*, *7*, 229–250. https://doi.org/10.1080/15440478.2010.504043.

Corrales-Ureña, Y. R., Villalobos-Bermúdez, C., Pereira, R., Camacho, M., Estrada, E., Argüello-Miranda, O., & Vega-Baudrit, J. R. (2018). Biogenic silica-based microparticles obtained as a sub-product of the nanocellulose extraction process from pineapple peels. *Scientific Reports*, *8*(1), 10417.

Costa, C. S. M. F., Fonseca, A. C., Serra, A. C., Coelho, J. F. J., Fonseca, A. C., Serra, A. C., & Dynamic, J. F. J. C. (2016, May). Dynamic mechanical thermal analysis of polymer composites reinforced with natural fibers *3724*. https://doi.org/10.1080/15583724.2015.1108334.

Cruz, J., & Fangueiro, R. (2016). Surface modification of natural fibers: A review. *Procedia Engineering*, *155*, 285–288. https://doi.org/10.1016/j.proeng.2016.08.030.

Dai, D., & Fan, M. (2014). Wood fibres as reinforcements in natural fibre composites: Structure, properties, processing and applications. In: Hodzic, A., & Shanks, R. (Eds.), *Natural Fibre Composites*. Elsevier, Amsterdam, pp. 3–65.

Di Schino, A., & Corradi, M. (2020, January 1). Construction and building materials. *AIMS Materials Science*. https://doi.org/10.3934/matersci.2020.2.157.

Elkhaoulani, A., Arrakhiz, F. Z., Benmoussa, K., Bouhfid, R., & Qaiss, A. (2013). Materials and design mechanical and thermal properties of polymer composite based on natural fibers: Moroccan hemp fibers/polypropylene. *Materials and Design*, *49*, 203–208. https://doi.org/10.1016/j.matdes.2013.01.063.

Eskandari, S., Zeraatkish, Y., Moghaddasi, R., & Azizi, J. (2021, September 14). Determinant factors energy efficiency and emission of pollutants Co2 & So2 in Iran's agricultural sector. *International Journal of Environmental Science and Technology*. https://doi.org/10.1007/s13762-021-03583-w.

Faruk, O., Bledzki, A. K., Fink, H. P., & Sain, M. (2012). Biocomposites reinforced with natural fibers: 2000–2010. *Progress in Polymer Science*, *37*(11), 1552–1596. https://doi.org/10.1016/j.progpolymsci.2012.04.003.

Gardetti, M. A. (2019). Introduction and the concept of circular economy. In: *Circular Economy in Textiles and Apparel*. Woodhead Publishing, pp. 1–11.

Gazzola, P., Pavione, E., Pezzetti, R., & Grechi, D. (2020). Trends in the fashion industry. The perception of sustainability and circular economy: A gender/generation quantitative approach. *Sustainability (Switzerland), 12*(7), 1–19. https://doi.org/10.3390/su12072809.

Gewert, B., Plassmann, M. M., & MacLeod, M. (2015). Pathways for degradation of plastic polymers floating in the marine environment. *Environmental Science: Processes & Impacts, 17*(9), 1513–1521.

Gholampour, A., & Ozbakkaloglu, T. (2020). A review of natural fiber composites: Properties, modification and processing techniques, characterization, applications. *Journal of Materials Science, 55*(3), 829–892.

Gowda, T. G. Y., Sanjay, M. R., Bhat, K. S., Madhu, P., Yogesha, B., Gowda, T. G. Y., Sanjay, M. R., Bhat, K. S., & Madhu, P. (2018). Polymer matrix-natural fiber composites: An overview polymer matrix-natural fiber composites: An overview. *Cogent Engineering, 47*. https://doi.org/10.1080/23311916.2018.1446667.

Islam, M. S., Kovalcik, A., Hasan, M., & Thakur, V. K. (2015). Natural fiber reinforced polymer composites. *International Journal of Polymer Science, 2015*. Hindawi Limited. https://doi.org/10.1155/2015/813568.

Jani, S. P., Senthil Kumar, A., Khan, M. A., & Sujin Jose, A. (2021). Design and optimization of unit production cost for AWJ process on machining hybrid natural fibre composite material. *International Journal of Lightweight Materials and Manufacture, 4*(4), 491–497. https://doi.org/10.1016/j.ijlmm.2021.07.002.

Kaushik, N., & Singhal, S. (2018). Hybrid combination of Taguchi-GRA-PCA for optimization of wear behavior in AA6063/SiCp matrix composite. *Production & Manufacturing Research, 6*(1), 171–189, DOI: 10.1080/21693277.2018.1479666.

Kerni, L., Singh, S., Patnaik, A., & Kumar, N. (2020). A review on natural fiber reinforced composites. *Materials Today: Proceedings, 28*, 1616–1621. https://doi.org/10.1016/j.matpr.2020.04.851.

Khan, M. A., Hassan, M. M., Taslima, R., & Mustafa, A. I. (2006). Role of pretreatment with potassium permanganate and urea on mechanical and degradable properties of photocured coir (cocos nucifera) fiber with 1,6-hexanediol diacrylate. *Journal of Applied Polymer Science, 100*, 4361–4368. https://doi.org/10.1002/app.23863.

Kim, N. K., Dutta, S., & Bhattacharyya, D. (2018, 2017 June). A review of flammability of natural fibre reinforced polymeric composites. *162*, 64–78. https://doi.org/10.1016/j.compscitech.2018.04.016.

Kozłowski, R., & Władyka-przybylak, M. (2008, April). Review flammability and fire resistance of composites reinforced by natural fibers. 446–453. https://doi.org/10.1002/pat.

Ku, H., Wang, H., Pattarachaiyakoop, N., & Trada, M. (2011). A review on the tensile properties of natural fiber reinforced polymer composites. *Composites Part B: Engineering, 42*(4), 856–873. https://doi.org/10.1016/j.compositesb.2011.01.010.

Lau, K., Hung, P., Zhu, M., & Hui, D. (2018, 2017 September). Properties of natural fibre composites for structural engineering applications. *Composites Part B, 136*, 222–233. https://doi.org/10.1016/j.compositesb.2017.10.038.

Le Duigou, A., Pillin, I., Bourmaud, A., Davies, P., & Baley, C. (2008). Effect of recycling on mechanical behaviour of biocompostable flax/poly (l-lactide) composites. *Composites Part A: Applied Science and Manufacturing, 39*(9), 1471–1478.

Li, X., Tabil, L. G., & Panigrahi, S. (2007). Chemical treatments of natural fiber for use in natural fiber-reinforced composites: A review. *Journal of Polymers and the Environment, 15*, 25–33. https://doi.org/10.1007/s10924-006-0042-3.

Liu, K., Takagi, H., Osugi, R., & Yang, Z. (2012). Effect of lumen size on the effective transverse thermal conductivity of unidirectional natural fiber composites. *Composites Science and Technology, 72*(5), 633–639. https://doi.org/10.1016/j.compscitech.2012.01.009.

Lopez, S., Akizu-Gardoki, O., & Lizundia, E. (2020). Comparative life cycle assessment of high performance lithium-sulfur battery cathodes. journal Of cleaner production, 124528. https://doi.org/10.1016/j.jclepro.2020.124528.

Lotfi, A., Li, H., Dao, D. V., & Prusty, G. (2021). Natural fiber – reinforced composites: A review on material, manufacturing, and machinability. *Journal of Thermoplastic Composite Materials*, *34*(2), 238–284. https://doi.org/10.1177/0892705719844546.

Luthada, P. (2019, November 13). Sustainable composites. *Addcomposites*. www.addcomposites.com/post/sustainable-composites-fiber-and-automated-manufacturing#:~:text=Many%20automotive%20components%20are%20already.

Manickam, P., & Duraisamy, G. (2019). 3Rs and circular economy. In: *Circular Economy in Textiles and Apparel*. Woodhead Publishing, pp. 77–93.

Manimaran, P., Senthamaraikannan, P., Sanjay, M. R., & Marichelvam, M. K. (2018, 2017 July). Study on characterization of Furcraea foetida new natural fiber as composite reinforcement for lightweight applications. *Carbohydrate Polymers*, *181*, 650–658. https://doi.org/10.1016/j.carbpol.2017.11.099.

Marsh, K., & Bugusu, B. (2007). Food packaging – roles, materials, and environmental issues. *Journal of Food Science*, *72*(3), R39–R55. Wiley. https://doi.org/10.1111/j.1750-3841.2007.00301.x.

McCracken, A., & Sadeghian, P. (2018). Corrugated cardboard core sandwich beams with bio-based flax fiber composite skins. *Journal of Building Engineering*, *20*, 114–122. https://doi.org/10.1016/j.jobe.2018.07.009.

Mugdha Bhat, K., Rajagopalan, J., Mallikarjunaiah, R., Nagaraj Rao, N., & Sharma, A. (2022). *Eco-Friendly and Biodegradable Green Composites*. IntechOpen. https://doi.org/10.5772/intechopen.98687.

Muhammad, A., Wardana Pratikto, N. G., & Irawan, Y. S. (2015). The morphology of coconut fiber surface under chemical treatment. *Matéria (Rio J.)*, *20*(1).

Muthukumar, K., Sabariraj, R. V., Kumar, S. D., & Sathish, T. (2020). Materials today: Proceedings investigation of thermal conductivity and thermal resistance analysis on different combination of natural fiber composites of banana, pineapple and jute. *Materials Today: Proceedings*, *21*, 976–980. https://doi.org/10.1016/j.matpr.2019.09.140.

Mwaikambo, L. Y., & Ansell, M. P. (2002). Chemical modification of hemp, sisal, jute, and kapok fibers by alkalization. *Journal of Applied Polymer Science*, *84*(12), 2222–2234. https://doi.org/10.1002/app.10460.

Pakravan, H. R., Asgharian Jeddi, A. A., Jamshidi, M., Memarian, F., & Saghafi, A. M. (2019). Properties of recycled carpet fiber reinforced concrete. *Use of Rrecycled Plastics in Eco-Efficient Concrete*, 411–425. https://doi.org/10.1016/b978-0-08-102676-2.00019-0.

Parbin, S., Waghmare, N. K., Singh, S. K., & Khan, S. (2019). Mechanical properties of natural fiber reinforced epoxy composites: A review. *Procedia Computer Science*, *152*, 375–379. https://doi.org/10.1016/j.procs.2019.05.003.

Park, S., Shou, W., Makatura, L., Matusik, W., & Fu, K. K. (2022). 3D printing of polymer composites: Materials, processes, and applications. *Matter*, *5*(1), 43–76. https://doi.org/10.1016/j.matt.2021.10.018.

Pornwannachai, W., Ebdon, J. R., & Kandola, B. K. (2018). Fire-resistant natural fibre-reinforced composites from flame retarded textiles. *Polymer Degradation and Stability*, *154*, 115–123. https://doi.org/10.1016/j.polymdegradstab.2018.05.019.

Prabhakar, M. N., Ur, A., Shah, R., & Song, J. (2015). A review on the flammability and flame retardant properties of natural fibers and polymer matrix based composites. *Composites Research, 28*(2), 29–39.

Puttegowda, M., Pulikkalparambil, H., & Rangappa, S. M. (2021 October–December). Trends and developments in natural fiber composites. *Applied Science and Engineering Progress*, *14*(4), 543–552. https://doi.org/10.14416/j.asep.2021.06.006.

Quilter, A. (2004). *StackPath*. www.aviationpros.com, www.aviationpros.com/engines-components/aircraft-airframe-accessories/article/10386441/composites-in-aerospace-applications.

Rahman, M. S. (2010). Jute-a versatile natural fibre. Cultivation, extraction and processing. In: Müssig, J. (Ed.), *Industrial Applications of Natural Fibres Industrial Applications of Natural Fibres: Structure, Properties and Technical Applications*. Wiley, 135–161. https://doi.org/10.1002/9780470660324.ch6.

Ramli, N., Mazlan, N., Ando, Y., Leman, Z., Abdan, K., Aziz, A. A., & Sairy, N. A. (2018, June). Natural fiber for green technology in automotive industry: A brief review. *IOP Conference Series: Materials Science and Engineering*, *368*, 012012. https://doi.org/10.1088/1757-899x/368/1/012012.

Salit, M. S. (2014). Biopolymer. In: *Tropical Natural Fibre Composites. Engineering Materials*. Springer, Singapore. pp. 39–52. https://doi.org/10.1007/978-981-287-155-8_3.

Sanjay, M. R., Madhu, P., Jawaid, M., Senthamaraikannan, P., Senthil, S., & Pradeep, S. (2018). Characterization and properties of natural fiber polymer composites: A comprehensive review. *Journal of Cleaner Production*, *172*, 566–581. Elsevier Ltd. https://doi.org/10.1016/j.jclepro.2017.10.101.

Sanjay, M. R., Siengchin, S., Parameswaranpillai, J., Jawaid, M., Pruncu, C. I., & Khan, A. (2019, 2018 October). A comprehensive review of techniques for natural fibers as reinforcement in composites: Preparation, processing and characterization. *Carbohydrate Polymers*, *207*, 108–121. https://doi.org/10.1016/j.carbpol.2018.11.083.

Sari, P. S., Thomas, S., Spatenka, P., Ghanam, Z., & Jenikova, Z. (2019). Effect of plasma modification of polyethylene on natural fibre composites prepared via rotational moulding. *Composites Part B: Engineering*, *177*, 107344. https://doi.org/10.1016/j.compositesb.2019.107344.

Shabir, I., Dash, K. K., Dar, A. H., Pandey, V. K., Fayaz, U., Srivastava, S., & R, N. (2023). Carbon footprints evaluation for sustainable food processing system development: A comprehensive review. *Future Foods*, *7*, 100215. sciencedirect. https://doi.org/10.1016/j.fufo.2023.100215.

Shahidi, S., Wiener, J., & Ghoranneviss, M. (2013). Surface modification methods for improving the dyeability of textile fabrics. *Eco-Friendly Textile Dyeing and Finishing*, *10*, 53911.

Shahzad, A. (2009). *Impact and Fatigue Properties of Natural Fibre Composites*. Thesis, Swansea University. http://cronfa.swan.ac.uk/Record/cronfa43056.

Shanmugam, V., Mensah, R. A., Försth, M., Sas, G., Restás, Á., Addy, C., Xu, Q., Jiang, L., Neisiany, R. E., Singha, S., George, G., Jose E, T., Berto, F., Hedenqvist, M. S., Das, O., & Ramakrishna, S. (2021, 2020 December). Circular economy in biocomposite development: State-of-the-art, challenges and emerging trends. *Composites Part C: Open Access*, *5*, 100138. https://doi.org/10.1016/j.jcomc.2021.100138.

Sinha, A., Narang, H., & Bhattacharya, S. (2017). Mechanical properties of natural fibre polymer composites. *Journal of Polymer Engineering*, *37*(9), 879–895. https://doi.org/10.1515/polyeng-2016-0362.

Srinivasan, V. S., Boopathy, S. R., Sangeetha, D., & Ramnath, B. V. (2014). Evaluation of mechanical and thermal properties of banana – flax based natural fibre composite. *Journal of Materials & Design*, *60*, 620–627. https://doi.org/10.1016/j.matdes.2014.03.014.

Staikos, T., Heath, R., Haworth, B., & Rahimifard, S. (2008). End-of-life management of shoes and the role of biodegradable materials. *Proceedings of the 13th CIRP International Conference on Life Cycle Engineering, LCE 2006, January*, 497–502.

Suharty, N. S., Ismail, H., Diharjo, K., Handayani, D. S., & Firdaus, M. (2016). Effect of kenaf fiber as a reinforcement on the tensile, flexural strength and impact toughness properties of recycled polypropylene/halloysite composites. *Procedia Chemistry*, *19*, 253–258. https://doi.org/10.1016/j.proche.2016.03.102.

Sydow, Z., & Bieńczak, K. (2018, April 3). The overview on the use of natural fibers reinforced composites for food packaging. *Journal of Natural Fibers*, *16*(8), 1189–1200. https://doi.org/10.1080/15440478.2018.1455621.

Thyavihalli Girijappa, Y. G., Mavinkere Rangappa, S., Parameswaranpillai, J., & Siengchin, S. (2019). Natural fibers as sustainable and renewable resource for development of eco-friendly composites: A comprehensive review. *Frontiers in Materials*, *6*. https://doi.org/10.3389/fmats.2019.00226.

Várdai, R., Lummerstorfer, T., Pretschuh, C., Jerabek, M., Gahleitner, M., Bartos, A., Móczó, J., Anggono, J., & Pukánszky, B. (2021). Improvement of the impact resistance of natural fiber – reinforced polypropylene composites through hybridization. *Polymers for Advanced Technologies*, *32*(6), 2499–2507. https://doi.org/10.1002/pat.5280.

Verma, D., Jain, R. K., Jain, S., & Jain, A. (2014). Natural fiber reinforced polymer composites: A review on potential for corrosion resistance. *Journal of Reinforced Plastics and Composites*, *33*(4), 342–354.

Vigneshwaran, S., John, K. M., Deepak Joel Johnson, R., Uthayakumar, M., Arumugaprabu, V., & Kumaran, S. T. (2021). Conventional and unconventional machining performance of natural fibre-reinforced polymer composites: A review. *Journal of Reinforced Plastics and Composites*, *40*(15–16). https://doi.org/10.1177/0731684420958103.

Wang, H., Zhang, J., Fu, H., Wang, W., & Wang, Q. (2020). Effect of an antioxidant on the life cycle of wood flour/polypropylene composites. *Journal of Forestry Research*, *31*(4), 1435–1443.

Wypych, G. (2018). *Functional Fillers: Chemical Composition, Morphology, Performance, Applications*. Elsevier. ISBN 978-1-927885-37-6.

Xu, J., & Varna, J. (2019). Matrix and interface microcracking in carbon fiber/polymer structural micro-battery. *Journal of Composite Materials*, *53*(25), 3615–3628. https://doi.org/10.1177/0021998319843616.

Yurduseven, S. (2023, March 30). *Tetkik*, 43–68. https://dergipark.org.tr/tr/download/article-file/3050423, https://doi.org/10.55709/tetkikdergisi.2023.3.56.

Zhou, J., Wang, B., Xu, C., Xu, Y., Tan, H., Zhang, X., & Zhang, Y. (2022). Performance of composite materials by wood fiber/polydopamine/silver modified PLA and the antibacterial property. *Journal of Materials Research and Technology*, *18*, 428–438.

3 PLA Synthesis and Applications

3.1 INTRODUCTION

The continued use of fossil fuel-based plastics poses multifaceted challenges that underscore the urgent need for sustainable alternatives. These challenges include limited supply of fossil fuels, ever rising costs associated with fossil fuel-based plastics, and sustainability (Mohla, 2023). Environmental pollution is a significant problem associated with conventional plastics, which contribute to waste management challenges, environmental pollution, and poor air quality due to their slow degradation (Barnes, 2019). This issue is further complicated by the expensive nature of recycling or incinerating conventional plastics. The high waste generated by industrial sectors such as packaging, textiles, and consumer products intensifies these environmental impacts (Ncube et al., 2021). These issues highlight the need for industries to shift toward more sustainable practices, including decreasing reliance on petroleum-based products and developing ecofriendly materials.

PLA emerges as a suitable solution to the pressing environmental and sustainability challenges posed by conventional fossil fuel-based plastics. Inherently biodegradable PLA degrades into carbon dioxide and water through microbial action under normal environmental conditions (Taib et al., 2023). This property of PLA directly addresses the critical issue of plastic waste accumulation and its adverse environmental impacts. Moreover, PLA is produced from renewable resources such as corn starch or sugarcane (Khan et al., 2023). The material's outstanding physical and mechanical properties, including strength, stiffness, biocompatibility, and thermo-plasticity, enable its application across various industries; applications of PLA range from packaging and automotive to biomedical devices without compromising performance (Domenek & Ducruet, 2016). The versatility of PLA is further enhanced by compatibility with a multitude of processing techniques, promising a broad spectrum of applications.

3.2 OVERVIEW OF PLA

PLA is an aliphatic polyester derived from renewable resources such as corn sugar, making it a sustainable alternative to fossil-based polymers (Sin, 2012). Its chemical structure includes the monomer lactic acid (2-hydroxypropionic acid), which can be produced through microbial fermentation, leading to optically pure L- or D-lactic acid (see Figure 3.1).

PLA's production process not only highlights environmental benefits but also its versatility (de Albuquerque et al., 2021). Properties of PLA can be tailored by

DOI: 10.1201/9781032678894-3

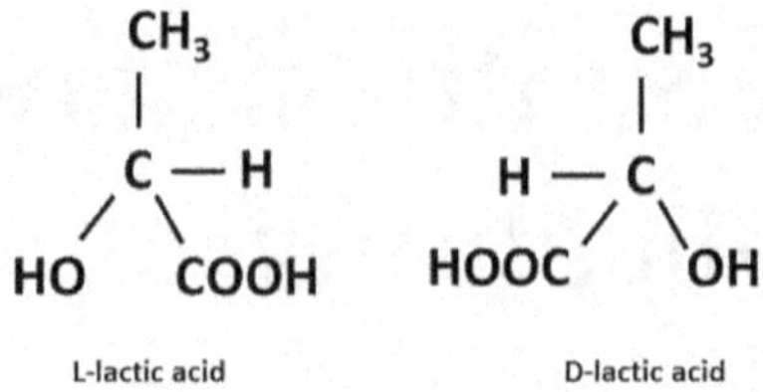

FIGURE 3.1 L-lactic acid and D-lactic acid

adjusting the ratio of the enantiomers, affecting its crystallinity and mechanical characteristics. PLA degrades primarily through hydrolysis of the ester bond, without requiring enzymes, which allows for its breakdown in compostable settings. The rate of degradation is influenced by the article's shape, size, isomer ratio, and hydrolysis temperature (Zhang et al., 2008). In addition to its biodegradability, PLA exhibits high strength, modulus, and thermal stability under specific conditions. However, it degrades above 200 °C, which limits its processing window (Fan et al., 2004).

PLA applications range as broadly as medical devices, due to its biocompatibility, to industrial packaging and textiles, supported by its ability to be processed using conventional plastic equipment. Modifying PLA's stereochemical structure to obtain high-molecular-weight polymers enhances its utility in food contact and medical applications (Cheng et al., 2009). PLA carries the US Food and Drug Administration's designation of generally recognized as safe. However, challenges such as its hydrophobic nature, comparatively low transition temperatures, and slow degradation rate at room conditions prompt ongoing research to enhance its properties through copolymerization or blending with other materials (Nampoothiri et al., 2010). Overall, PLA stands as a promising material in the quest for sustainable polymer solutions, evidenced by its significant market growth and broadening application range.

3.3 PLA SYNTHESIS

PLA is synthesized by three main methods: direct polycondensation, azeotropic dehydrative condensation, and ring-opening polymerization (see Figure 3.2). The synthesis process begins with lactic acid derived from sugar fermentation. Then the three main methods convert lactic acid to lactide, which is then polymerized to PLA. Each synthesis method affects the polymer's properties, such as molecular weight and crystallinity, catering to specific applications.

3.3.1 Direct Polycondensation

Synthesizing PLA via condensation has seen significant advancements and variations in methodologies. The focus of all the methodologies is producing high-molecular-weight PLA through polycondensation without byproducts. The existing studies present a variety of approaches and optimizations to overcome inherent limitations of traditional polycondensation reactions (Södergård et al., 2022).

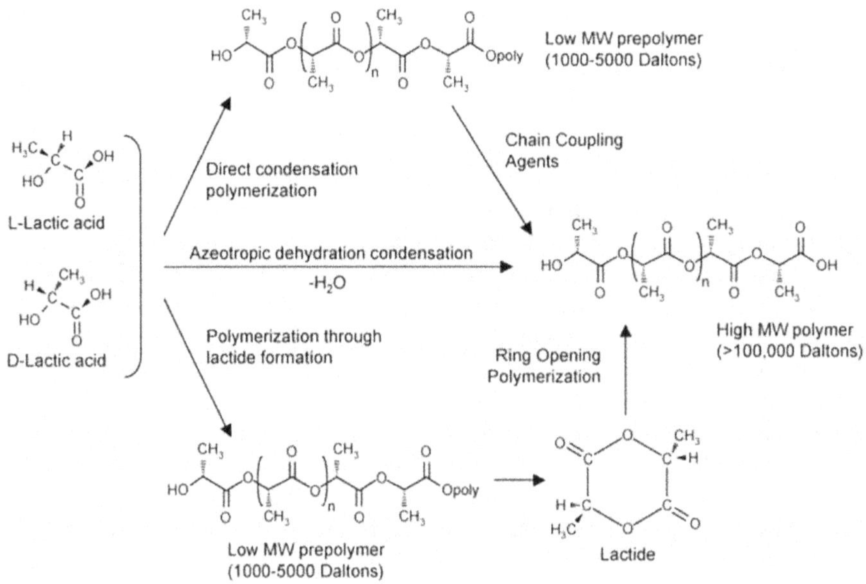

FIGURE 3.2 Pathways for synthesizing PLA from lactic acid.

Source: Espinoza-Montero et al. (2023).

Achmad et al. (2009) showcased a novel approach to PLA synthesis through direct polycondensation under vacuum without the use of catalysts, solvents, or initiators and achieved PLA with a molecular weight of up to 90 kDa. This simple method presented an attractive pathway for on-site production despite the higher activation energy than that with catalyzed processes and the thermal degradation encountered above 200 °C. Then Hu et al. (2016) provided a comprehensive review of azeotropic polycondensation and solid-state ring-opening polymerization for producing higher-molecular-weight PLA. These methods, by effectively removing byproducts and controlling the heating, demonstrated significant potential in overcoming the drawbacks of direct polycondensation.

Theodorou et al. (2023) introduced catalytic polycondensation complemented by kinetic modeling that accurately predicted molecular weight increase over time. This method, optimized for high-molecular-weight PLA production, underscores the importance of catalysts and controlled conditions in achieving desirable polymer properties. These studies highlight the challenges and drawbacks of using chain-extending agents and esterification-promoting adjuvants and in turn the need for innovations that minimize impurities, complexity, and costs associated with these additives.

In summary, the evolution of PLA synthesis through condensation reflects a balance between innovative chemical engineering approaches and the practical considerations of production costs, environmental impact, and application-specific requirements. The shift toward methods that allow for producing high-molecular-weight PLA without compromising purity or increasing operational complexity marks a significant stride in the field, promising broader applicability and sustainability of PLA as a biodegradable polymer.

3.3.2 Azeotropic Dehydrative Condensation

Azeotropic polycondensation (AP) is a specialized method of polycondensation aimed at overcoming the limitations of direct polycondensation in polymer production, particularly for achieving higher molecular weight polymers (Desai et al., 2023). In direct polycondensation, the byproducts (e.g., water) generated during the reaction can be difficult to remove completely from the reaction mixture, often leading to polymers of low molecular weight and quality. AP addresses this issue by using appropriate azeotropic solvents to efficiently remove water or other byproducts from the reaction mixture (Hu et al., 2016). This process manipulates the equilibrium between the monomer and polymer in an organic solvent, enabling the production of polymers with relatively high molecular weights in a single step. Additionally, AP allows for the reaction temperature to be lower than the polymer's melting point, which helps avoid impurities caused by depolymerization and racemization. The choice of solvent is critical in AP, as it significantly influences the conditions under which the polymerization occurs and the properties of the resulting polymer. High-molecular-weight PLA synthesis through AP represents a significant advancement in the field of polymer chemistry (Montané et al., 2020).

AP is part of a broader set of improved polycondensation techniques developed over the recent decade to overcome the limitations of traditional polycondensation methods. These new methods aim to produce PLA with higher molecular weights and improved properties by efficiently removing byproducts and optimizing reaction conditions. One innovative method involves using a Soxhlet extractor equipped with a molecular sieve (3 Å) to remove trace water from the refluxed solvent during AP. This setup allows for the efficient removal of byproducts, facilitating the synthesis of PLA with a molecular weight exceeding 30,000 $g{\cdot}mol^{-1}$. This approach highlights the importance of removing water to shift the equilibrium toward polymers with higher molecular weights (Kim & Woo, 2002).

Maryanty et al. (2021) highlighted the use of xylene as an azeotropic solvent and $SnCl_2$ as a catalyst in AP and achieved PLA of approximately 57,066.9 g/mol. This study underscored the importance of operating conditions, revealing that optimal results were obtained with specific amounts of lactic acid and solvent, alongside a 30-hour polymerization time. Mitsui Chemicals' commercial approach to AP in Japan achieved high-molecular-weight PLA (Mw ~300,000) by azeotropically dehydrating lactic acid in a high-boiling aprotic solvent under reduced pressure (Avérous, 2008). While effective, this method raises concerns about catalyst residue, necessitating additional steps to mitigate toxicity for biomedical applications.

The synthesis of PLA through AP has been pivotal in developing medical-grade polymers. The ability to tailor PLA's properties by manipulating the ratio of lactic acid isomers has led to materials suitable for surgical sutures, bone implants, and drug-delivery systems, with regulatory approval from bodies such as the FDA. Despite its advantages, AP for PLA synthesis is not without its challenges. The presence of catalyst residues remains a significant concern, particularly for medical applications, due to potential toxicity and non-biodegradability (Avérous, 2008). Research efforts are therefore directed toward optimizing the process to reduce or eliminate these residues without compromising the polymer's quality or increasing production costs.

3.3.3 Ring-Opening Polymerization

Ring-opening polymerization (ROP) is the cornerstone method for synthesizing high-molecular-weight PLA with the purities necessary for advanced applications (Kayan, 2020). Initially demonstrated by Carothers in the 1930s, ROP's success in creating PLA with molecular weights exceeding 100,000 wasn't realized until the 1950s after DuPont enhanced lactide purification methods (Kricheldorf, 2014). This process is characterized by its versatility, employing various methods like solution, bulk, melt, or suspension processes and different mechanisms, ionic (anionic or cationic) or coordination–insertion, depending on the choice of catalyst.

Trifluoromethanesulfonic acid and its methyl ester stand out as unique cationic initiators for lactide polymerization, promoting a nucleophilic attack on the carbonyl group and acyl–oxygen bond cleavage (Byers et al., 2018). Anionic polymerizations, on the other hand, are advanced by nucleophiles like potassium methoxide, minimizing racemization and leading to well-defined polymers. However, both methods, despite their high reactivity, are sensitive to impurities and side reactions like racemization and transesterification. Industrial-scale PLA synthesis favors bulk and melt polymerization due to their compatibility with low-toxicity catalysts and lower impurity levels. Among various catalysts, tin-based compounds, particularly tin(II) bis-2-ethylhexanoate (stannous octoate), are preferred for their efficiency and low toxicity. Stannous octoate has emerged as a frontrunner due to its high catalytic efficiency, approval for food and drug contact, and its ability to synthesize PLA with minimal racemization. This has made it a catalyst of choice in commercial-scale PLA production, where bulk and melt polymerization are preferred (Balla et al., 2021). These methods are favored over solution or suspension processes due to their lower toxicity and fewer impurities.

However, ROP is not without drawbacks. The process is susceptible to racemization and transesterification, which can compromise the polymer's mechanical properties and purity (Li et al., 2020). Additionally, the use of metal catalysts raises concerns about potential toxicity, particularly when PLA is intended for medical or food packaging applications. In response to these challenges, advancements have been made to refine the ROP process. For instance, Jacobsen et al. (2000) developed a one-stage continuous process that significantly reduces polymerization time while maintaining the quality of PLA. This process exemplifies the ongoing improvements aimed at overcoming the limitations of ROP, ensuring its viability as a production method for high-quality PLA. The continued evolution of this method reflects the balance of its inherent benefits, such as the ability to achieve high molecular weights, with its drawbacks, like sensitivity to impurities and the need for careful control of reaction conditions.

3.4 PLA DEGRADATION

The shift toward PLA reflects an urgent need to address the environmental challenges posed by traditional plastics. Deriving as it does from renewable resources, PLA offers significant advantages, including biocompatibility and comparable mechanical properties (see Figure 3.3). However, its degradation, encompassing hydrolytic, enzymatic, microbial, and photodegradative pathways, presents both opportunities and challenges in effectively managing PLA waste. In this section, we critically examine PLA's degradation mechanisms, crucial for enhancing its environmental sustainability and application scope.

3.4.1 Hydrolysis

The breakdown of PLA through hydrolysis plays a pivotal role in its role as an ecofriendly alternative to traditional plastics. This process results from moisture interacting with PLA's ester bonds, leading to decreased molecular weight and the release of soluble monomers and oligomers that are soluble. The hydrolytic reaction is self-catalyzing, as the formation of acidic groups accelerates degradation by lowering the environment's pH and increasing proton concentrations. Research indicates that the amorphous sections of PLA are primarily affected by hydrolysis, which in turn boosts the crystallinity of the polymer (Elsawy et al., 2017).

Four principal factors influence the rate of PLA hydrolysis: the amount of water absorbed, diffusion rate and coefficient of chain fragments within the polymer, and how soluble the degradation products are (Zaaba & Jaafar, 2020). Hydrolysis can be heterogeneous (surface reactions) or homogeneous (bulk erosion), with heterogeneous degradation occurring when hydrolysis surpasses the rate at which water diffuses into the particle, causing the particle to reduce in size. Conversely, during homogeneous degradation, the particle's shape is preserved, but its volume diminishes over time.

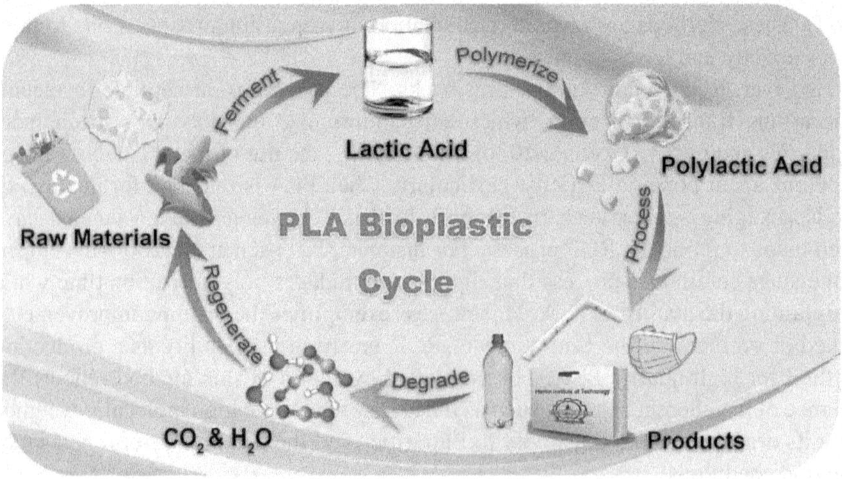

FIGURE 3.3 PLA degradation cycle.

Source: Yu et al. (2023).

Investigations into PLA's hydrolytic degradation have considered the impact of variables such as water-ethanol mixes, titanium dioxide (TiO_2) nanoparticles, chain extenders based on epoxy, varying temperatures, alkaline solutions, and organic modifiers (Luo et al., 2012). For example, the presence of ethanol at 50% concentration quickened PLA's degradation rate due to ethanol's more rapid diffusion through the polymer matrix (Khajavi et al., 2022). Similarly, TiO_2 nanofillers sped up the degradation process by serving as nucleating agents that modified the crystallization behavior of PLA during hydrolysis.

Additionally, the surrounding environment's pH and temperature significantly affect hydrolytic degradation, with studies showing accelerated degradation at both very high and very low pH levels. This is attributed to different mechanisms of degradation, including chain-end cleavage in acidic environments and random ester bond cleavage under basic conditions. Notably, the degradation shifts from surface to bulk as temperatures exceed the glass transition temperature, underscoring the intricate nature of PLA degradation and its sensitivity to external conditions. Understanding the dynamics of hydrolytic degradation is vital for creating PLA materials with designed degradation rates suitable for specific uses, such as drug delivery systems with controlled release (Da Silva et al., 2018). By comprehending the determinants of hydrolytic degradation, it is possible to engineer PLA-based materials that decompose at desired rates, thereby reducing environmental impact and enhancing the utility of PLA products.

3.4.2 Enzymes

The breakdown of PLA through enzymes such as lipase, esterase, and alcalase is influenced by several factors including pH levels, temperatures, the stereochemistry of the chains, and the crystallinity of the PLA material. The effectiveness of PLA biodegradation in natural settings hinges on optimal environmental conditions and the presence of specific microorganisms. Research shows that while alkaline proteases can effectively break down PLA into lactic acid, the effectiveness of acid and neutral proteases is significantly lower (He et al., 2008).

This limitation restricts the use of commercial proteases for degrading PLA in standard environmental settings. To overcome this, sophisticated molecular biology methods are employed to identify and characterize extracellular enzymes from microbes that can degrade PLA, with findings pointing to proteases (specifically serine proteases), lipases, and cutinases as key to this process. These enzymes latch onto and proliferate on the PLA surface, initiating the hydrolysis of its ester bonds and releasing lactic acid oligomers and monomers. This degradation process begins on the exterior and moves inward, ensuring thorough material breakdown. Enhancing enzymatic degradation involves adding specific enzymes such as proteinase K, which primarily acts on the PLA surface, differing from autocatalysis which occurs more rapidly internally (Cui et al., 2022). Proteinase K shows a particular preference for degrading L-lactic acid, and its activity is greatly affected by the polymer's stereochemical composition and crystallinity, underscoring the need for precise enzyme selection to optimize PLA degradation.

3.4.3 Microbes

Microbial degradation, a crucial process for recycling organic compounds into less harmful forms, plays a significant role in the breakdown of PLA in the environment. This degradation occurs when microorganisms secrete enzymes, breaking PLA down into smaller molecules after hydrolysis. Inducers like gelatin and amino acids are necessary to stimulate this enzymatic activity, leading to the eventual conversion of PLA into carbon dioxide, water, or methane (Kawai, 2010). Despite PLA's biodegradable nature, its microbial degradation in soil is notably slow due to its inherent resistance to microbial attack, presenting a challenge for its use in sustainable applications.

To address this, research has focused on enhancing PLA biodegradation through the introduction of microbial consortia, including mesophilic bacteria and actinomycetes, into the soil (Kim & Park, 2010). These consortia have shown potential in accelerating the degradation process, especially when combined with organic waste or dairy manure, which can provide a more conducive environment for microbial activity. Furthermore, specific bacterial species such as *Stenotrophomonas pavanii* and *P. geniculata* have been identified as beneficial for PLA degradation under both aerobic and anaerobic conditions, showcasing the adaptability of certain microbes to diverse environmental conditions (Bubpachat et al., 2018).

On a broader scale, understanding PLA's environmental degradation is essential given its widespread use in consumer products and packaging, with a significant portion destined for landfill disposal. The biodegradation of PLA is influenced by various factors, including material properties like molecular weight and crystallinity, as well as external conditions such as moisture, temperature, and oxygen availability (Qi et al., 2017). Aerobic and anaerobic pathways offer different rates of degradation, with aerobic conditions facilitating faster breakdown. This discrepancy highlights the importance of selecting appropriate disposal methods to ensure the effective degradation of PLA, avoiding long-term environmental accumulation.

Studies comparing the biodegradation rates of PLA to other biopolymers under different conditions reveal PLA's relatively slow degradation, underscoring the need for improved understanding and optimization of conditions that can enhance its breakdown. Real-world composting experiments, for instance, demonstrate the potential for PLA to degrade efficiently in controlled environments, emphasizing the role of temperature and microbial action in accelerating the process. These findings suggest the possibility of designing better waste management practices that can accommodate the specific needs of PLA degradation, contributing to more sustainable use of biodegradable polymers.

3.4.4 Photodegradation

The degradation of PLA upon exposure to UV light, known as photodegradation, entails the polymer absorbing UV radiation, triggering photochemical transformations. This process starts with photoionization (classified as Norrish Type I) and progresses to the breaking of polymer chains (Norrish Type II), leading to various outcomes like Norrish reactions, crosslinking, and oxidation (Sakai & Tsutsumi, 2022). Research has highlighted that additives such as oleic acid serve as photosensitizers, quickening the rate at which chain scissions occur and thereby enhancing the crystallinity of PLA without impeding its biodegradation. Moreover, molybdenum disulfide (MoS_2) integrated into PLA compounds heightened PLA's sensitivity to visible light, influencing the photodegradation pace and contributing to more significant weight loss in films (Han et al., 2018).

The impact of photodegradation on PLA, especially in outdoor settings, is profound, with UV light and moisture being the primary degradative agents. This exposure leads to significant alterations in the chemical makeup of PLA, characterized by chain breaks and the formation of novel functional groups. Certain compounds, such as TMPD (N,N,N',N'-tetramethyl-1,4-phenylenediamine), are known to accelerate PLA's photodegradation by releasing free radicals that target the polymer's backbone in a manner akin to the Norrish II reaction (More et al., 2023).

Researchers incorporated nanosilver and titanium dioxide (TiO_2) into PLA to adjust how it degrades under light exposure. The PLA composites infused with nanosilver delay photodegradation longer than unmodified PLA; nanosilver extended the material's photostability. Conversely, TiO2 nanoparticles sped up PLA photodegradation due to their optimal dispersion (Mucha et al., 2014).

These findings emphasize the nuanced nature of PLA's photodegradation and how different additives and composites formulations can significantly influence the material's response to UV light. Developing PLA materials with specific degradation profiles for targeted uses, particularly those requiring improved resilience against UV light, hinges on a deep understanding of these photodegradation mechanisms.

3.5 MULTIFARIOUS PLA APPLICATIONS

3.5.1 Surgical Sutures

Multifarious advantages of PLA in medical devices have been reported in the literature. PLA medical devices slowly degrade after fulfilling their function, avoiding the need for secondary surgery and reducing long-term complications such as immune rejection and bacterial infection. PLA has a history of safety in medical applications, with approval from regulatory agencies. PLA-based sutures help internal wound healing to avoid secondary removal (Liu et al., 2020).

The polymers used for biodegradable sutures include PLA and its copolymer, poly(lactic-co-glycolic acid) (PLGA). PLA sutures are noted for their advantages, such as prolonged tensile strength retention and suitability for specific applications like bone repair (see Figure 3.4). Antibacterial sutures can be developed from Vircryl Plus, and PLA-based fibers are used in medical nanofibers and mats. Examples include using solution blow spinning to generate PLGA nanofibers for wound healing and implanting PLA nanofibrous mats loaded with anticancer drugs to suppress tumor growth in mice (Böncü & Ozdemir, 2022).

3.5.2 Medical Devices

PLA has been applied in orthopedic fixation devices, emphasizing its advantages over non-biodegradable metallic materials. It is highlighted for its ability to avoid

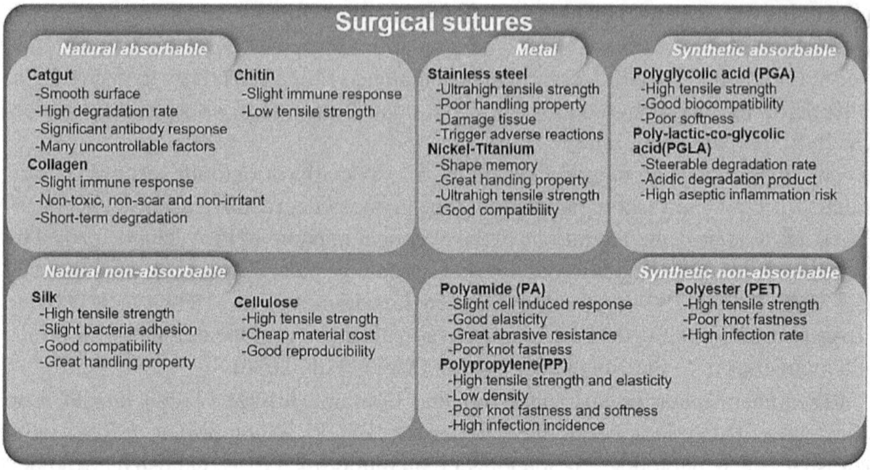

FIGURE 3.4 Materials commonly used as surgical sutures.

Source: Li et al. (2023).

stress shielding, a concern in orthopedics, where metallic devices can reduce stress on bones to the point of causing complications. The PLA plates and screws used in fracture stabilization provide support until the bone heals without causing stress shielding (Pina & Ferreira, 2012).

Additionally, the mechanical strength of semicrystalline PLA makes it useful in orthopedic fixation. Self-reinforced PLA screws are noted for their improved mechanical properties and successful use in fracture internal fixation. PLA has potential for orthopedic devices to incorporate pharmaceutical ingredients for targeted drug delivery applications. Examples include controlled release of bone morphogenic protein and antibiotics from PLA implants (Chou et al., 2022). Some challenges with PLA use, such as acidic degradation products and difficulties in adjusting degradation rates, are acknowledged. Overall, PLA has a longstanding history in orthopedics, presenting advantages but also facing certain reported issues.

3.5.3 Coronary Stents

The challenges associated with metallic coronary stents, such as ions leaching and restenosis, prompt the need for alternative solutions such as PLA. Biodegradable coronary stents, which degrade without causing long-term immune reactions, are considered an ideal option.

PLLA, a strong biodegradable polymer, is utilized in manufacturing these stents. The Igaki-Tamai stent, based on PLLA scaffold, has shown satisfactory performance in treating peripheral arteries, with a 10-year follow-up indicating complete degradation without hyperplasia development (Kuwabara et al., 2020). Abbott's biodegradable vascular scaffold everolimus-eluting PLLA coronary stents also demonstrated promising results in clinical trials, with zero stent thrombosis reported in a trial of

130 patients (Nef et al., 2016). Various fabrication methods including laser machining, rapid prototyping, and knitting technologies have been explored for biodegradable stents. Despite the potential of PLLA stents, limited clinical trial experience exists, with only Abbott's scaffold and Kyoto Medical's Igaki-Tamai stent completing trials and publishing findings. Larger trials are underway to further evaluate the safety and effectiveness of biodegradable stents, emphasizing the need for long-term follow-up.

3.5.4 Tissue Engineering

The use of PLA scaffolds in tissue engineering aims to restore, maintain, or improve tissue functions through biological substitutes or tissue reconstruction. The traditional regenerative strategy is transplantation, but tissue engineering provides an alternative by using patients' own cells seeded into a three-dimensional porous scaffold. PLA-based scaffolds are attractive due to their high surface-to-volume ratio, porosity, biodegradability, and mechanical properties (Sartore et al., 2019). However, the hydrophobic nature of PLA limits its use in tissue engineering, affecting cell adhesion. To overcome this limitation, blends of PLA with bioactive polymers such as collagen or chitosan are commonly used. These blends enhance cell viability, attachment, and proliferation. For bone tissue engineering, reinforced PLA-based scaffolds with starch or silk have been developed to match bone properties and provide proper load-bearing capacity (Büyük et al., 2023).

Various fabrication methods for PLA-based scaffolds include solvent casting, emulsion freeze drying, thermally induced phase separation, electrospinning, 3D printing, and gas foaming. While PLA offers advantages, its degradation can create a local acidic environment, causing adverse tissue responses and limiting clinical use. Ongoing research aims to improve PLA functionality and expand its applications in tissue engineering.

3.5.5 Drug Delivery

Drug delivery aims to administer pharmaceutical compounds to achieve therapeutic effects, and biodegradable polymers like PLA are widely used due to their biocompatibility, biodegradability, mechanical strength, and ability to encapsulate various drugs. PLA-based polymers are employed in different dosage forms, including pellets, microcapsules, microparticles, and nanoparticles. PLA's long degradation period makes it suitable for long-term release applications, while PLA-based PLGA is used for short-term drug delivery (Kapoor et al., 2015).

These biodegradable carriers have been extensively studied for controlled delivery of proteins, vaccines, genes, antigens, and growth factors (see Figure 3.5). PLA-based systems offer continuous drug release, and targeted therapy is crucial, particularly in cancer treatment, to reduce side effects and enhance therapeutic effects. For instance, PEG-PLA nanoparticles for siRNA delivery in breast cancer and folate-PEG-PLA carrying curcumin-loaded micelles demonstrated successful targeted drug delivery (Liu et al., 2014). Moreover, the responsive nature of PLA allows it to adapt to environmental conditions, leading to responses such as degradation, drug release, and

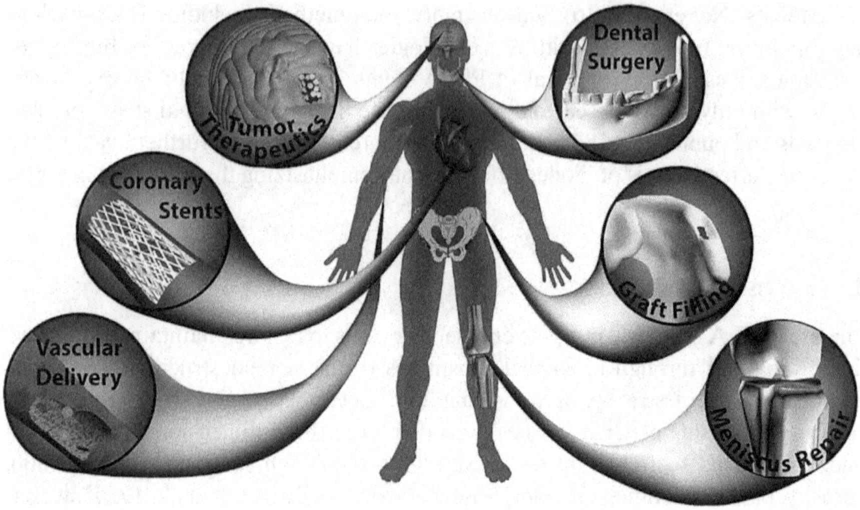

FIGURE 3.5 PLA-controlled drug delivery carriers.

Source: Tyler et al. (2016).

swelling. Examples include PLA-based carriers with near-infrared-sensitive hollow gold nanoshells for photo triggered drug release (Rajan et al., 2019).

3.5.6 Textiles

Over the last two decades, nylon, PET, and polypropylene have been dominant synthetic polymers in the fiber/textile industry (Jahandideh et al., 2021). However, concerns about the heavy reliance on petroleum-based polymers and challenges in solid waste management have driven the development of biodegradable materials in the sector. PLA as a biodegradable alternative has gained widespread adoption in various applications, with fiber/textiles being a significant area of utilization, including short-cut fibers, bulked continuous filaments, multifilaments, staple fibers, and spun bond fabrics. PLA fibers/textiles exhibit not only biodegradability but also high functionality. NatureWorks LLC conducted a comparison of PLA fiber with other common fibers and demonstrated PLA's potential to replace traditional fibers in textiles (Farrington et al., 2008). PLA fibers offer superior properties including hydrophilicity, good dyeability, low flammability, antibacterial characteristics, and excellent weather resistance. These qualities position PLA as a promising and environmentally friendly alternative in the textile industry.

3.5.7 Packaging

In the past decade, PLA has seen significant development for various primary packaging applications, particularly in oriented and flexible films. Oriented PLA film, created through extrusion casting with drawing-induced crystallization, offers transparency alongside improved strength, thermal and impact resistance, and elongation.

This makes it widely utilized in packaging such as bags, bottles, disposable beverage cups, and cutlery (Cheng et al., 2015). However, a drawback of oriented PLA film is its brittleness, which can lead to folding lines and breakage in applications like bags.

To address this limitation, flexible PLA film has been developed by blending PLA with plasticizers or other ductile materials. While this enhances flexibility, it comes at the sacrifice of some of PLA's inherent strength. Thermoforming is an alternative method of manufacturing unoriented or flexible PLA film. Efforts to improve PLA's brittleness include blending it with other polymers like polycaprolactone (PCL) or poly (butylene-adipate-co-terephthalate) (PBAT). Blending with PCL enhanced fracture energy through stress relaxation and energy dissipation, while PLA/PBAT reduced brittleness but also reduced tensile modulus compared with pure PLA (Gigante et al., 2019). Additionally, the brittleness of PLA could be mitigated by incorporating acetonitrile, which improved flexibility and thermal stability (Tripathi et al., 2021). Flexible PLA films are increasingly used as environmentally friendly and compostable alternatives to traditional plastics like LDPE and HDPE.

Commercially available PLA packaging exhibits superior mechanical properties to those of polystyrene and comparable properties to those of polyethylene terephthalate (PET). PLA has gained prominence in the market for biodegradable packaging due to its economic feasibility and current high consumption (Rai et al., 2021). Initially utilized in high-value films, rigid thermoforms, and food and beverage containers, PLA has found applications in additional areas such as fresh products across Europe, Japan, and the United Staes. Danone, a French company, was among the pioneers in using PLA in yoghurt cups for the German market (Kostic et al., 2022) (see Figure 3.6).

Over the last decade, PLA's use in packaging has expanded globally, particularly in the fresh products category, where it serves as food packaging for short shelf-life items like fruits and vegetables. PLA applications encompass containers, drinking cups, sundae and salad cups, wrappers for sweets, lamination films, blister packages, water bottles, and compostable yard bags for national or regional composting programs. Efforts are underway to explore new applications such as cardboard or paper coatings, particularly in the fast-food market for items like cups and plates. However, to address PLA's limitations, including limited mechanical and barrier properties and heat resistance, and to meet market expectations, there is a need to substantially increase global PLA production.

FIGURE 3.6 Cups made from PLA.

Source: Kostic et al. (2022).

3.5.8 Composites

The adoption of PLA in the composites sector represents a significant stride toward sustainable material science. PLA's biodegradability and renewable source base counter the environmental drawbacks of conventional composites. With its application spanning from lightweight and acoustic insulation materials in aviation to everyday household items, PLA-based composites emerge as a compelling alternative (Bajpai et al., 2012).

PLA not only reduces ecological footprints but also enhances mechanical properties through the integration of reinforcing agents. This strategic amalgamation with fibers, micro- and nanofillers, and specific additives is meticulously engineered to rectify PLA's inherent shortcomings such as brittleness, thereby unlocking new potentials in durability and performance (Murariu & Dubois, 2016). Such advancements make PLA composites key players in green technology, extending their utility beyond traditional domains to include long-term applications in automotive and electronics. This evolution reflects a broader industry shift toward materials that combine environmental responsibility with technological innovation, paving the way for PLA composites to challenge the status quo of petroleum-based counterparts (Ilyas et al., 2021). Through focused research on synthesis, modification, and application, PLA composites not only exemplify the potential for greener alternatives but also underscore the critical role of material science in fostering a sustainable future.

3.6 CHALLENGES WITH PLA

Biodegradable polymers like PLA offer potential solutions but come with their own challenges, including low toughness, high cost, and slow biodegradation, that limit their effectiveness (Jem & Tan, 2020). Additionally, PLA's permanency in soil can lead to extensive environmental pollution, similar to that caused by conventional plastics, necessitating further research into its degradation behavior and long-term impact (Zaaba & Jaafar, 2020). Table 3.1 lists some of the existing notable challenges with PLA that have been identified in the literature.

TABLE 3.1
Challenges with PLA

Criteria	Challenge
Cost and Mechanical Properties	PLA is more expensive than traditional petroleum-based plastics, limiting its widespread adoption. It exhibits inferior mechanical and physical properties, posing challenges in high-strength applications.
Heat Stability and Water Barrier Properties	PLA has poor heat stability and water barrier properties compared with conventional thermoplastics, limiting its use in high-temperature scenarios.
Development of Petroleum Industry	The reliance on crude oil for the chemical and plastic industries contributes to environmental issues like greenhouse gas emissions. Transitioning to biobased plastics like PLA addresses resource sustainability and pollution.
Regulatory and Public Awareness	The future of bioplastics is influenced by government regulations and public awareness. Infrastructure for garbage collection, composting, and recycling is essential for the adoption of bioplastics.

3.7 CONCLUSIONS

PLA offers a promising solution to the environmental challenges posed by conventional plastics owing to its biodegradability, renewable sourcing, and versatile applications across industries. However, challenges such as cost, mechanical properties, and degradation behavior underscore the need for ongoing research and innovation to enhance its performance and sustainability. The development of efficient synthesis methods and degradation mechanisms, alongside expanding applications, demonstrates PLA's potential in contributing to a more sustainable future. Addressing its limitations through technological advancements and increased public and regulatory support can further facilitate PLA's adoption and impact as an ecofriendly alternative to traditional plastics.

REFERENCES

Achmad, F., Yamane, K., Quan, S., & Kokugan, T. (2009). Synthesis of polylactic acid by direct polycondensation under vacuum without catalysts, solvents and initiators. *Chemical Engineering Journal*, *151*(1), 342–350. https://doi.org/10.1016/j.cej.2009.04.014

Avérous, L. (2008). Polylactic acid: Synthesis, properties and applications. In *Monomers, polymers and composites from renewable resources* (pp. 433–450). Elsevier.

Bajpai, P. K., Singh, I., & Madaan, J. (2012). Development and characterization of PLA-based green composites: A review. *Journal of Thermoplastic Composite Materials*, *27*(1), 52–81. https://doi.org/10.1177/0892705712439571

Balla, E., Daniilidis, V., Karlioti, G., Kalamas, T., Stefanidou, M., Bikiaris, N. D., Vlachopoulos, A., Koumentakou, I., & Bikiaris, D. N. (2021). Poly (lactic acid): A versatile biobased polymer for the future with multifunctional properties – from monomer synthesis, polymerization techniques and molecular weight increase to PLA applications. *Polymers*, *13*(11), 1822.

Barnes, S. J. (2019). Understanding plastics pollution: The role of economic development and technological research. *Environmental Pollution*, *249*, 812–821.

Böncü, T. E., & Ozdemir, N. (2022). Effects of drug concentration and PLGA addition on the properties of electrospun ampicillin trihydrate-loaded PLA nanofibers. *Beilstein Journal of Nanotechnology*, *13*(1), 245–254.

Bubpachat, T., Sombatsompop, N., & Prapagdee, B. (2018). Isolation and role of polylactic acid-degrading bacteria on degrading enzymes productions and PLA biodegradability at mesophilic conditions. *Polymer Degradation and Stability*, *152*, 75–85.

Büyük, N. İ., Aksu, D., & Torun Köse, G. (2023). Effect of different pore sizes of 3D printed PLA-based scaffold in bone tissue engineering. *International Journal of Polymeric Materials and Polymeric Biomaterials*, *72*(13), 1021–1031.

Byers, J. A., Biernesser, A. B., Delle Chiaie, K. R., Kaur, A., & Kehl, J. A. (2018). Catalytic systems for the production of poly (lactic acid). In *Synthesis, structure and properties of poly (lactic acid)* (pp. 67–118). Cham: Springer.

Cheng, H., Yang, Y., Li, S., Hong, J., & Jang, G. (2015). Modification and extrusion coating of polylactic acid films. *Journal of Applied Polymer Science*, *132*(35).

Cheng, Y., Deng, S., Chen, P., & Ruan, R. (2009). Polylactic acid (PLA) synthesis and modifications: A review. *Frontiers of Chemistry in China*, *4*, 259–264.

Chou, P.-Y., Lee, D., Weng, C.-C., Wu, R.-C., Liao, C.-T., & Liu, S.-J. (2022). Bone morphogenetic protein-, antimicrobial agent-, and analgesic-incorporated nanofibrous scaffolds for the therapy of alveolar clefts. *Pharmaceutics*, *14*(2), 374.

Cui, L., Wang, X., Szarka, G., Hegyesi, N., Wang, Y., Sui, X., & Pukánszky, B. (2022). Quantitative analysis of factors determining the enzymatic degradation of poly (lactic acid). *International Journal of Biological Macromolecules*, *209*, 1703–1709.

Da Silva, D., Kaduri, M., Poley, M., Adir, O., Krinsky, N., Shainsky-Roitman, J., & Schroeder, A. (2018). Biocompatibility, biodegradation and excretion of polylactic acid (PLA) in medical implants and theranostic systems. *Chemical Engineering Journal*, *340*, 9–14.

de Albuquerque, T. L., Júnior, J. E. M., de Queiroz, L. P., Ricardo, A. D. S., & Rocha, M. V. P. (2021). Polylactic acid production from biotechnological routes: A review. *International Journal of Biological Macromolecules*, *186*, 933–951.

Desai, H., Mehta, T., & Shah, N. (2023). Azeotropic dehydrative (solution) polycondensation of lactic acid to polylactic acid (PLA): A in-depth review of an overlooked method for manufacturing PLA. *Polymer-Plastics Technology and Materials*, 1–9.

Domenek, S., & Ducruet, V. (2016). Characteristics and applications of PLA. *Biodegradable and Biobased Polymers for Environmental and Biomedical Applications*, *10*, 9781119117360.

Elsawy, M. A., Kim, K.-H., Park, J.-W., & Deep, A. (2017). Hydrolytic degradation of polylactic acid (PLA) and its composites. *Renewable and Sustainable Energy Reviews*, *79*, 1346–1352.

Espinoza-Montero, P. J., Montero-Jiménez, M., Rojas-Quishpe, S., Alcívar León, C. D., Heredia-Moya, J., Rosero-Chanalata, A., Orbea-Hinojosa, C., & Piñeiros, J. L. (2023). Nude and modified eelectrospun nanofibers, application to air purification. *Nanomaterials*, *13*(3). https://doi.org/10.3390/nano13030593

Fan, Y., Nishida, H., Shirai, Y., Tokiwa, Y., & Endo, T. (2004). Thermal degradation behaviour of poly (lactic acid) stereocomplex. *Polymer Degradation and Stability*, *86*(2), 197–208.

Farrington, D. W., Lunt, J., Davies, S., & Blackburn, R. S. (2008). Poly (lactic acid) fibers (PLA). In *Polyesters and polyamides* (pp. 140–170). Elsevier.

Gigante, V., Canesi, I., Cinelli, P., Coltelli, M. B., & Lazzeri, A. (2019). Rubber toughening of polylactic acid (PLA) with poly (butylene adipate-co-terephthalate)(PBAT): Mechanical properties, fracture mechanics and analysis of ductile-to-brittle behavior while varying temperature and test speed. *European Polymer Journal*, *115*, 125–137.

Han, W., Luo, C., Yang, Y., Ren, J., Xuan, H., & Ge, L. (2018). Free-standing polylactic acid/chitosan/molybdenum disulfide films with controllable visible-light photodegradation. *Colloids and Surfaces A: Physicochemical and Engineering Aspects*, *558*, 488–494.

He, Y., Wu, T., Wei, J., Fan, Z., & Li, S. (2008). Morphological investigation on melt crystallized polylactide homo-and stereocopolymers by enzymatic degradation with proteinase K. *Journal of Polymer Science Part B: Polymer Physics*, *46*(10), 959–970.

Hu, Y., Daoud, W. A., Cheuk, K. K. L., & Lin, C. S. K. (2016). Newly developed techniques on polycondensation, ring-opening polymerization and polymer modification: Focus on poly(lactic acid). *Materials*, *9*(3). https://doi.org/10.3390/ma9030133

Ilyas, R. A., Sapuan, S. M., Harussani, M. M., Hakimi, M. Y. A. Y., Haziq, M. Z. M., Atikah, M. S. N., Asyraf, M. R. M., Ishak, M. R., Razman, M. R., Nurazzi, N. M., Norrrahim, M. N. F., Abral, H., & Asrofi, M. (2021). Polylactic acid (PLA) biocomposite: Processing, additive manufacturing and advanced applications. *Polymers*, *13*(8). https://doi.org/10.3390/polym13081326

Jacobsen, S., Fritz, H.-G., Degée, P., Dubois, P., & Jérôme, R. (2000). New developments on the ring opening polymerisation of polylactide. *Industrial Crops and Products*, *11*(2–3), 265–275.

Jahandideh, A., Ashkani, M., & Moini, N. (2021). Biopolymers in textile industries. In *Biopolymers and their industrial applications* (pp. 193–218). Elsevier.

Jem, K. J., & Tan, B. (2020). The development and challenges of poly (lactic acid) and poly (glycolic acid). *Advanced Industrial and Engineering Polymer Research*, *3*(2), 60–70.

Kapoor, D. N., Bhatia, A., Kaur, R., Sharma, R., Kaur, G., & Dhawan, S. (2015). PLGA: A unique polymer for drug delivery. *Therapeutic Delivery*, *6*(1), 41–58.

Kawai, F. (2010). Polylactic acid (PLA)-degrading microorganisms and PLA depolymerases. In *Green polymer chemistry: Biocatalysis and biomaterials* (pp. 405–414). ACS Publications.

Kayan, A. (2020). Recent studies on single site metal alkoxide complexes as catalysts for ring opening polymerization of cyclic compounds. *Catalysis Surveys from Asia*, *24*, 87–103.

Khajavi, M. Z., Ebrahimi, A., Mortazavian, A. M., Farhoodi, M., & Ahmadi, S. (2022). Hydrolytic degradation mechanism of modified polylactic acid in different food simulants. *Food Packaging and Shelf Life*, *34*, 100956.

Khan, A., Sapuan, S. M., Siddiqui, V. U., Zainudin, E. S., Zuhri, M. Y. M., & Harussani, M. M. (2023). A review of recent developments in kenaf fiber/polylactic acid composites research. *International Journal of Biological Macromolecules*, 127119.

Kim, K. W., & Woo, S. I. (2002). Synthesis of high-molecular-weight poly (L-lactic acid) by direct polycondensation. *Macromolecular Chemistry and Physics*, *203*(15), 2245–2250.

Kim, M. N., & Park, S. T. (2010). Degradation of poly (L-lactide) by a mesophilic bacterium. *Journal of Applied Polymer Science*, *117*(1), 67–74.

Kostic, D., Hoffmann, U., Fürtauer, S., Fell, T., Yilmaz, C., Burth, D., & Sängerlaub, S. (2022). Packaging weight, filling ratio and filling efficiency of yogurt and relevant packagings depending on commercial packaging design, material, packaging type and filling quantity. *Dairy*, *3*(3), 668–692.

Kricheldorf, H. (2014). Polycondensation. In *History and new results*. Heidelberg: Springer.

Kuwabara, K., Zen, K., Yashige, M., Ito, N., Kadoya, Y., Wakana, N., Yanishi, K., & Matoba, S. (2020). Comparative analysis of the paclitaxel-eluting peripheral igaki-tamai stent and the drug-free Igaki-Tamai stent using optical coherence tomography and histological analysis in a porcine iliac artery model. *Circulation Journal*, *84*(5), 799–805.

Li, G., Zhao, M., Xu, F., Yang, B., Li, X., Meng, X., Teng, L., Sun, F., & Li, Y. (2020). Synthesis and biological application of polylactic acid. *Molecules*, *25*(21), 5023.

Li, Y., Meng, Q., Chen, S., Ling, P., Kuss, M. A., Duan, B., & Wu, S. (2023). Advances, challenges, and prospects for surgical suture materials. *Acta Biomaterialia*, *168*, 78–112. https://doi.org/10.1016/j.actbio.2023.07.041

Liu, S., Yu, J., Li, H., Wang, K., Wu, G., Wang, B., Liu, M., Zhang, Y., Wang, P., & Zhang, J. (2020). Controllable drug release behavior of polylactic acid (PLA) surgical suture coating with ciprofloxacin (CPFX) – polycaprolactone (PCL)/polyglycolide (PGA). *Polymers*, *12*(2), 288.

Liu, Y., Zhu, Y.-H., Mao, C.-Q., Dou, S., Shen, S., Tan, Z.-B., & Wang, J. (2014). Triple negative breast cancer therapy with CDK1 siRNA delivered by cationic lipid assisted PEG-PLA nanoparticles. *Journal of Controlled Release*, *192*, 114–121.

Luo, Y.-B., Wang, X.-L., & Wang, Y.-Z. (2012). Effect of TiO2 nanoparticles on the long-term hydrolytic degradation behavior of PLA. *Polymer Degradation and Stability*, *97*(5), 721–728.

Maryanty, Y., Hadiantoro, S., Widjajanti, K., Putri, D. I. K., & Nikmah, L. (2021). Poly lactic acid (PLA) development using lactic acid product from rice straw fermentation with azeotropic polycondensation process. *IOP Conference Series: Materials Science and Engineering*, *1053*(1), 12043. https://doi.org/10.1088/1757-899X/1053/1/012043

Mohla, A. (2023). *Plastics & fossil fuels-processing & seeking alternatives*. BFC Publications.

Montané, X., Montornes, J. M., Nogalska, A., Olkiewicz, M., Giamberini, M., Garcia-Valls, R., Badia-Fabregat, M., Jubany, I., & Tylkowski, B. (2020). Synthesis and synthetic mechanism of polylactic acid. *Physical Sciences Reviews*, *5*(12), 20190102.

More, N., Avhad, M., Utekar, S., & More, A. (2023). Polylactic acid (PLA) membrane – significance, synthesis, and applications: A review. *Polymer Bulletin*, *80*(2), 1117–1153.

Mucha, M., Bialas, S., & Kaczmarek, H. (2014). Effect of nanosilver on the photodegradation of poly (lactic acid). *Journal of Applied Polymer Science*, *131*(8).

Murariu, M., & Dubois, P. (2016). PLA composites: From production to properties. *Advanced Drug Delivery Reviews*, *107*, 17–46. https://doi.org/10.1016/j.addr.2016.04.003

Nampoothiri, K. M., Nair, N. R., & John, R. P. (2010). An overview of the recent developments in polylactide (PLA) research. *Bioresource Technology*, *101*(22), 8493–8501.

Ncube, L. K., Ude, A. U., Ogunmuyiwa, E. N., Zulkifli, R., & Beas, I. N. (2021). An overview of plastic waste generation and management in food packaging industries. *Recycling*, *6*(1), 12.

Nef, H., Wiebe, J., Achenbach, S., Münzel, T., Naber, C., Richardt, G., Mehilli, J., Wöhrle, J., Neumann, T., & Biermann, J. (2016). Evaluation of the short-and long-term safety and therapy outcomes of the everolimus-eluting bioresorbable vascular scaffold system in patients with coronary artery stenosis: Rationale and design of the German – Austrian ABSORB RegIstRy (GABI-R). *Cardiovascular Revascularization Medicine*, *17*(1), 34–37.

Pina, S., & Ferreira, J. M. F. (2012). Bioresorbable plates and screws for clinical applications: A review. *Journal of Healthcare Engineering*, *3*, 243–260.

Qi, X., Ren, Y., & Wang, X. (2017). New advances in the biodegradation of poly (lactic) acid. *International Biodeterioration & Biodegradation*, *117*, 215–223.

Rai, P., Mehrotra, S., Priya, S., Gnansounou, E., & Sharma, S. K. (2021). Recent advances in the sustainable design and applications of biodegradable polymers. *Bioresource Technology*, *325*, 124739.

Rajan, M., Praphakar, R. A., & Pradeepkumar, P. (2019). Polymer composite strategies in cancer therapy, augment stem cell osteogenesis, diagnostics in the central nervous System, and drug delivery. In *Polymer nanocomposites in biomedical engineering* (pp. 235–270). Cham: Springer.

Sakai, W., & Tsutsumi, N. (2022). Photodegradation and radiation degradation. In *Poly (lactic acid) synthesis, structures, properties, processing, applications, and end of life* (pp. 439–454). John Wiley & Sons, Inc.

Sartore, L., Inverardi, N., Pandini, S., Bignotti, F., & Chiellini, F. (2019). PLA/PCL-based foams as scaffolds for tissue engineering applications. *Materials Today: Proceedings*, *7*, 410–417.

Sin, L. T. (2012). *Polylactic acid: PLA biopolymer technology and applications*. William Andrew.

Södergård, A., Stolt, M., & Inkinen, S. (2022). Industrial production of high-molecular-weight poly (lactic acid). In *Poly (lactic acid) synthesis, structures, properties, processing, applications, and end of life* (pp. 29–44). John Wiley & Sons, Inc.

Taib, N.-A. A. B., Rahman, M. R., Huda, D., Kuok, K. K., Hamdan, S., Bakri, M. K. B., Julaihi, M. R. M. B., & Khan, A. (2023). A review on poly lactic acid (PLA) as a biodegradable polymer. *Polymer Bulletin*, *80*(2), 1179–1213.

Theodorou, A., Raptis, V., Baltzaki, C. I. M., Manios, T., Harmandaris, V., & Velonia, K. (2023). Synthesis and modeling of poly(L-lactic acid) via polycondensation of L-lactic acid. *Polymers*, *15*(23). https://doi.org/10.3390/polym15234569

Tripathi, N., Misra, M., & Mohanty, A. K. (2021). Durable polylactic acid (PLA)-based sustainable engineered blends and biocomposites: Recent developments, challenges, and opportunities. *ACS Engineering Au*, *1*(1), 7–38.

Tyler, B., Gullotti, D., Mangraviti, A., Utsuki, T., & Brem, H. (2016). Polylactic acid (PLA) controlled delivery carriers for biomedical applications. *Advanced Drug Delivery Reviews*, *107*, 163–175. https://doi.org/10.1016/j.addr.2016.06.018

Yu, J., Xu, S., Liu, B., Wang, H., Qiao, F., Ren, X., & Wei, Q. (2023). PLA bioplastic production: From monomer to the polymer. *European Polymer Journal*, *193*, 112076. https://doi.org/10.1016/j.eurpolymj.2023.112076

Zaaba, N. F., & Jaafar, M. (2020). A review on degradation mechanisms of polylactic acid: Hydrolytic, photodegradative, microbial, and enzymatic degradation. *Polymer Engineering & Science*, *60*(9), 2061–2075.

Zhang, X., Espiritu, M., Bilyk, A., & Kurniawan, L. (2008). Morphological behaviour of poly (lactic acid) during hydrolytic degradation. *Polymer Degradation and Stability*, *93*(10), 1964–1970.

4 Natural Fiber-Reinforced PLA Biocomposites

4.1 INTRODUCTION

Biocomposites are composites materials that are formed by a matrix (resin) reinforced with natural fibers (Gholampour & Ozbakkaloglu, 2020). These materials are increasingly gaining approval at the industrial scale due to their high adaptability and superior performance. The matrix phase is formed by polymers derived from renewable and nonrenewable resources and plays a crucial role in holding the fibers together, transferring stresses onto them, and safeguarding them from mechanical harm and environmental deterioration (Faheed, 2024).

Biocomposites are defined as biocompatible and/or ecofriendly composites. They consist of a large variety of organic and/or inorganic components, such as natural and synthetic polymers, polysaccharides, proteins, sugars, ceramics, metals, and nanocarbons (Jaafar et al., 2019). The difference with this class of composites is that they are biodegradable and pollute the environment less. Natural fibers as reinforcements in composites offer advantages such as reducing reliance on nonrenewable resources and reducing effects on the environment. These substances have been effectively utilized in the fields of tissue engineering, wound care, drug delivery, and nanotechnology as hydrogels, scaffolding, matrices, and implants (Sharma et al., 2024).

In recent decades, natural fiber composites have emerged as promising structural materials in packaging, furniture, and private housing construction. Researchers have explored natural fibers for use in these composites including flax, sisal, banana, kenaf, hemp, bamboo, jute, feather, and wool. These natural fibers, when combined with polymer resins, offer a sustainable and environmentally friendly alternative to traditional synthetic materials. One advantage of natural fibers is their low density, which results in higher specific tensile strength and stiffness than glass fibers, in addition to their lower manufacturing costs. As such, biocomposites could be a viable ecological alternative to carbon, glass, and manmade fiber composites (Prakash et al., 2022).

PLA is a biodegradable biopolymer derived from renewable resources. It offers several desirable properties, including excellent transparency, processability, glossy appearance, and high rigidity. However, PLA also has some limitations, such as brittleness, high rate of crystallization, low damage resistance (less than 10% elongation at break), and slow degradation (hydrolysis of the backbone ester groups) limit its widespread use (Qin et al., 2011; Zhao et al., 2011).

Biocomposites are fabricated by combining natural fibers in a matrix material. The matrix material can be biodegradable, nonbiodegradable, or synthetic. Synthetic matrix materials, along with natural fibers, are used to form hybrid biocomposites. Indeed, biocomposites are an innovative class of materials that combine the benefits

DOI: 10.1201/9781032678894-4

of natural fibers and polymers to create ecofriendly, biodegradable, and high-performance materials. They offer a sustainable and environmentally friendly alternative to traditional composites, with a wide range of potential applications in various industries. PLA-based natural fiber composites are fully organic and combine the biodegradability of PLA with the mechanical properties of natural fibers. These composites have been extensively studied as potential replacements for petroleum-based products. Researchers have investigated the properties and applications of PLA-based natural fiber composites to assess their suitability for various industries. Environmental Benefits of PLA and Natural Fiber Composites includes Reduced greenhouse gas emissions: PLA's manufacturing process requires less energy, resulting in lower emissions compared to traditional polymers (Mukherjee & Kao, 2011).

Biodegradable composites have lower processing and waste disposal costs than synthetic materials, and they are biodegradable because they derive from renewable sources, reducing environmental impact. Their life cycle assessments are also favorable, with lower transportation requirements and reduced greenhouse gas emissions (Shalwan & Yousif, 2013). In this chapter, we discuss natural fiber-reinforced PLA biocomposite processing techniques (such as surface treatment, properties, and applications are discussed in detail. This chapter provides the overview of PLA biocomposites.

4.2 PROCESSING NATURAL FIBER-REINFORCED PLA BIOCOMPOSITES

The processing techniques for natural fiber-reinforced PLA biocomposites are crucial in determining the final properties of the composites. These techniques involve combining PLA matrices with natural fibers to form composite material. The choice of processing technique depends on factors including the type of natural fiber, the desired properties of the composite, and the intended application.

The processing techniques can be broadly classified into two categories: thermomechanical and solution-based methods. Thermomechanical methods such as extrusion and injection molding involve applying heat and mechanical force to mix the PLA and natural fibers. On the other hand, solution-based methods such as solution casting involve dissolving the PLA and natural fibers in a suitable solvent, followed by evaporating the solvent to form the composite. Each processing technique has its advantages and disadvantages. For instance, thermomechanical methods produce composites with high mechanical strength and dimensional stability, but they can cause thermal degradation of the natural fibers, whereas solution-based methods produce composites with excellent interfacial adhesion between the PLA and natural fibers, but they require the use of toxic solvents.

The processing technique also affects the dispersion of the natural fibers in the PLA matrix. Good dispersion of the fibers is essential for achieving high mechanical strength and other desirable properties in the composite. Therefore, it is important to optimize the processing conditions, such as temperature, pressure, and mixing time, to ensure good dispersion of the fibers. In addition to the abovementioned techniques, processing methods such as compression molding and 3D printing, are also being explored for producing natural fiber-reinforced PLA biocomposites. These methods offer unique advantages such as the ability to produce complex shapes and structures and the potential for rapid prototyping and mass production. Paunonen et al. used epoxidized linseed oil as a coupling agent and studied the effect of surface modifying PLA with pulp fiber composites at 30 wt%. They also observed aging under hydrothermal conditions (Paunonen et al., 2020).

4.2.1 Melt Blending

Melt blending is a thermomechanical processing technique that is commonly used to produce natural fiber-reinforced PLA biocomposites. This technique involves melting the PLA and then mixing the molten PLA with the natural fibers. The main advantage of melt blending is its simplicity and cost-effectiveness. It does not require the use of solvents, which makes it environmentally friendly, and it can produce composites with good mechanical properties and dimensional stability.

However, melt blending has some limitations. The high temperatures can cause thermal degradation of the natural fibers that reduce the mechanical strength of the composite. Moreover, it can be difficult to achieve good dispersion of the fibers in the PLA matrix, especially when high fiber loadings are used. To overcome these limitations, various strategies have been proposed. For instance, compatibilizers and coupling agents can improve the interfacial adhesion between the PLA and natural fibers, thereby enhancing the mechanical properties of the composite, twin-screw extruders also improve fiber dispersion in the PLA matrix. In conclusion, melt blending is a versatile and cost-effective technique for producing natural fiber-reinforced PLA biocomposites. However, careful optimization of the processing conditions is required to achieve the desired properties in the composite.

4.2.2 Solution Casting

Solution casting is a simple and cost-effective method for fabricating natural fiber-reinforced PLA biocomposites. In this process, the PLA and natural fibers are dissolved in a suitable solvent. The solution is then cast into a mold and allowed to dry, causing the solvent to evaporate and leaving behind a composite material. The main advantage of solution casting is that it does not require high temperatures or pressures, which can degrade the natural fibers, but it requires toxic solvents can be a disadvantage, and the drying process can be time-consuming. Solution casting allows for producing thin films and coatings, but it can be challenging to achieve uniform dispersion of the fibers in the PLA matrix, which can affect the mechanical properties of the composite (Milenkovic et al., 2021). Despite these challenges, solution casting remains a popular method for fabricating natural fiber reinforced PLA biocomposites due to its simplicity, low cost, and the ability to produce composites with excellent interfacial adhesion between the PLA and natural fibers.

4.2.3 Compression Molding

Compression molding is a commonly used method for producing natural fiber-reinforced PLA biocomposites. In this process, the PLA and natural fibers are placed in a heated mold, and pressure is applied to form the composite. This method is advantageous because it can produce composites with high mechanical strength and dimensional stability. However, the high temperatures used in this process can cause thermal degradation of the natural fibers.

Compression molding is suitable for producing large and complex parts, and it offers reductions in weight, cost, and carbon dioxide emission. However, it requires precise control of the molding temperature, compression pressure, and duration to ensure the quality of the composite. Dong et al. (2014) investigated the effects of varying coir weight percentages (5 wt%–30 wt%) on PLA biocomposites. They found that while tensile and flexural modulus increased, tensile and flexural strength decreased. Faludi et al. (2013) also studied the impact of reinforcement on PLA biocomposites using maize cob and wood fibers and found that compression molding improved mechanical properties due to the reinforcement.

4.2.4 Injection Molding and Extrusion

Injection molding is a popular method for producing natural fiber-reinforced PLA biocomposites. In this process, the PLA and natural fibers are melted and injected into a mold under high pressure. Extrusion is a continuous process in which the PLA and natural fibers are melted and forced through a die to form a continuous profile.

Injection molding produces composites with excellent surface finish and complex shapes. However, the high temperatures and pressures used in this process can degrade the natural fibers. Extrusion is advantageous because it can produce composites with consistent cross-sectional profiles. However, the high temperatures used in this process can cause thermal degradation of the natural fibers.

Despite these challenges, injection molding is widely used in fabricating natural fiber-reinforced PLA biocomposites due to its ability to produce composites with high mechanical strength and dimensional stability and extrusion is widely used due to its ability to produce composites with high mechanical strength and dimensional stability. Orue et al. (2016) and Komal et al. (2020) investigated the effects of fiber treatment and fabrication methods on PLA-based biocomposites reinforced with sisal and banana fibers and found that while fabrication methods (injection molding, extrusion injection molding, and compression molding) did not significantly impact thermal properties, glass transition temperature, hardness, or chemical composition and the orientation and distribution of fibers influenced crystallinity, mechanical properties, and dynamic properties. These findings highlight the importance of fiber arrangement in optimizing the performance of biocomposites

4.3 SURFACE TREATMENT OF NATURAL FIBERS

Surface treatment of natural fibers is crucial in preparing natural fiber-reinforced PLA biocomposites. The main objective of surface treatment is to improve the interfacial adhesion between the fibers and the PLA matrix. This is achieved by modifying the surface properties of the fibers, such as their chemical composition, surface roughness, and wettability. Surface treatment entails chemical, physical, and biological treatments. Each of these methods has its advantages and disadvantages, and the choice of method depends on the type of natural fiber, the desired properties of the composite, and the intended application (Table 4.1).

TABLE 4.1
Chemical, Physical, and Biological Treatments of Natural Fibers

Treatment	Effect and Results
Chemical	
Alkali (Mercerization)	Reduces fiber diameter by breaking the fiber bundle, thereby increasing the fiber surface area, which results in good adhesion with the matrix and improves mechanical and thermal behaviors of the composite.
Silane	One of the most effective coupling agents for natural fibers surface modification. It is a multifunctional molecule that deposits on the fiber surface, enhancing the linkage with the matrix through a siloxane bridge. Silane improves the fiber matrix adhesion and stabilizes the composite properties.
Acetylation	This treatment is known as esterification and plasticizes natural fibers. The reaction of the acetyl group $\left(CH_3CO\right)$ with the hydroxyl groups $\left(\text{-}OH\right)$ reduces the hydrophilicity of natural fiber and improves the dimensional stability of the composites.
Benzoylation	Benzoyl chloride decreases the hydrophilic nature of the natural fibers and improves their compatibility with the matrix, which thereby enhances the thermal stability and strength of the composite.
Peroxide	Improves the interfacial adhesion and thermal stability and reduces the moisture absorption of the fiber and matrix.
Malleated coupling agents	Give proficient interactions with the functional surface of the fiber and matrix, which reduces the melting temperature and lowers the stiffness of the fibers.
Sodium chlorite $\left(NaClO_2\right)$	Used to bleach fibers in acid solution. $NaClO_2$ is acidified and releases choleric acid $\left(HClO_2\right)$, which undergoes an oxidation reaction and forms chlorine dioxide $\left(ClO_2\right)$. ClO_2 reacts with lignin constituents and removes it from the fiber, improving the adhesive properties of the fiber.
Acrylation and acrylonitrile grafting	Acrylic acid reacts with the hydroxyl groups of the fiber to generate additional reactive cellulose macro-radicals, which helps to create good interfacial bonding. It reduces hydroxyl groups from the fiber and improves moisture resistance.
Calcium hydroxide	Helps to degrade the amorphous materials present in the fiber structure, increasing the crystallinity index of cellulose and improving thermal stability.
Isocyanate	Acts as a coupling agent in fiber surface modification, improves fiber moisture resistance, and provides better bonding with the matrix to improve the composite properties.
Triazine	Reacts with the hydroxyl groups of cellulose and lignin in the natural fibers to improve their moisture resistance properties.
Stearic acid	In ethyl alcohol solution, is used to treat natural fiber surfaces to facilitate interfacial bonding of fiber and matrix.
Permanganate	In acetone solution, treats fiber surfaces to enhance interfacial adhesion between natural fibers and matrix by reducing fiber hydrophilicity and increasing fiber thermal stability.
Fatty acid derivate (oleoyl chloride)	Fatty acid derivative act as a coupling agent to alter natural fiber surface to improves its adhesion with the matrix and wettability properties of natural fiber in composites.

(Continued)

TABLE 4.1 (*Continued*)
Chemical, Physical, and Biological Treatments of Natural Fibers

Treatment	Effect and Results
Physical	
Plasma	Offers a unique approach of modifying the physio-chemical structures of fiber surfaces without altering the bulk structures and characteristics of the composites. Plasma has been regarded as a clean and dry method for fiber treatments that increases surface roughness, which facilitates dispersion within the matrices.
Corona discharge	A green technique for modifying fibers and reinforcing them into composites that tend to show significant improvements in mechanical properties.
Steam explosion	Involves heating natural fiber materials at high temperatures and pressures followed by mechanical disruption of the pretreated material by violent discharge (explosion) into a collecting tank. This improves fiber properties, giving them smoother surfaces, reduced stiffness, and better bending and dispersion in the matrices.
Biological	
Enzymatic	Used to treat natural fibers to improve the interface between the fibers and the matrix, leading to improved mechanical properties of the composites. However, it is expensive and limited to pilot scale only.
Fungal	Removes noncellulosic components and lignin from the fiber surface by the action of specific enzymes as well as increases hemicellulose solubility, thereby reducing hydrophobicity.

Source: Siakeng et al. (2019).

4.3.1 Chemical Treatments

Chemical treatments involve the use of chemicals to modify the surface properties of the fibers. These treatments can remove impurities from the fiber surface, introduce functional groups that can form chemical bonds with the PLA matrix, and increase the surface roughness of the fibers.

4.3.1.1 Alkaline Treatment

Alkaline treatment, also known as mercerization, is one of the most common chemical treatments used for natural fibers. This treatment involves soaking the fibers in an alkaline solution, usually sodium hydroxide, for a certain period of time. The alkaline treatment removes impurities such as hemicellulose, lignin, and waxes from the fiber surface, exposing the cellulose microfibrils. This increases the surface roughness of the fibers, leading to better mechanical interlocking with the PLA matrix. Moreover, the alkaline treatment introduces hydroxyl groups onto the fiber surface. These hydroxyl groups can form hydrogen bonds with the PLA matrix, improving the interfacial adhesion between the fibers and the matrix. However, excessive alkaline treatment can damage the cellulose microfibrils and reduce the mechanical properties of the fibers. Therefore, the conditions of the alkaline treatment, such as the concentration of the alkaline solution and the treatment time, need to be carefully optimized.

4.3.1.2 Silane Treatment

Silane treatment is another common chemical treatment used for natural fibers. This process involves the application of silane coupling agents to the fiber surface. Silane coupling agents are molecules that have two different functional groups: one that can form a bond with the fiber surface, and another that can form a bond with the PLA matrix. The silane treatment can improve the interfacial adhesion between the fibers and the PLA matrix by forming chemical bonds at the fiber/matrix interface. This can lead to significant improvements in the mechanical properties of the composite. Moreover, the silane treatment can increase the hydrophobicity of the fibers, which can reduce the water absorption of the composite. This is particularly important for applications where the composite is exposed to moisture. However, the effectiveness of the silane treatment depends on several factors, including the type of silane coupling agent used, the concentration of the silane solution, and the treatment time. Therefore, these parameters need to be carefully optimized to achieve the desired properties in the composite.

4.3.1.3 Acetylation

Acetylation is a chemical reaction of the natural fibers with acetic anhydride. This process introduces acetyl groups onto the fiber surface, which increases the hydrophobicity of the fibers and improves their compatibility with the PLA matrix. Acetylation can significantly improve the mechanical properties of the composite by enhancing the interfacial adhesion between the fibers and the PLA matrix. Moreover, it can reduce the water absorption of the composite, which is beneficial for applications where the composite is exposed to moisture. However, similar to other chemical treatments, the effectiveness of acetylation depends on factors such as the concentration of the acetic anhydride solution, the treatment time, and the temperature. These parameters need to be carefully optimized to achieve the desired properties in the composite.

4.3.2 Physical Treatments

Physical treatments are nonchemical methods used to modify the surface properties of natural fibers. These treatments increase the surface roughness of the fibers, improve their wettability, and create functional groups on the fiber surface that can form physical bonds with the PLA matrix. Physical treatments include methods such as plasma treatment and corona treatment.

4.3.2.1 Plasma Treatment

Plasma treatment involves using ionized gas, or plasma, to modify the surface properties of natural fibers. Plasma removes impurities from the fiber surface, creates functional groups that can form physical bonds with the PLA matrix, and increases the surface roughness of the fibers. The main advantage of plasma treatment is that it does not require chemicals, which makes it environmentally friendly. Moreover, it produces composites with good mechanical properties and dimensional stability. However, plasma treatment requires specialized equipment and can be energy intensive. Moreover, the effects of plasma treatment can be temporary, as the functional groups created on the fiber surface can reorient over time, reducing the effectiveness of the treatment.

4.3.2.2 Corona Treatment

Corona treatment is another physical method used to modify the surface properties of natural fibers. This process involves applying a high-voltage electric field to the fiber surface, creating a corona discharge that can modify the fiber surface. Corona treatment can increase the surface energy of the fibers, improve their wettability, and create functional groups on the fiber surface that can form physical bonds with the PLA matrix. This can lead to improvements in the mechanical properties of the composite. However, like plasma treatment, corona treatment requires specialized equipment and can be energy intensive. Moreover, the effects of corona treatment can be temporary, and the treatment may need to be repeated periodically to maintain the improved surface properties of the fibers.

4.3.3 Biological Treatments

Biological treatments involve the use of biological agents, such as enzymes and microorganisms, to modify the surface properties of natural fibers. These treatments can remove impurities from the fiber surface, introduce functional groups that can form chemical bonds with the PLA matrix, and increase the surface roughness of the fibers. Biological treatments are environmentally friendly, as they do not require the use of toxic chemicals. Moreover, they can be performed under mild conditions, which prevents thermal degradation of the natural fibers. However, biological treatments can be time-consuming, and their effectiveness can depend on the type of biological agent used, the treatment time, and the environmental conditions.

4.3.3.1 Enzymatic Treatment

Enzymatic treatment uses of enzymes to modify the surface properties of natural fibers. Enzymes are proteins that can catalyze specific chemical reactions. In the context of natural fiber surface treatment, enzymes can break down certain components of the fiber surface, such as lignin and hemicellulose, thereby exposing the cellulose microfibrils. The main advantage of enzymatic treatment is that it is highly specific, meaning that it can selectively modify the fiber surface without causing significant damage to the fibers. Moreover, enzymatic treatment is environmentally friendly, as it does not require the use of toxic chemicals. However, enzymatic treatment can be time-consuming, and the effectiveness of the treatment can depend on several factors, including the type of enzyme used, the concentration of the enzyme solution, the treatment time, and the temperature and pH of the treatment conditions.

4.4 PROPERTIES OF NATURAL FIBER-REINFORCED PLA BIOCOMPOSITES

The properties of natural fiber-reinforced PLA biocomposites are influenced by several factors, including the type of natural fiber, the PLA matrix, the fiber/matrix interface, and the processing techniques. These properties determine the performance of the composites in various applications.

4.4.1 Mechanical Properties

The mechanical properties of natural fiber-reinforced PLA biocomposites, such as tensile strength, flexural strength, and impact strength, are particularly important as they determine the load-bearing capacity of the composites. Tensile strength is a measure of the resistance of a material to breaking under tension. It is one of the most important mechanical properties of natural fiber-reinforced PLA biocomposites. The tensile strength of these composites is influenced by the fiber/matrix adhesion, the dispersion of the fibers in the PLA matrix, and the orientation and volume fraction of the fibers. Impact strength is a measure of the resistance of a material to fracture under sudden impact. It is a crucial mechanical property of natural fiber-reinforced PLA biocomposites, especially for applications where the composites are subjected to sudden loads or impacts. The impact strength of these composites is influenced by the fiber/matrix adhesion, the dispersion of the fibers in the PLA matrix, the orientation and volume fraction of the fibers, and the toughness of the fibers and the PLA matrix. Table 4.2 shows the different effects of natural fibers on PLA biocomposites.

TABLE 4.2
Mechanical Characteristics of PLA Biocomposites Reinforced with Natural Fibers

Fiber	Manufacturing Methods	Fibers wt%	Tensile Strength (MPa)	Tensile Modulus (GPa)	Flexural Strength (MPa)	Flexural Modulus (GPa)
Kenaf	Injection molding	20			115	7.6
Banana	Extrusion injection molding	20	57	1.61	104	5.6
Banana	Injection molding	20	52	1.43	104	4.8
Sisal	Compression molding	20	80.6	5.3	249.8	9.75
Rice straw	Solvent casting	30	22.27	2.59	26	
Rice husk	Extrusion compression molding	20			90	4.5
Hemp	Compression molding	40 (vol%)	54.6	8.5	112.7	
Jute	Compression molding	30 (vol%)	64.13	3.39	97.74	7.36
Jute	Compression molding	50	32.3	2.11	41.8	
Banana and sisal	Injection molding	30	79	4.1	125	–
Pine wood flour	Twin screw extruder	50 wt%	66.2	5.4	98	
Okra	Twin screw extruder	30 wt%	58.4	4.6		
Coir	Twin screw extruder	10 wt%	57.9	4.0	107.1	
Flax	Compression molding	50 wt%	151	18.5	215	
Ramie	Compression molding	30 wt%	53	4.3	104	
Oil seed filler	Twin screw extruder	8 wt%	62.6	1.43		
PBAT & office wastepaper	Injection molding	20 wt% & 10 wt%	49	2.9	73	

Source: Trivedi et al. (2023).

4.4.2 Physical Properties

The physical properties of natural fiber-reinforced PLA biocomposites, such as density and water absorption, are important as they influence the performance of the composites in various applications.

4.4.2.1 Density

Density is a measure of the mass per unit volume of a material. The density of natural fiber-reinforced PLA biocomposites is influenced by the densities of the PLA matrix and the natural fibers, as well as the volume fraction of the fibers. One of the advantages of using natural fibers as reinforcements in PLA biocomposites is that they have lower densities than many synthetic fibers. This can result in composites with lower densities, which is beneficial for applications where weight reduction is important. However, the density of the composites can also be influenced by the techniques used to fabricate the composites. For instance, techniques that involve pressure, such as compression molding and injection molding, can result in composites with higher densities.

4.4.2.2 Water Absorption

Water absorption is a measure of the amount of water that a material can absorb. The water absorption of natural fiber-reinforced PLA biocomposites is influenced by the hydrophilicity of the natural fibers and the PLA matrix, as well as the fiber/matrix adhesion. Natural fibers are generally more hydrophilic than PLA, which means that they can absorb more water; this increases the water absorption of the composites, which can affect their mechanical properties and dimensional stability. However, the water absorption of composites can be reduced by using surface treatments to increase the hydrophobicity of the fibers and improve the fiber/matrix adhesion. For instance, chemical treatments, such as silane treatment and acetylation, can introduce hydrophobic groups onto the fiber surface, thereby reducing the water absorption of the composites.

4.4.3 Thermal Properties

The thermal properties of natural fiber reinforced PLA biocomposites, such as thermal stability and thermal conductivity, are crucial as they influence the performance of the composites in various applications.

4.4.3.1 Thermal Stability

Thermal stability is a measure of the ability of a material to maintain its properties under high temperatures. It is an important property for natural fiber-reinforced PLA biocomposites that will be exposed to high temperatures. The thermal stability of these composites is influenced by the thermal stability of the natural fibers and the PLA matrix, as well as the fiber/matrix adhesion. Natural fibers have lower thermal stability than PLA, which means they degrade at lower temperatures. The thermal stability of PLA biocomposites and their individual components (PLA and fibers) can be assessed using thermogravimetric analysis and derivative thermogravimetry (Gond & Gupta, 2020; Sun et al., 2017).

PLA biocomposites undergo three stages of heat degradation. The first is moisture evaporation, when the initial weight loss of 5%–10% occurs around 100 °C due to the evaporation of moisture content in the fibers. The second stage involves the breakdown of hemicellulose and cellulose within the fibers; this coincides with the thermal degradation of the PLA matrix. Hemicellulose degrades before cellulose due to its lower heat stability. The final stage occurs at higher temperatures and involves the degradation of lignin and other noncellulosic components. Lignin's high thermal stability requires elevated temperatures for breakdown. Pérez-Fonseca et al. (2016) demonstrated that annealing significantly improves the thermal stability of PLA biocomposites by altering their crystallinity.

4.4.3.2 Thermal Conductivity

Thermal conductivity is a measure of the ability of a material to conduct heat. It is another important property for natural-fiber reinforced PLA biocomposites, especially for applications where the composites are exposed to heat.

The thermal conductivity of these composites is influenced by the thermal conductivity of the natural fibers and the PLA matrix, as well as the fiber/matrix adhesion and the orientation and volume fraction of the fibers. Natural fibers generally have lower thermal conductivity than PLA, which means that they provide thermal insulation. The thermal degradation of PLA biocomposites reinforced with pineapple fibers (30 wt%) began at 325 °C and peaked at 415 °C, resulting in an 80% weight loss and leaving behind 18% residue (Kaewpirom & Worrarat, 2014). PLA biocomposites reinforced with hemp fibers exhibited a slightly lower starting degradation temperature of 303 °C and a maximum degradation temperature of 398 °C (Masirek et al., 2007). During deterioration, there was a 95% weight loss with 1.2% residue remaining. Boubekeur et al. (2020) investigated the impact of incorporating waste wood flour into PLA biocomposites and found that adding waste wood flour did not significantly alter the thermal characteristics of the biocomposites. However, their thermal conductivity can be increased by aligning the fibers in the direction of heat flow and increasing the volume fraction of the fibers.

4.5 APPLICATIONS OF NATURAL FIBER-REINFORCED PLA BIOCOMPOSITES

Fiber-reinforced PLA biocomposites have found a wide range of applications due to their unique properties such as high strength, light weight, biodegradability, and cost-effectiveness. These applications span across various industries, including automotive, construction, consumer goods, and medical and healthcare (Keya et al., 2019).

4.5.1 Automotive Industry

In the automotive industry, natural fiber-reinforced PLA biocomposites are increasingly being used as alternatives to traditional materials such as metals and synthetic polymers including in interior panels, door panels, seat backs, engine components, and trunk liners (Gupta et al., 2016).

These biocomposites offer multiple advantages. First, they are lighter than traditional materials, which reduces vehicle weight and improves fuel efficiency. Second, they are biodegradable, which reduces the environmental impact of the vehicle at the end of its life cycle. Lastly, they are cost-effective, which reduces the manufacturing cost of the vehicle (Singh et al., 2023).

However, despite these advantages, the use of natural fiber-reinforced PLA biocomposites in the automotive industry also presents some challenges. The variable properties of natural fibers can affect the consistency of the composites, which can be a concern for automotive applications where consistent performance is required. Moreover, the hydrophilic nature of natural fibers can lead to moisture absorption, which affects the durability of the composites. Various surface treatments and manufacturing techniques have been developed to address these challenges.

4.5.2 Construction Industry

Natural fiber-reinforced PLA biocomposites are increasingly being used in the construction industry due to their high strength, light weight, biodegradability, and cost-effectiveness. These biocomposites are used in various parts of a building, including interior panels, door panels, insulation materials, and structural components (Lila et al., 2013).

The use of these biocomposites in the construction industry offers several advantages. First, they are lighter than traditional materials, which improves a building's energy efficiency. Secondly, they are biodegradable, which can help to reduce the environmental impact of the building at the end of its life cycle. Lastly, they are cost-effective, which reduces the construction cost of the building.

4.5.3 Consumer Goods

Natural fiber-reinforced PLA biocomposites are also used in the production of various consumer goods, including furniture, packaging materials, and sporting goods. These biocomposites offer several advantages over traditional materials, including lower weight, higher strength, and biodegradability. Moreover, natural fiber-reinforced PLA biocomposites can be processed using conventional manufacturing techniques, which facilitates their adoption in the consumer goods industry.

4.5.4 Medical and Health Care Applications

In the medical and health care sector, natural fiber-reinforced PLA biocomposites are used in surgical sutures, implants, and drug delivery systems. These biocomposites offer several advantages over traditional materials, including biocompatibility, biodegradability, and the ability to be processed into complex shapes (Gupta et al., 2016). Moreover, natural fibers improve the mechanical properties of the biocomposites, making them suitable for load-bearing applications. However, the use of natural fibers in medical and health care applications also presents some challenges, including potential immune responses and the need for sterilization.

4.6 CHALLENGES AND FUTURE PROSPECTS

The development of natural fiber-reinforced PLA biocomposites faces several challenges that need to be addressed to realize their full potential. However, these challenges also open avenues for future research and innovation. One of the primary challenges is the cost and availability of high-quality natural fibers; while natural fibers are generally less expensive than synthetic ones, the cost can vary significantly depending on the type of fiber and its source. Additionally, ensuring a consistent supply of natural fibers can be challenging due to factors such as agricultural production cycles and regional availability. Another challenge is the compatibility between natural fibers and PLA matrix. Due to the hydrophilic nature of natural fibers and the hydrophobic characteristics of PLA, there can be issues with interfacial bonding that affect the mechanical properties of the composites.

Future research is looking into advanced surface treatments for natural fibers to improve their compatibility with PLA matrices. These treatments aim to modify the surface properties of the fibers to enhance bonding and, consequently, the mechanical performance of the composites. Innovative processing techniques are also being explored to optimize the manufacturing of natural fiber-reinforced PLA biocomposites. These techniques aim to improve efficiency, reduce costs, and enhance the properties of the final product. Research is ongoing to enhance the performance properties of these biocomposites, such as strength, durability, and thermal stability. This involves exploring different fiber treatments, composite formulations, and reinforcement strategies.

4.7 CONCLUSIONS

The growing demand for renewable and recyclable materials, coupled with concerns about energy consumption, environmental regulations, and plastic pollution, has driven the development of biodegradable composites. PLA and natural fibers offer promising alternatives to traditional synthetic fibers and polymers, which pose significant challenges in terms of recycling and sustainability. PLA composites reinforced with natural fibers exhibit great potential for various applications. These biocomposite materials offer a range of attractive properties, positioning them as viable substitutes for current synthetic materials derived from fossil fuels. Research has explored the synergistic properties of PLA and natural fibers, highlighting their compatibility in composite materials. However, some studies have shown a decrease in certain properties, indicating a need for further optimization. The growing availability of biopolymers, the unique characteristics of natural fibers, and the environmental benefits of biodegradable composites compared with synthetic materials all justify continued research in this area.

REFERENCES

Boubekeur, B., Belhaneche-Bensemra, N., & Massardier, V. (2020). Low-density polyethylene/poly (lactic acid) blends reinforced by waste wood flour. *Journal of Vinyl and Additive Technology*, *26*(4), 443–451.

Dong, Y., Ghataura, A., Takagi, H., Haroosh, H. J., Nakagaito, A. N., & Lau, K.-T. (2014). Polylactic acid (PLA) biocomposites reinforced with coir fibres: Evaluation of mechanical performance and multifunctional properties. *Composites Part A: Applied Science and Manufacturing*, *63*, 76–84.

Faheed, N. K. (2024). Advantages of natural fiber composites for biomedical applications: A review of recent advances. *Emergent Materials*, *7*(1), 63–75. https://doi.org/10.1007/s42247-023-00620-x

Faludi, G., Dora, G., Renner, K., Móczó, J., & Pukánszky, B. (2013). Biocomposite from polylactic acid and lignocellulosic fibers: Structure – property correlations. *Carbohydrate Polymers*, *92*(2), 1767–1775.

Gholampour, A., & Ozbakkaloglu, T. (2020). A review of natural fiber composites: Properties, modification and processing techniques, characterization, applications. *Journal of Materials Science*, *55*(3), 829–892. https://doi.org/10.1007/s10853-019-03990-y

Gond, R. K., & Gupta, M. K. (2020). A novel approach for isolation of nanofibers from sugarcane bagasse and its characterization for packaging applications. *Polymer Composites*, *41*(12), 5216–5226.

Gupta, G., Kumar, A., Tyagi, R., & Kumar, S. (2016). Application and future of composite materials: A review. *International Journal of Innovative Research in Science, Engineering and Technology*, *5*(5), 6907–6911.

Jaafar, J., Siregar, J. P., Mohd Salleh, S., Mohd Hamdan, M. H., Cionita, T., & Rihayat, T. (2019). Important considerations in manufacturing of natural fiber composites: A review. *International Journal of Precision Engineering and Manufacturing – Green Technology*, *6*(3), 647–664. https://doi.org/10.1007/s40684-019-00097-2

Kaewpirom, S., & Worrarat, C. (2014). Preparation and properties of pineapple leaf fiber reinforced poly (lactic acid) green composites. *Fibers and Polymers*, *15*, 1469–1477.

Keya, K. N., Kona, N. A., Koly, F. A., Maraz, K. M., Islam, M. N., & Khan, R. A. (2019). Natural fiber reinforced polymer composites: History, types, advantages, and applications. *Materials Engineering Research*, *1*(2), 69–87. https://doi.org/10.25082/mer.2019.02.006

Komal, U. K., Lila, M. K., & Singh, I. (2020). PLA/banana fiber based sustainable biocomposites: A manufacturing perspective. *Composites Part B: Engineering*, *180*, 107535.

Lila, M., Kumar, F., & Sharma, S. (2013). Composites from waste for civil engineering applications. *I-Manager's Journal on Material Science*, *1*(3), 1.

Masirek, R., Kulinski, Z., Chionna, D., Piorkowska, E., & Pracella, M. (2007). Composites of poly (L-lactide) with hemp fibers: Morphology and thermal and mechanical properties. *Journal of Applied Polymer Science*, *105*(1), 255–268.

Milenkovic, S., Slavkovic, V., Fragassa, C., Grujovic, N., Palic, N., & Zivic, F. (2021). Effect of the raster orientation on strength of the continuous fiber reinforced PVDF/PLA composites, fabricated by hand-layup and fused deposition modeling. *Composite Structures*, *270*, 114063.

Mukherjee, T., & Kao, N. (2011). PLA based biopolymer reinforced with natural fibre: A review. *Journal of Polymers and the Environment*, *19*, 714–725.

Orue, A., Jauregi, A., Unsuain, U., Labidi, J., Eceiza, A., & Arbelaiz, A. (2016). The effect of alkaline and silane treatments on mechanical properties and breakage of sisal fibers and poly (lactic acid)/sisal fiber composites. *Composites Part A: Applied Science and Manufacturing*, *84*, 186–195.

Paunonen, S., Berthold, F., & Immonen, K. (2020). Poly(lactic acid)/pulp fiber composites: The effect of fiber surface modification and hydrothermal aging on viscoelastic and strength properties. *Journal of Applied Polymer Science*, *137*(42). https://doi.org/10.1002/app.49617

Pérez-Fonseca, A. A., Robledo-Ortíz, J. R., González-Núñez, R., & Rodrigue, D. (2016). Effect of thermal annealing on the mechanical and thermal properties of polylactic acid – cellulosic fiber biocomposites. *Journal of Applied Polymer Science*, *133*(31).

Prakash, S. O., Sahu, P., Madhan, M., & Johnson Santhosh, A. (2022). A review on natural fibre-reinforced biopolymer composites: Properties and applications. *International Journal of Polymer Science, 2022*. Hindawi Limited. https://doi.org/10.1155/2022/7820731

Qin, L., Qiu, J., Liu, M., Ding, S., Shao, L., Lü, S., Zhang, G., Zhao, Y., & Fu, X. (2011). Mechanical and thermal properties of poly (lactic acid) composites with rice straw fiber modified by poly (butyl acrylate). *Chemical Engineering Journal, 166*(2), 772–778.

Shalwan, A., & Yousif, B. F. (2013). In state of art: Mechanical and tribological behaviour of polymeric composites based on natural fibres. *Materials & Design, 48*, 14–24.

Sharma, M., Verma, S., Chauhan, G., Arya, M., & Kumari, A. (2024). Mycelium-based biocomposites: Synthesis and applications. *Environmental Sustainability*. https://doi.org/10.1007/s42398-024-00305-z

Siakeng, R., Jawaid, M., Ariffin, H., Sapuan, S. M., Asim, M., & Saba, N. (2019). Natural fiber reinforced polylactic acid composites: A review. *Polymer Composites, 40*(2), 446–463. John Wiley and Sons Inc. https://doi.org/10.1002/pc.24747

Singh, B., Ahmad, K. A., Manikandan, M., Pai, R., Ng, E. Y. K., & Yidris, N. (2023). Natural fiber reinforced composites and their role in aerospace engineering. In T. Khan & M. Jawaid (Eds.), *Green hybrid composite in engineering and non-engineering applications* (pp. 61–76). Springer Nature. https://doi.org/10.1007/978-981-99-1583-5_5

Sun, Z., Zhang, L., Liang, D., Xiao, W., & Lin, J. (2017). Mechanical and thermal properties of PLA biocomposites reinforced by coir fibers. *International Journal of Polymer Science, 2017*(1), 2178329.

Trivedi, A. K., Gupta, M. K., & Singh, H. (2023). PLA based biocomposites for sustainable products: A review. *Advanced Industrial and Engineering Polymer Research, 6*(4), 382–395. https://doi.org/10.1016/j.aiepr.2023.02.002

Zhao, Y., Qiu, J., Feng, H., Zhang, M., Lei, L., & Wu, X. (2011). Improvement of tensile and thermal properties of poly (lactic acid) composites with admicellar-treated rice straw fiber. *Chemical Engineering Journal, 173*(2), 659–666.

5 Natural Fiber-Reinforced PLA Hybrid Biocomposites

5.1 INTRODUCTION

In recent years, the push toward sustainable materials has gained momentum as environmental concerns continue to rise. Biocomposites, which consist of natural fibers embedded within a biopolymer matrix, represent a significant step forward in the pursuit of sustainability. These materials combine the environmental benefits of natural fibers such as biodegradability, renewability, and minimal environmental impact with the functional advantages of synthetic polymers (Nabilah Haris et al. 2022). As a result, biocomposites have found applications in various industries, including automotive, aviation, and construction, due to their enhanced mechanical properties, reduced weight, and improved environmental sustainability.

Natural fibers, sourced from plants like flax, hemp, jute, and bamboo, offer several advantages over traditional synthetic fibers. They are renewable, have low energy consumption during production, and are often less harmful to the environment. When these fibers are combined with biopolymers, the resulting biocomposites not only perform well but also contribute to reducing the ecological footprint of the end product (Amiandamhen, Meincken, and Tyhoda 2020). This makes natural fiber-based biocomposites a promising alternative to conventional composite materials, particularly in applications where environmental impact is a significant consideration.

The importance of hybrid composites lies in their ability to combine two or more types of fibers or matrices to create materials with enhanced properties compared to their individual components (Alias et al. 2021). The concept of hybridization in biocomposites involves leveraging the strengths of each component while mitigating their weaknesses (Suriani, Ilyas, et al. 2021). In the context of natural fiber-reinforced composites, hybridization can lead to materials with improved mechanical performance, better thermal stability, and enhanced impact resistance (Suriani, Radzi, et al. 2021). This approach is especially valuable in applications where a balance of strength, weight, and flexibility is required. By integrating different natural fibers or combining natural fibers with synthetic fibers or biopolymers, hybrid composites can be tailored to meet specific performance requirements, making them a versatile and valuable option for various industrial applications.

PLA plays a crucial role in hybrid biocomposites as a widely used biopolymer derived from renewable resources such as corn starch or sugarcane. PLA is well-known for its biodegradability, compostability, and relatively low environmental

DOI: 10.1201/9781032678894-5

impact compared with traditional petroleum-based polymers (Ilyas et al. 2022). Its role in hybrid biocomposites is particularly significant due to its favorable processing characteristics and compatibility with natural fibers. PLA offers several advantages when used as a matrix material in biocomposites. It has a relatively low melting temperature, which facilitates processing using conventional polymer techniques. Additionally, PLA exhibits a good balance of strength and rigidity, which when combined with natural fibers results in composites with enhanced mechanical properties. However, PLA also has some limitations, such as lower impact resistance and thermal stability than other polymers. By creating hybrid composites with PLA, these limitations can be addressed while retaining the environmental benefits of using natural fibers, making PLA a critical component in the development of sustainable hybrid biocomposites (Raquez et al. 2013).

This chapter provides a comprehensive review of natural fiber-reinforced PLA hybrid biocomposites, focusing on their development, properties, and applications. The chapter is structured to cover several key aspects. First, we offer an overview of natural fibers and PLA, detailing the types of natural fibers commonly used in PLA-based composites, their properties, and their interactions with PLA. We also discuss the characteristics of PLA as a matrix material and its suitability for hybrid composites.

Second, we explore the various techniques used to produce natural fiber-reinforced PLA hybrid biocomposites, including extrusion, injection molding, and compression molding, highlighting the advantages and challenges associated with each method. Third, we analyze the mechanical and thermal properties of hybrid composites, covering aspects such as strength, stiffness, impact resistance, and thermal stability, along with comparisons with pure PLA and other types of biocomposites. We then discuss the practical applications of these hybrid composites, with examples from various industries such as automotive, packaging, and construction, as well as potential future applications and innovations. Finally, we address the current challenges in developing and utilizing PLA-based hybrid biocomposites, including issues related to fiber-matrix adhesion, durability, and cost, and suggests possible future research directions to overcome these challenges and enhance the performance and sustainability of these materials.

5.2 NATURAL FIBERS IN HYBRID COMPOSITES

Natural fibers have gained considerable attention in recent years as environmentally friendly alternatives to synthetic fibers in composite materials. These fibers, derived from renewable plant sources, are integral to the development of hybrid composites due to their sustainability, low cost, and favorable mechanical properties. The use of natural fibers in hybrid composites not only reduces the reliance on nonrenewable resources but also enhances the overall environmental profile of the resulting materials. Figure 5.1 shows natural fiber-reinforced hybrid composites (Suriani, Ilyas, et al. 2021). In this section, we provide an in-depth overview of natural fibers focusing on the most commonly used, their properties, and their advantages in hybrid composites.

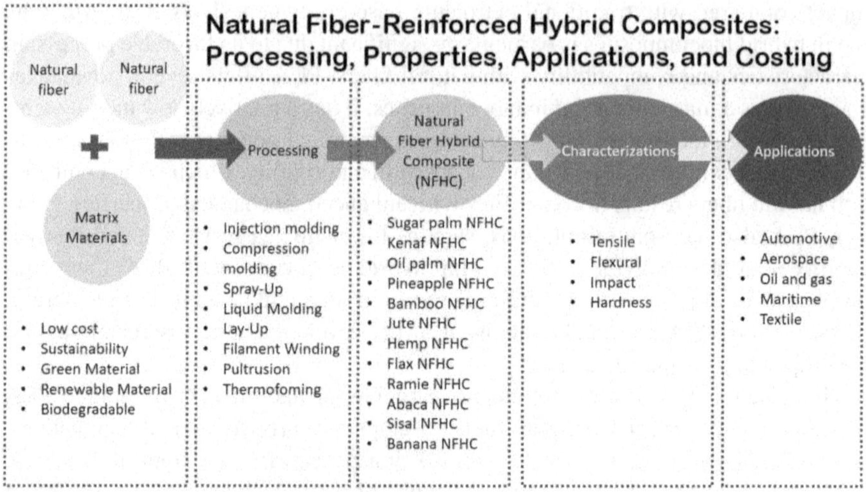

FIGURE 5.1 Natural fiber-reinforced hybrid composites.

Source: Reprinted with permission from Suriani, Ilyas, et al. (2021).

Natural fibers can be broadly categorized into three main types: bast fibers, leaf fibers, and seed fibers. Bast fibers, such as jute, flax, and hemp, are extracted from the outer layer of the plant stem and are known for their high strength and stiffness. Leaf fibers, like sisal, are obtained from the leaves of the plant and are recognized for their durability and resilience. Seed fibers, including cotton, are derived from the seeds of the plant and are valued for their softness and flexibility (Veerasimman et al. 2022).

Among these, bast fibers are the most commonly used in hybrid composites due to their superior mechanical properties and ease of processing. Jute, flax, hemp, and sisal are some of the most widely used natural fibers in hybrid composites. Jute is a bast fiber that is abundant in tropical regions and is known for its high tensile strength and modulus (Mahbubul Islam 2019). It is widely used in the production of bags, ropes, and carpets, and its inclusion in hybrid composites can significantly improve their mechanical properties. Flax, another bast fiber, is prized for its high specific stiffness and low density, making it an ideal reinforcement in lightweight composites. Hemp, which is also a bast fiber, offers excellent mechanical properties, including high tensile strength, stiffness, and durability. Its use in hybrid composites can enhance the material's performance in demanding applications. Sisal, a leaf fiber, is known for its toughness and resistance to abrasion. It is commonly used in the automotive industry for producing interior components, and its inclusion in hybrid composites can improve their impact resistance and durability.

The properties of natural fibers make them particularly suitable for use in hybrid composites. One of the most significant advantages of natural fibers is their high specific strength, which refers to the strength-to-weight ratio of the material (Sanjay et al. 2018). This property is crucial in applications where weight reduction is essential, such as in the automotive and aviation industries. Natural fibers also have low density, which further contributes to the lightweight nature of the composites. Additionally, natural fibers exhibit good thermal and acoustic insulation properties, making them suitable for use in construction and building materials. Another notable property of natural fibers is their biodegradability, which ensures that the composites can be disposed of in an environmentally friendly manner at the end of their life cycle.

The inclusion of natural fibers in hybrid composites offers several advantages. First, natural fibers are renewable resources that are widely available and can be cultivated with low environmental impact. This makes them attractive for sustainable composite materials. Second, natural fibers are biodegradable, meaning that they can break down naturally over time without causing harm to the environment. This is a significant advantage over synthetic fibers, which can persist in the environment for decades. Third, natural fibers have low abrasiveness, which reduces wear and tear on processing equipment, leading to lower maintenance costs and longer tool life. Lastly, natural fibers are often less expensive than synthetic fibers, making them a cost-effective option for composite materials (Dhand et al. 2015).

Despite these advantages, there are also some challenges associated with the use of natural fibers in hybrid composites. One of the main challenges is the variability in the properties of natural fibers, which can affect the consistency and quality of the composites. Factors such as the growing conditions, harvesting methods, and processing techniques can all influence the properties of natural fibers. Additionally, natural fibers are hydrophilic, meaning that they absorb moisture, which can lead to swelling, degradation, and reduced mechanical properties in the composites. To address these challenges, various treatments and surface modifications can be applied to natural fibers to improve their compatibility with the matrix material and enhance their performance in hybrid composites (Omran et al. 2021; Kamaruddin et al. 2022).

5.3 PLA AS A MATRIX MATERIAL IN HYBRIDS

PLA is one of the most widely used biopolymers in sustainable materials, particularly in hybrid composites. Derived from renewable resources such as corn starch or sugarcane, PLA has gained popularity due to its biodegradability, compostability, and lower environmental impact than that of traditional petroleum-based polymers. As a matrix material in hybrid composites, PLA offers several advantages including favorable mechanical properties, ease of processing, and compatibility with natural fibers (Ilyas et al. 2021).

The characteristics of PLA make it a suitable matrix material for hybrid composites. PLA is a thermoplastic polymer, which means it can be melted and reprocessed multiple times without significant degradation of its properties. This characteristic makes PLA highly versatile and compatible with various manufacturing techniques, such as injection molding, compression molding, and extrusion. Additionally, PLA has a relatively low melting temperature, typically ranging between 150 °C and 190 °C, which allows for energy-efficient processing. PLA also exhibits good transparency, which is an important attribute in applications where aesthetic considerations are important. Furthermore, PLA has a good balance of mechanical properties, including tensile strength, modulus, and impact resistance, making it suitable for use in a wide range of applications (Siakeng et al. 2019).

The benefits of using PLA as a matrix material in hybrid composites are manifold. One of the most significant benefits is PLA's biodegradability. Unlike conventional petroleum-based polymers, PLA can break down into natural components under industrial composting conditions, reducing the environmental impact of the material at the end of its life cycle. This biodegradability is particularly advantageous in applications where the composite materials are intended for single-use or short-term use, such as in packaging and disposable products. Another benefit of PLA is its renewability. Since PLA is derived from renewable resources, its production contributes to reducing the dependence on finite fossil fuels and decreases the carbon footprint of the materials. PLA also offers good mechanical properties, which can be further enhanced through hybridization with natural fibers. The inclusion of natural fibers in PLA-based composites can improve the stiffness, strength, and impact resistance of the material, making it suitable for demanding applications in industries such as automotive, construction, and consumer goods (Trivedi, Gupta, and Singh 2023).

Despite its many benefits, PLA also presents certain limitations and challenges that need to be addressed to optimize its use in hybrid composites. One of the main limitations of PLA is its low thermal stability. PLA softens and deforms at temperatures above its glass transition temperature, around 60 °C. This limits the use of PLA-based composites in applications that require high thermal resistance, such as in automotive engine components or electrical housings. Additionally, PLA is prone to hydrolytic degradation, meaning that it can break down in the presence of moisture; this decreases the mechanical properties of the composite over time, particularly in humid environments. Another challenge associated with PLA is its brittleness: It has low elongation at break, which means that it is prone to cracking or fracturing under stress. To overcome this limitation, plasticizers or toughening agents can be added to the PLA matrix to improve its flexibility and impact resistance (Surya et al. 2020).

PLA is a promising matrix material for hybrid composites due to its biodegradability, renewability, and favorable mechanical properties. Its use in hybrid composites enhances the sustainability and performance of the materials, making them suitable for a wide range of applications. However, the limitations and challenges associated with PLA, such as its low thermal stability, brittleness, and cost, must be carefully considered and addressed to fully realize its potential in hybrid composites. Through ongoing research and development, it is possible to optimize the use of PLA in hybrid composites, improving their performance and expanding their application potential.

5.4 FABRICATING NATURAL FIBER-REINFORCED PLA HYBRID BIOCOMPOSITES

5.4.1 Processing Hybrid Biocomposites

Fabricating natural fiber-reinforced PLA hybrid biocomposites involves a series of processes that are critical to achieving the desired material properties and performance. These processes include preparing and treating the fibers, selecting appropriate composite fabrication methods, and optimizing the fiber–matrix interface. Each of these steps plays a crucial role in determining the quality, consistency, and effectiveness of the final biocomposite material.

5.4.2 Preparing and Treating Natural Fibers

The preparation and treatment of natural fibers are essential steps in the fabrication process, as they significantly influence the mechanical properties, durability, and overall performance of the resulting hybrid biocomposites. Natural fibers, being hydrophilic, tend to absorb moisture from the environment, which leads to swelling, degradation, and poor adhesion with the hydrophobic PLA matrix. To mitigate these issues, several treatment methods are employed to modify the surface characteristics of the fibers, enhance their compatibility with the matrix, and improve the fiber–matrix bonding (Alsubari et al. 2021).

One common method of fiber treatment is chemical modification, which involves the use of chemical agents such as alkalis, silanes, and maleic anhydride to alter the surface properties of the fibers (Bangar et al. 2022). Alkali treatment, also known as mercerization, is widely used to remove natural impurities such as lignin, hemicellulose, and waxes from the fiber surface, thereby increasing the surface roughness and enhancing the mechanical interlocking with the PLA matrix (Ilyas et al. 2017). This treatment also reduces the hydrophilicity of the fibers, making them more compatible with the hydrophobic matrix. Silane treatment is another effective method that involves the use of silane coupling agents to form a covalent bond between the fiber surface and the PLA matrix, thereby improving the adhesion and mechanical properties of the composite. Additionally, maleic anhydride grafting can be used to modify the PLA matrix, enhancing its compatibility with the natural fibers and improving the overall performance of the hybrid composite.

In addition to chemical treatments, physical methods such as plasma treatment, corona discharge, and thermal treatment are also used to modify the surface properties of natural fibers (Jang et al. 2012). Plasma treatment involves the use of ionized gas to etch the fiber surface, increasing its roughness and enhancing the adhesion with the PLA matrix. Corona discharge treatment is similar to plasma treatment but involves the use of high-voltage electrical discharge to modify the fiber surface. Thermal treatment, on the other hand, involves the controlled heating of fibers to remove moisture and reduce their hydrophilicity. Each of these methods has its advantages and limitations, and the choice of treatment depends on the specific requirements of the composite application.

5.4.3 Composites Fabrication Methods

Several techniques can be used to fabricate hybrid biocomposites, each with its advantages and challenges. Table 5.1 shows techniques for PLA hybrid composites. Among the most commonly used methods are injection molding, compression molding, and extrusion. Injection molding is widely used in producing hybrid biocomposites due to its ability to produce complex shapes with high precision and repeatability. In this process, the PLA matrix and treated natural fibers are mixed and melted together,

TABLE 5.1
Fabricating PLA Hybrid Composites

Fibers	Processing Technique	References
Banana/sisal fiber	Injection molding	(Asaithambi, Ganesan, and Ananda Kumar 2014)
Flax/jute	Compression molding	(Manral, Ahmad, and Chaudhary 2019)
Polycaprolactone/oil palm mesocarp	Melt blending	(Eng et al. 2014)
Montmorillonite nanoclay/short kenaf	Double extrusion	(Kaiser et al. 2013)
Corn stover/wheat straw/soy stalks	Extrusion + injection molding	(Sanchez-Garcia and Lagaron 2010; Nyambo, Mohanty, and Misra 2010)
Hemp/sisal	Injection molding	(Pappu, Pickering, and Thakur 2019)
Coir/pineapple leaf	Melt mixing	(Siakeng et al. 2019b)
Banana/kenaf	Molding	(Thiruchitrambalam et al. 2009)
Cotton gin waste/flax	Extrusion + melt blending	(Bajracharya, Bajwa, and Bajwa 2017)
Bamboo/microfibrillated cellulose	Milling	(Okubo, Fujii, and Thostenson 2009)
Hemp/yarn	Compression molding/prepreg	(Baghaei, Skrifvars, and Berglin 2013)
MMT clay/aloe vera	Extrusion	(Ramesh, Prasad, and Narayana 2020)
Softwood flour/cellulose	Injection molding	(Bledzki, Franciszczak, and Meljon 2015)

and the molten mixture is then injected into a mold cavity under high pressure. The mold is then cooled, and the solidified composite part is ejected.

Injection molding is particularly suitable for mass production and can be used to produce parts with intricate geometries and fine details. However, achieving uniform dispersion of fibers within the matrix and preventing fiber breakage during processing are some of the challenges associated with this method. Compression molding is another common technique used in the fabrication of hybrid biocomposites. In this process, the PLA matrix and natural fibers are placed into an open mold cavity, and pressure and heat are applied to consolidate the materials and form the composite part. Compression molding is advantageous for producing large and thick parts with good surface finish and mechanical properties. It is also suitable for processing composites with high fiber content, which can enhance the strength and stiffness of the material.

However, compression molding requires careful control of processing parameters such as temperature, pressure, and time to avoid defects such as voids, incomplete filling, or fiber misalignment. Extrusion is a continuous processing method that is used to produce long and uniform composite profiles, such as sheets, rods, and pipes. In this process, the PLA matrix and natural fibers are mixed and melted in an extruder, and the molten mixture is then forced through a die to form the desired shape. The extruded composite is then cooled and cut to the required length. Extrusion is advantageous for producing composite materials with consistent cross-sectional dimensions and is suitable for high-volume production. However, similar to injection molding, achieving uniform fiber dispersion and maintaining fiber integrity during processing are challenges that need to be addressed. In addition to these primary fabrication methods, other techniques such as resin transfer molding, vacuum-assisted resin infusion, and additive manufacturing (3D printing) are also being explored for producing natural fiber-reinforced PLA hybrid biocomposites. Each of these methods offers unique advantages and can be tailored to specific applications and material requirements.

The optimization of the fiber–matrix interface is a critical aspect of hybrid biocomposite fabrication, as it directly impacts the mechanical performance, durability, and overall quality of the material. The interface between the natural fibers and the PLA matrix plays a crucial role in stress transfer and load distribution within the composite. A strong and well-bonded interface ensures efficient load transfer from the matrix to the fibers, resulting in improved mechanical properties such as tensile strength, flexural strength, and impact resistance. To optimize the fiber-matrix interface, several strategies can be employed. One approach is to use coupling agents or compatibilizers that enhance the chemical bonding between the fibers and the matrix (Pasquini et al. 2008). Silane coupling agents (Atiqah et al. 2018) and maleic anhydride (Shah et al. 2016) are commonly used to improve fiber–matrix adhesion. Another approach is to modify the surface of the PLA matrix to increase its affinity for natural fibers. This can be achieved through chemical grafting, plasma treatment, or the addition of functional fillers that promote interfacial bonding.

Moreover, the processing parameters, such as temperature, pressure, and mixing time, play a significant role in determining the quality of the fiber–matrix interface. For example, optimizing the processing temperature can help prevent the degradation of natural fibers and PLA during fabrication, while controlling the pressure and mixing time can ensure uniform fiber dispersion and prevent the formation of voids or defects in the composite (Kaiser et al. 2013). The fabrication of natural fiber-reinforced PLA hybrid biocomposites involves a series of carefully controlled processes, including the preparation and treatment of natural fibers, the selection of appropriate fabrication methods, and the optimization of the fiber-matrix interface. Each of these steps is crucial in achieving the desired material properties and performance, making it essential to consider the specific requirements of the composite application when selecting and optimizing the fabrication process. Through ongoing research and development, advancements in processing techniques and interface optimization will continue to enhance the performance and sustainability of natural fiber-reinforced PLA hybrid biocomposites, expanding their potential applications in various industries.

5.5 MECHANICAL PROPERTIES OF PLA HYBRID BIOCOMPOSITES

The mechanical properties of natural fiber-reinforced PLA hybrid biocomposites are of importance in determining their suitability for various applications. These properties, including tensile strength, flexural strength, impact resistance, and overall durability, are influenced by factors such as the type and content of natural fibers, the PLA matrix, the fiber–matrix interface, and the fabrication process. Understanding and optimizing the mechanical properties of these hybrid biocomposites are essential for developing materials that meet the performance requirements of specific applications. One of the key advantages of hybridization in biocomposites is the synergistic effect that can be achieved by combining different types of fibers or fibers with the PLA matrix. This synergy results in improved mechanical properties that are greater than the sum of the individual components. For instance, the combination of high-strength natural fibers with the PLA matrix can lead to significant enhancements in tensile strength, stiffness, and impact resistance, making the hybrid composite suitable for structural and load-bearing applications.

Tensile properties are among the most critical mechanical characteristics of hybrid biocomposites, as they determine the material's ability to withstand tensile loads without failure. The tensile strength and modulus of a hybrid composite are influenced by the type and orientation of the natural fibers, the fiber–matrix interface, and the overall fiber content. Natural fibers with high tensile strength, such as flax and hemp, can significantly enhance the tensile properties of the PLA matrix when properly dispersed and aligned within the composite. The fiber-matrix interface also plays a crucial role in tensile performance, as a strong and well-bonded interface ensures efficient load transfer from the matrix to the fibers. Additionally, the fabrication process, including the processing temperature, pressure, and fiber co ntent, must be carefully controlled to optimize the tensile properties of the hybrid composite. Unlike its thermal properties, PLA's mechanical properties and crystallization behavior are highly dependent on Mw and stereochemical make-up backbone (Garlotta 1978). For instance, when Mw increased from 50 to 100 kDa (Södergård, Stolt, and Inkinen 2022), the tensile modulus of PLLA increased by a factor of two, and tensile strengths of 15.5, 80, and 150 MPa were obtained for Mw = 50, 150, and 200 k Da, respectively (Velde and Kiekens 2002).

Tensile properties: tensile strength (MPa), tensile modulus (GPa), ultimate strain (percent), and polymer density (g/cm^3) are the mechanical properties of PLA that have been studied the most in relation to biopolymers. Tensile properties are clearly best for the densest recorded polymers, especially PGA. PCL, on the other hand, appears to be the most fragile polymer, with a heavy strain at failure. Data on flexural properties were insufficient for comparisons. However, since flexural and tensile properties are largely associated, the tendencies observed here are likely to be the same as those observed when comparing flexural properties (Velde and Kiekens 2002).

Figure 5.2 shows the graphs of tensile tests of PLA and composites that were conducted by Eselini et al. (2019). The stress–strain analysis showed that including flax (FF) in hybrid form with basalt fibers (BF) reduced tensile strength, while adding BF alone enhanced it. The hybrid composites, particularly those with higher BF content, show improved tensile strength and elongation, with the PLA/30 BF composite achieving the highest tensile strength, about 30% higher than unfilled PLA. Elastic modulus results follow a similar trend, with BF increasing modulus values and FF decreasing them, suggesting that BF contributes positively to the mechanical properties of PLA composites.

Flexural properties, including flexural strength and flexural modulus, are also important indicators of the mechanical performance of hybrid biocomposites. These properties reflect the material's ability to resist bending or flexing under load, which

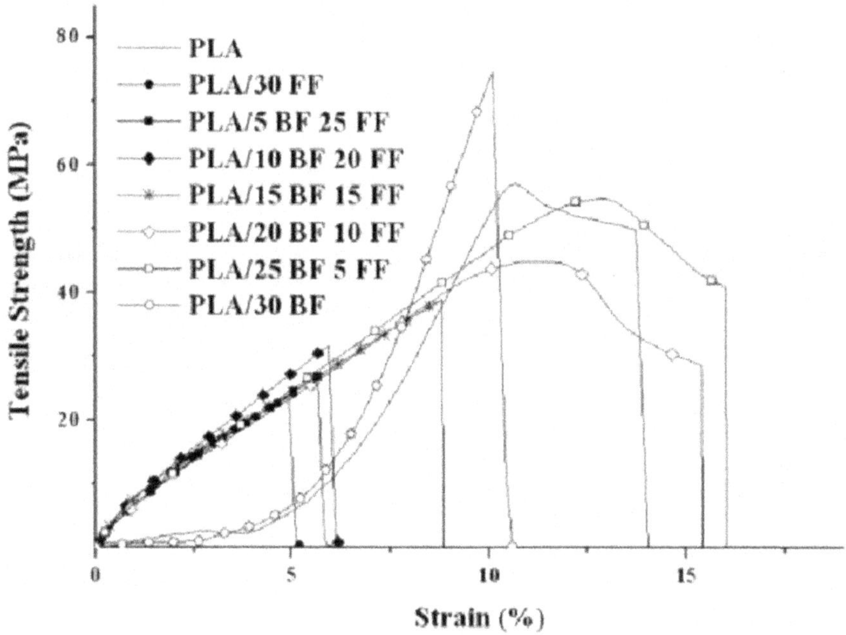

FIGURE 5.2 Tensile test graphs of PLA and composites. BF: basalt fiber; FF: flax fiber.

Source: Eselini et al. (2019).

is essential for applications such as structural panels, beams, and automotive components. Similar to tensile properties, the flexural performance of hybrid biocomposites is influenced by the type and content of natural fibers, the fiber–matrix interface, and the fabrication process. Natural fibers with high flexural strength, such as jute and sisal, can significantly improve the flexural properties of the PLA matrix, particularly when the fibers are well-dispersed and aligned. The fiber–matrix interface must be optimized to ensure efficient load transfer and prevent premature failure under flexural loads.

Impact resistance is another critical mechanical property of hybrid biocomposites, particularly for applications that involve dynamic or impact loading. The ability of a composite material to absorb and dissipate energy during impact is crucial for preventing catastrophic failure and ensuring the safety and durability of the material. The impact resistance of hybrid biocomposites is influenced by the toughness of the natural fibers, the ductility of the PLA matrix, and the quality of the fiber–matrix interface. Hybrid composites that combine tough natural fibers with a ductile PLA matrix tend to exhibit enhanced impact resistance, making them suitable for applications such as automotive interior components, protective gear, and packaging materials. The fiber–matrix interface must also be carefully engineered to optimize energy absorption and prevent fiber pull-out or matrix cracking during impact. Techniques such as surface treatment of fibers, the use of compatibilizers, and the careful selection of processing parameters can all contribute to enhancing the impact resistance of PLA hybrid biocomposites. Table 5.2 shows mechanical properties of various PLA hybrid composites.

TABLE 5.2
Mechanical Properties of Natural Fiber-Reinforced PLA Hybrid Composites

	Property					
Fiber	Tensile Strength (MPa)	Tensile Modulus (GPa)	Flexural Strength (MPa)	Flexural Modulus (GPa)	Impact	Ref.
Banana/sisal	79.00	4.10	125.00	5.60	47.80 kJ/m^2	(Asaithambi, Ganesan, and Ananda Kumar 2014)
Flax/jute	49.35	2.80	80.50	2.25	61.46 J/m	(Manral, Ahmad, and Chaudhary 2019)
Polycaprolactone/oil palm mesocarp	33.48	0.88	21.45	2.43	95.44 J/m	(Eng et al. 2014)
Montmorillonite nanoclay/short kenaf	37.00	2.80	50.00	7.50	82.00 kJ/m^2	(Kaiser et al. 2013)
Corn stover/ wheat straw/ soy stalks	58.00	5.55	80.00	6.90	23 J/m	(Sanchez-Garcia and Lagaron 2010; Nyambo, Mohanty, and Misra 2010)
Hemp/sisal	46.25	6.10	94.83	6.04	10.29 kJ/m^2	(Pappu, Pickering, and Thakur 2019)
Coir/pineapple leaf	18.00	5.00	33.00	5.00	4.3 kJ/m^2	(Siakeng et al. 2019)
Banana/kenaf	50.00	–	61.00	–	16.00 kJ/m^2	(Thiruchitrambalam et al. 2009)
Cotton gin waste/flax	–	–	13.99	3.97	–	(Bajracharya, Bajwa, and Bajwa 2017)
Bamboo/ microfibrillated cellulose	–	4.81	53.80	–	–	(Okubo, Fujii, and Thostenson 2009)
Hemp/yarn	62.00	6.50	122.00	9.00	25.00 kJ/m^2	(Baghaei, Skrifvars, and Berglin 2013)
MMT clay/aloe vera	56.00	3.20	100.00	6.70	55.00 kJ/m^2	(Ramesh, Prasad, and Narayana 2020)
Softwood flour/ cellulose	70.00	56.00	–	–	–	(Bledzki, Franciszczak, and Meljon 2015)

Table 5.2 shows the highest tensile strength for PLA/banana/sisal fiber at 79.00 MPa, while PLA/hemp/sisal showed the lowest tensile strength at 46.25 MPa. However, PLA/hemp/sisal achieved the highest tensile modulus at 6.10 GPa, whereas PLA/polycaprolactone/oil palm mesocarp showed the lowest at 0.88 GPa.

Asaithambi, Ganesan, and Ananda Kumar (2014) found that adding untreated hybrid fiber to virgin PLA provided support by acting as a stress bearer for the PLA matrix. Treating the fiber with alkali improved the adhesive characteristics of the fiber surface by eliminating natural and artificial impurities. Pappu, Pickering, and Thakur (2019) reported that fiber length/orientation, matrix properties (kinetics and crystallinity), and processing techniques all influence the properties of natural fiber-reinforced PLA composites. It is also worth noting that fiber surface roughness aids mechanical interlocking with matrices.

In terms of flexural strength, PLA/banana/sisal fiber obtained the highest value at 125.00 MPa, and PLA/polycaprolactone/oil palm mesocarp showed the lowest flexural strength at 21.45 MPa. Surprisingly, as with flexural modulus, PLA/montmorillonite nanoclay/short kenaf achieved the highest value at 7.50 GPa, while PLA/flax/jute achieved the lowest value at 2.25 GPa. In the study conducted by Asaithambi et al. (2014), the flexural properties of PLA composites improved with mercerization, which transformed crude cellulose structure I into refined cellulose structure II with reactive groups, resulting in short crystallites capable of making intimate bonds with PLA matrix. Alkali treatment increased the strength of chemical bonding between fiber and PLA matrix in composites. The tensile and flexural properties of PLA hybrid biocomposites typically show significant improvement over single-fiber PLA composites. For example, incorporating a second fiber with complementary properties can lead to a composite with better tensile strength and flexural modulus than a single-fiber composite. This is due to the ability of different fibers to distribute and carry the applied loads more effectively. Additionally, fiber hybridization tailors the properties of the composite to meet specific application requirements, thereby broadening the potential uses of these materials.

The mechanical properties of PLA hybrid biocomposites are also influenced by the fiber volume fraction, fiber orientation, and processing method. Higher fiber content generally enhances the strength and stiffness of the composite, but there is a critical point beyond which the properties may degrade due to poor fiber dispersion or excessive matrix depletion. Fiber orientation plays a crucial role in determining the direction-dependent properties of the composite, with aligned fibers providing superior tensile and flexural strength in the direction of alignment. However, random fiber orientation may be advantageous in applications where isotropic properties are desired. The processing method, whether it is injection molding, compression molding, or extrusion, impacts the distribution and orientation of fibers, the quality of the fiber-matrix interface, and ultimately the mechanical performance of the composite.

Overall, the mechanical properties of PLA hybrid biocomposites can be tailored and optimized through careful selection of fiber types, fiber content, fiber orientation, and processing techniques. These materials offer a promising alternative to traditional composites, particularly in applications where a balance between mechanical performance, environmental sustainability, and cost is desired. As research and development in this field continue to advance, further improvements in the mechanical properties of hybrid biocomposites are expected, paving the way for their increased adoption in various industries.

5.6 THERMAL PROPERTIES OF PLA HYBRID BIOCOMPOSITES

The thermal properties of natural fiber-reinforced PLA hybrid biocomposites are critical in determining their performance and stability under different temperature conditions. These properties include thermal stability, degradation behavior, and the influence of hybridization on the composite's thermal characteristics. Understanding and optimizing these thermal properties are essential for ensuring the suitability of hybrid biocomposites in applications where temperature fluctuations and thermal resistance are important factors (Asyraf et al. 2021).

Thermal stability refers to the ability of a material to maintain its structural integrity and mechanical properties when exposed to elevated temperatures. For PLA hybrid biocomposites, thermal stability is influenced by the inherent thermal properties of both the PLA matrix and the natural fibers used in the composite. PLA, being a biodegradable thermoplastic, has a relatively low melting point and glass transition temperature compared to conventional thermoplastics like polyethylene or polypropylene. This makes it essential to carefully consider the thermal stability of the hybrid composite, especially in applications where high-temperature resistance is required.

The incorporation of natural fibers into the PLA matrix can affect the thermal stability of the composite in various ways. Natural fibers, being composed of cellulose, hemicellulose, and lignin, have their degradation temperatures, which can influence the overall thermal behavior of the composite (Raquez et al. 2013). For instance, lignocellulosic fibers like flax and jute typically begin to degrade at temperatures around 200 °C, which is lower than the degradation temperature of PLA. This means that the thermal stability of the hybrid composite may be limited by the thermal degradation of the natural fibers, especially if the fibers are not properly treated or if the processing temperatures are not carefully controlled.

One of the ways to improve the thermal stability of PLA hybrid biocomposites is through the use of thermal stabilizers or flame retardants. These additives can be incorporated into the PLA matrix or applied as surface treatments to the natural fibers to enhance their resistance to thermal degradation. Additionally, the surface treatment of fibers plays a role in improving the thermal stability of the composite by reducing the moisture content and improving the fiber-matrix adhesion, which can help to prevent the premature degradation of the fibers during processing.

Degradation behavior is another important aspect of the thermal properties of PLA hybrid biocomposites. The degradation of these materials typically occurs through thermal, hydrolytic, and oxidative mechanisms, depending on the environmental conditions to which the composite is exposed. Thermal degradation is particularly relevant in applications where the composite is subjected to high temperatures over extended periods. The degradation behavior of PLA hybrid biocomposites is influenced by factors such as the type and content of natural fibers, the presence of additives or stabilizers, and the processing conditions. PLA, being a biodegradable polymer, undergoes thermal degradation primarily through chain scission mechanisms, leading to a reduction in molecular weight and mechanical properties over time. The incorporation of natural fibers can influence the degradation behavior of the composite by providing additional pathways for thermal degradation, particularly if the fibers contain residual moisture or other impurities. To mitigate this, fiber treatment methods such as drying, chemical modification, and the use of compatibilizers can be employed to enhance the thermal stability and degradation resistance of the hybrid composite (Norrrahim et al. 2021).

Hybridization can have a significant impact on the thermal properties of PLA composites, often resulting in improved thermal performance compared to single-fiber composites. The presence of different types of natural fibers in the composite can lead to a more complex thermal degradation profile, where the degradation temperatures of individual components may complement each other. This can result in a hybrid composite with a broader temperature range of stability, making it suitable for applications that require resistance to both high and low temperatures.

The influence of hybridization on the thermal properties of PLA biocomposites also extends to their thermal conductivity and heat deflection temperature. Natural fibers, depending on their type and orientation, can affect the thermal conductivity of the composite, which is an important consideration in applications where heat dissipation or insulation is required. The heat deflection temperature, which is the temperature at which the composite begins to deform under load, can also be influenced by the choice of natural fibers and the overall fiber content. Hybrid composites that combine fibers with different thermal properties can be engineered to achieve the desired thermal conductivity and heat deflection temperature, thereby enhancing the performance of the composite in specific applications.

The thermal properties of PLA are very reliant on molecular weight, molecular weight distribution, and composition. PLA with different physical properties can be easily obtained due to its stereochemical composition and thermal history, which decided either PLA is amorphous or semicrystalline in solid state (Pang, Shanks, and Daver 2015). "PLA's thermal properties depend on its molecular weight, distribution, and composition. Different physical properties of PLA can be achieved based on its stereochemical composition and thermal history, determining whether it is amorphous or semicrystalline." Glass transition temperature (T_g) is very important for amorphous PLA because T_g zone is a place where polymer chain mobility occurs. On the other hand, both T_g and melting temperature (T_m) are important parameters for semicrystalline PLA (Duc et al. 2014). Adding hybrid fibers as reinforcement alters both the T_g and T_m of PLA composites with a suitable ratio of fiber reinforcement to the polymer matrix. Table 5.3 shows a few PLA hybrid composites along with their thermal properties.

Table 5.3 shows that the highest T_g is PLA/coir/pineapple leaf fiber at 290.07 °C, while the lowest T_g is for PLA/hydroxyapatite/membrane mat at 153.60 °C.

TABLE 5.3
Thermal Properties of PLA Hybrid Composites

Fibers	Processing Technique	Property T_m (°C)	T_g (°C)	Reference
Coir/pineapple leaf fiber	Melt blending	290.07	-	(Siakeng et al. 2018)
PBSA/starch	Extrusion	165.35	54.01	(Meng et al. 2019)
Clay/RCF	Mold blending	176.30	-	(Lee, Wang, and Teramoto 2008)
Hydroxyapatite/membrane mat	Air jet spinning	153.60	51.29	(Abdal-hay, Sheikh, and Lim 2013)
Chitosan/basalt	Reactive blending + injection molding	158.01	63.32	(Kumar & Prakash 2020)

Several factors can affect T_g like molecular adhesion, chain modification ability, steric effect, molecular weight, bonding nature, and cross-link density. However, the hybrid composite that achieved the highest T_m was PLA/chitosan/basalt at 63.32 °C, while PLA/hydroxyapatite/membrane mat achieved the lowest T_m of 51.29 °C (Bandaru et al. 2016).

Researchers have conducted an array of studies on coir/pineapple leaf fiber reinforced with polymer (Siakeng et al. 2020; Rigolin et al. 2019; Munawar et al. 2015). Siakeng et al. (2018) investigated the thermal properties of coir fiber (CF) and pineapple leaf fiber (PALF) as fiber reinforcement and fabricated PLA hybrid composites using melt blending. They found that CF increased the thermal degradation of pure PLA while PALF slowed the degradation. However, PALF combined with PLA as a reinforcing fiber showed better thermal stability than the CF-reinforced composite.

The thermal properties of PLA hybrid biocomposites are influenced by various factors, including the type and content of natural fibers, the presence of additives or stabilizers, and the processing conditions. Thermal stability, degradation behavior, and the influence of hybridization are key considerations in the design and development of these materials for different applications. By optimizing the thermal properties of PLA hybrid biocomposites, it is possible to create materials that are not only environmentally sustainable but also capable of performing well in a wide range of temperature conditions. As research in this area continues to advance, further improvements in the thermal properties of hybrid biocomposites are expected, making them increasingly viable for use in demanding applications across various industries.

5.7 MORPHOLOGICAL ANALYSIS OF HYBRID COMPOSITES

The morphological analysis of natural fiber-reinforced PLA hybrid biocomposites is crucial for understanding the microstructural characteristics that influence the material's mechanical, thermal, and overall performance. Morphological studies involve the examination of fiber–matrix interactions, fiber dispersion and distribution, and the presence of any defects or voids within the composite structure. Scanning electron microscopy (SEM), an advanced technique, is commonly used to perform these analyses, providing detailed insights into the internal structure, surface morphology, and fracture surfaces of hybrid biocomposites. SEM provides high-resolution images of the composite's surface, allowing researchers to analyze the fiber–matrix interface, the dispersion of fibers within the matrix, and the presence of any defects such as voids, cracks, or delamination. The quality of the interface is a critical factor in the mechanical properties of the composite; a strong and well-bonded interface ensures efficient load transfer from the matrix to the fibers, resulting in enhanced tensile, flexural, and impact properties. SEM analysis can reveal the extent of fiber pull-out, fiber breakage, and matrix cracking, which are indicative of the quality of the interface and the overall performance of the composite.

The interaction between the natural fibers and the PLA matrix is a key factor in determining the performance of hybrid biocomposites. The fiber–matrix interface is where the load transfer occurs, and any weaknesses in this region can lead to premature failure of the composite. Morphological analysis using SEM can reveal critical information about the nature of the fiber-matrix interface, such as the presence of voids or gaps that can weaken the composite. A strong interface is typically characterized by minimal gaps between the fibers and the matrix, indicating good adhesion and efficient load transfer. This can be achieved through various fiber treatment methods, such as chemical modifications or the use of coupling agents, which enhance the compatibility between the hydrophilic natural fibers and the hydrophobic PLA matrix.

Azlin et al. (2022) performed SEM analysis of fractured tensile specimens and identified various failure mechanisms in the kenaf/PLA and polyester/PLA layers, including matrix cracking, fiber breakage, and fiber pull-out, as shown in Figure 5.3. The S5 laminate, with a higher kenaf content, exhibited significant fiber pull-out due to its inability to withstand tensile forces, resulting in lower tensile strength.

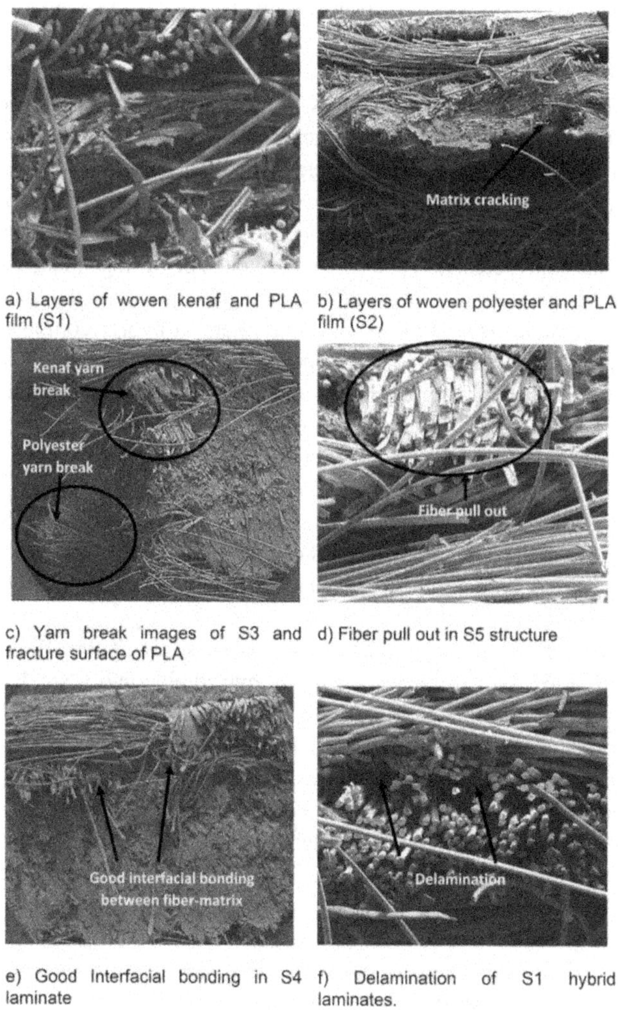

FIGURE 5.3 Morphological micrograph of hybrid laminated composites of S1, S2, S3, S4, and S5 tensile fracture.

Source: Azlin et al. (2022).

However, good interfacial bonding in the S4 hybrid laminate led to better stress transfer and higher tensile strength, while delamination was observed in the S1 laminate, indicating poor load transfer between layers.

Fiber dispersion and distribution within the PLA matrix are also important factors that influence the overall properties of hybrid biocomposites. Uniform dispersion of fibers ensures that the load is evenly distributed throughout the composite, leading to improved mechanical properties. However, achieving uniform dispersion can be challenging, especially with natural fibers, which tend to agglomerate due to their inherent variability in size, shape, and surface characteristics. SEM analysis can help identify regions of fiber agglomeration or clustering that can serve as stress concentrators and reduce the mechanical performance of the composite. Proper processing techniques, such as optimizing the mixing parameters and using compatibilizers, can help achieve better fiber dispersion and distribution.

In addition to fiber–matrix interaction and fiber dispersion, the morphological analysis of hybrid biocomposites also involves the examination of any defects or voids within the composite structure. Voids are small air pockets that can form during the fabrication process, especially in composites that involve natural fibers, which can retain moisture. The presence of voids can negatively impact the mechanical properties of the composite, as they act as stress concentrators and weaken the material. SEM analysis can be used to detect and quantify these voids, providing valuable information for improving the fabrication process and enhancing the quality of the final composite.

The dispersion and distribution of multiple fibers in hybrid composites are particularly important in achieving the desired balance of properties. In hybrid composites, different types of natural fibers are combined to take advantage of their complementary properties. For example, one type of fiber may provide high tensile strength, while another offers better flexibility or impact resistance. The success of this approach depends on the even distribution of both types of fibers within the matrix, ensuring that the composite benefits from the synergistic effects of hybridization. Morphological analysis using SEM can provide detailed insights into the distribution of each fiber type and help identify any issues related to fiber compatibility or processing conditions.

Another aspect of morphological analysis involves studying the effects of various processing techniques on the microstructure of hybrid biocomposites. Different fabrication methods, such as injection molding, compression molding, or extrusion, can lead to variations in the microstructure, affecting properties like fiber orientation, fiber length distribution, and defects. For example, processing methods that involve high shear forces, such as extrusion, can lead to better fiber dispersion but can also cause fiber breakage, reducing the overall length of the fibers and potentially impacting the composite's mechanical properties. SEM analysis can be used to evaluate the effects of different processing techniques on the microstructure, providing guidance for optimizing the fabrication process to achieve the desired properties.

Morphological analysis plays a critical role in understanding and optimizing the properties of natural fiber-reinforced PLA hybrid biocomposites. Techniques such as SEM provide detailed insights into the fiber–matrix interaction, fiber dispersion and distribution, and the presence of defects or voids within the composite structure. By analyzing these microstructural characteristics, researchers can identify potential issues and make informed decisions on how to improve the performance and quality of hybrid biocomposites. As the field of hybrid composites continues to evolve, advances in morphological analysis will play a key role in developing new materials with tailored properties for a wide range of applications.

5.8 APPLICATIONS OF PLA HYBRID BIOCOMPOSITES

The unique combination of mechanical properties, environmental sustainability, and cost-effectiveness makes PLA hybrid biocomposites suitable for a wide range of applications across various industries (Asyraf et al. 2022). These materials offer a compelling alternative to traditional composites, particularly in sectors where the balance between performance and environmental impact is critical. Some of the most promising applications of PLA hybrid biocomposites include the automotive sector, construction and building materials, consumer goods, and biomedical applications (Figure 5.4) (Khomenkova and Larysa 2019).

In the automotive sector, PLA hybrid biocomposites are increasingly being considered as a sustainable alternative to conventional synthetic composites for interior and exterior components. The automotive industry is under growing pressure to reduce the environmental impact of vehicles, not only through improved fuel efficiency but also by incorporating sustainable materials into vehicle design. PLA hybrid biocomposites offer the potential to reduce the overall weight of vehicles,

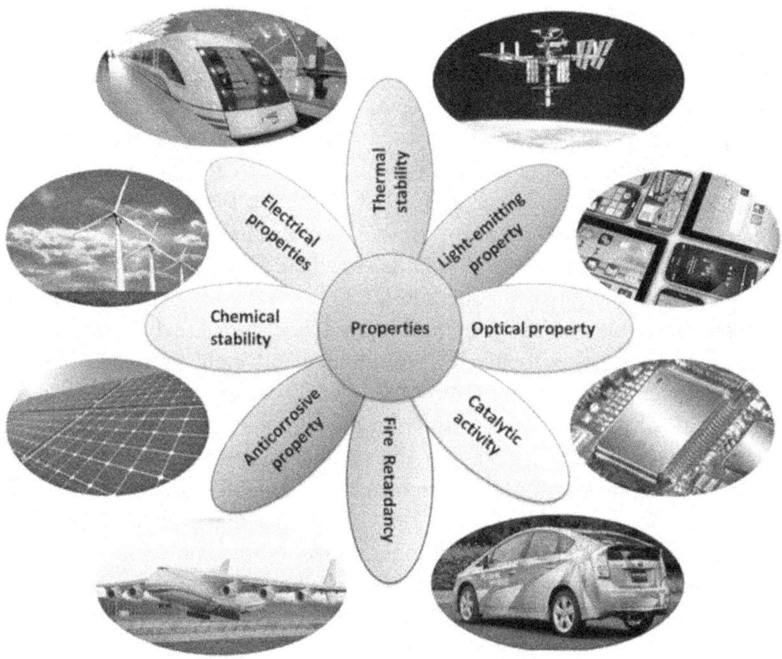

FIGURE 5.4 Composites and hybrid materials: properties and applications.

Source: Khomenkova and Larysa (2019).

which can contribute to improved fuel efficiency and reduced carbon emissions. Additionally, the use of natural fibers in these composites aligns with the industry's goals of increasing the use of renewable materials.

Components such as door panels, dashboards, and interior trim can be manufactured using PLA hybrid biocomposites, offering both mechanical performance and aesthetic appeal. The ability to mold these materials into complex shapes further enhances their suitability for automotive applications. Furthermore, due to the excellent mechanical strength of natural fiber-reinforced PLA composites, they are favored to be used in the automotive sector. Natural fiber-reinforced PLA showed comparable mechanical strength with that of normal laminated glass fiber-reinforced plastics, and the specific strength was three times higher than mild steel's (Sukmawan, Takagi, and Nakagaito 2016).

Jute fiber composite showed greater damping behavior, so jute fiber composites can be practical options as automotive parts since they possessed low vibration and noise (Furtado et al. 2014). Due to its relatively cheap cost, good properties, environmental friendliness, and ease of manufacturing, natural fiber-reinforced PLA composite can be applied in a wide range of applications, including the aviation and automotive industries (Alkbir et al. 2016). In the automotive industry, coconut fiber is used to make furniture in cars, while cotton is used for noise cancellation. Sometimes, wood fiber is also used as furniture and accessory in vehicles while flax, sisal, and hemp are applied in seatback linings and floor panels (Holbery and Houston 2006).

In the construction and building materials industry, PLA hybrid biocomposites are being used in insulation panels, flooring, and structural components. Natural fibers and their reinforcement in cement-based materials are being researched (Faruk et al. 2012). PLA hybrid biocomposites, with their favorable mechanical properties and biodegradability, are well-suited for use in sustainable construction projects. For example, insulation panels made from these composites can provide effective thermal insulation while being lighter and easier to handle than traditional materials. Additionally, the use of natural fibers in these composites can improve indoor air quality by reducing the emission of volatile organic compounds (VOCs), which are commonly associated with synthetic materials.

The consumer goods industry is another area where PLA hybrid biocomposites are finding applications, particularly in durable, lightweight products. Items such as packaging materials, furniture (Mazani et al. 2019), disposable cutlery, and consumer electronics housings can benefit from the use of PLA hybrid biocomposites. The demand for sustainable packaging solutions has grown significantly in recent years, driven by both consumer preference and regulatory pressure. PLA hybrid biocomposites offer an attractive option for packaging materials that are not only strong and lightweight but also compostable, reducing the environmental impact of single-use products. In the electronics industry, the use of these composites in the housings of devices such as laptops and mobile phones can help reduce the overall weight of the products while providing the necessary strength and durability.

Biomedical applications represent a rapidly growing area for PLA hybrid biocomposites, thanks to the biocompatibility and biodegradability of PLA. These materials are being used in medical devices, implants, and drug delivery systems. In particular, the use of natural fibers in combination with PLA can enhance the mechanical properties of biomedical implants, making them more suitable for load-bearing applications. For example, PLA hybrid biocomposites can be used to create bone fixation devices such as screws, plates, and pins that provide the necessary strength while gradually degrading in the body, eliminating the need for a second surgery to remove the implant. Additionally, the use of PLA in advanced drug-delivery systems allows for the controlled release of medications over time.

The versatility of PLA hybrid biocomposites extends to other sectors as well, including the aviation industry, where lightweight and high-strength materials are critical, and the sports and leisure industry, where the demand for ecofriendly products is growing. In aviation, these composites are used in lightweight components for aircraft interiors, contributing to overall weight reduction and improved fuel efficiency. In the sports industry, PLA hybrid biocomposites are used to manufacture products such as bicycles, helmets, and sporting equipment that offer a combination of performance, durability, and environmental sustainability.

PLA hybrid biocomposites offer a wide range of applications across multiple industries, thanks to their unique combination of mechanical properties, environmental sustainability, and cost-effectiveness. From automotive components to construction materials and biomedical devices, these composites provide a viable alternative to traditional materials, with the potential to significantly reduce the environmental impact of various products and processes. As research and development in this field continue to advance, the scope of applications for PLA hybrid biocomposites is expected to broaden, further solidifying their role in the transition toward more sustainable and ecofriendly materials.

5.9 ADVANTAGES AND CHALLENGES OF HYBRID BIOCOMPOSITES

PLA hybrid biocomposites present a compelling combination of benefits and challenges that must be carefully considered to fully leverage their potential. These materials, which combine natural fibers with PLA, offer enhanced properties through hybridization, but they also face challenges related to moisture absorption, fiber uniformity, quality control, and balancing mechanical performance with environmental sustainability. One of the primary advantages of PLA hybrid biocomposites is the enhanced properties that can be achieved through hybridization. By combining different natural fibers, it is possible to tailor the mechanical, thermal, and even aesthetic properties of the composite to meet specific application requirements. For example, the combination of stiff fibers such as flax with more flexible fibers such as hemp can result in a composite that offers both high strength and flexibility. This synergistic effect allows for the development of materials that outperform single-fiber composites in terms of tensile strength, impact resistance, and durability. Additionally, the use of natural fibers in these composites contributes to their biodegradability, making them an environmentally friendly alternative to traditional synthetic composites.

Another significant advantage of PLA hybrid biocomposites is that they reduce the environmental impact of various products and processes. Incorporating natural fibers into PLA hybrid biocomposites reduces the reliance on nonrenewable resources and lowers the overall carbon footprint of the materials. This is particularly important in industries such as automotive and construction, where the environmental impact of materials is a growing concern. The use of natural fibers not only reduces the weight of the composite, leading to potential energy savings during transportation and usage, but also aligns with the global shift toward more sustainable and ecofriendly materials. Moreover, at the end of their life cycle, these composites can be easily recycled or biodegraded, in contrast with conventional composites, further reducing their environmental impact.

However, despite these advantages, PLA hybrid biocomposites also face several challenges that must be addressed to fully realize their potential. One of the primary challenges is moisture absorption. Natural fibers are inherently hydrophilic, meaning they absorb moisture from the environment. This moisture absorption can lead to swelling, degradation of the fiber-matrix interface, and a reduction in mechanical properties over time. The presence of moisture can also accelerate the degradation of the PLA matrix, which is susceptible to hydrolysis under certain conditions. To mitigate these issues, various fiber treatment methods, such as chemical modification and moisture-resistant coatings, are being explored. However, these treatments can add complexity and cost to the manufacturing process, and their long-term effectiveness needs to be thoroughly evaluated.

Fiber uniformity and quality control represent another significant challenge in the production of PLA hybrid biocomposites. Natural fibers are inherently variable in terms of their size, shape, and mechanical properties, which can lead to inconsistencies in the final composite material. This variability can make it difficult to achieve consistent mechanical performance and quality in large-scale production. Moreover, the processing of natural fibers, including their extraction, treatment, and incorporation into the PLA matrix, can introduce further variability. Achieving uniform dispersion of fibers within the matrix is crucial for ensuring optimal load transfer and mechanical properties. However, this can be difficult to achieve, especially when multiple fibers are used. Advanced processing techniques, such as the use of compatibilizers or controlled processing environments, are being developed to address these issues, but they require further refinement.

Balancing mechanical performance and environmental sustainability is another key challenge in the development of PLA hybrid biocomposites. While natural fibers and PLA offers clear environmental benefits, there can be trade-offs in terms of mechanical performance. For example, while natural fibers can provide good tensile strength, they may not offer the same level of impact resistance or durability as synthetic fibers. Similarly, while PLA is biodegradable, it may not offer the same level of thermal or mechanical stability as conventional thermoplastics. These trade-offs must be carefully managed to ensure that the final composite material meets the performance requirements for its intended application while still providing environmental benefits. In some cases, this may involve the use of hybridization strategies that combine natural and synthetic fibers or the incorporation of additives to enhance the properties of the PLA matrix.

In addition to these technical challenges, there are also economic considerations that must be taken into account. The cost of natural fibers can vary significantly depending on factors such as availability, processing requirements, and market demand. Similarly, the production of PLA is currently more expensive than conventional petroleum-based plastics, which limits its adoption in cost-sensitive industries. To address these economic challenges, ongoing research is focused on developing more cost-effective production methods for both natural fibers and PLA, as well as exploring the use of alternative, lower-cost natural fibers that can still provide the desired properties.

Despite these challenges, the future of PLA hybrid biocomposites remains promising thanks to ongoing advances in material science and processing technologies. Innovations in fiber treatment, hybridization, and composite fabrication are expected to lead to new generations of materials that offer improved performance, consistency, and sustainability. For example, advanced coupling agents that enhance the adhesion between natural fibers and the PLA matrix significantly improve the mechanical properties and moisture resistance of the composites. Similarly, novel processing techniques, such as additive manufacturing or nano-reinforcement, enable producing composites with tailored properties for specific applications (Norfarhana, Ilyas, and Ngadi 2022).

While PLA hybrid biocomposites offer significant advantages in terms of enhanced properties and environmental sustainability, they also face several challenges related to moisture absorption, fiber uniformity, quality control, and balancing performance with sustainability. Addressing these challenges requires a multidisciplinary approach that combines advances in material science, processing technology, and economic analysis. As research in this field continues to evolve, it is likely that new solutions will emerge that enable the broader adoption of PLA hybrid biocomposites across a wide range of industries. By carefully managing the trade-offs between performance and sustainability, these materials have the potential to play a key role in the transition toward more ecofriendly and sustainable manufacturing practices.

5.10 FUTURE TRENDS AND RESEARCH DIRECTIONS

The future of PLA hybrid biocomposites is shaped by ongoing innovations in material processing, sustainable practices, and improvements in biodegradability and performance. As the demand for sustainable materials continues to grow, research and development efforts are increasingly focused on overcoming the current limitations of PLA hybrid biocomposites and expanding their range of applications. This section explores some of the key trends and research directions that are expected to drive the future of PLA hybrid biocomposites.

One of the most promising trends in the field of PLA hybrid biocomposites is new material processing techniques that enhance the properties of these composites while maintaining their environmental benefits. Traditional processing methods, such as injection molding and compression molding, have been widely used for fabricating PLA hybrid biocomposites, but they come with limitations such as the need for high temperatures and pressures, which can affect the thermal stability of the PLA matrix and the integrity of natural fibers. To address these challenges, researchers are exploring alternative processing methods, such as additive manufacturing

(3D printing), which allows for the precise control of fiber orientation and distribution within the composite. Additive manufacturing also offers the potential for creating complex geometries and customized components, which can be particularly valuable in applications such as biomedical devices and consumer goods.

Another area of innovation is the use of nano-reinforcements in PLA hybrid biocomposites. Incorporating nanomaterials such as cellulose nanocrystals into the PLA matrix can significantly enhance the mechanical, thermal, and barrier properties of the composites. These nano-reinforcements can improve the dispersion of natural fibers within the matrix, reduce moisture absorption, and increase the overall strength and durability of the composite. Additionally, nanomaterials enable the development of multifunctional composites with properties such as improved electrical conductivity or fire resistance, expanding the range of potential applications.

Advances in sustainable practices are also expected to play a key role in the future of PLA hybrid biocomposites. As industries and consumers become more environmentally conscious, there is increasing interest in developing composites that are not only biodegradable but also produced using sustainable and renewable resources. This includes the use of biobased additives and compatibilizers that enhance the properties of the PLA matrix without compromising its biodegradability. Researchers are also exploring the use of alternative natural fibers that are more sustainable or have lower environmental impacts, such as agricultural waste fibers or fibers derived from invasive plant species. These efforts are aligned with the broader goals of the circular economy, which seeks to minimize waste and make the most of available resources.

In terms of biodegradability, one of the key research directions is PLA hybrid biocomposites with improved degradation profiles. While PLA is biodegradable, its degradation can be slow under certain environmental conditions, such as in marine environments or landfills. To address this, researchers are investigating ways to accelerate the degradation process, such as by incorporating enzymes or biodegradable additives into the PLA matrix. Additionally, there is interest in developing composites that can degrade in a controlled manner, allowing for the release of embedded nutrients or other beneficial substances during the degradation process. This could be particularly useful in applications such as agricultural films or biomedical implants.

The future of PLA hybrid biocomposites also involves exploring new applications and expanding the use of these materials in industries that have traditionally relied on synthetic composites. For example, the aviation industry, which has stringent requirements for material performance, is beginning to explore the use of PLA hybrid biocomposites for nonstructural components, such as interior panels or insulation materials. Similarly, the renewable energy sector is interested in using these composites for components such as wind turbine blades or solar panel frames, where the combination of lightweight, strength, and environmental sustainability is highly desirable. As the field of PLA hybrid biocomposites continues to evolve, there will be a growing need for interdisciplinary research that integrates material science, engineering, environmental science, and economics. This includes the development of life cycle assessment methodologies that can accurately evaluate the environmental impact of PLA hybrid biocomposites across their entire life cycle, from raw material extraction to end-of-life disposal. Such assessments are crucial for identifying areas where improvements can be made and for guiding the development of more sustainable materials.

The future of PLA hybrid biocomposites is characterized by exciting trends and research directions that promise to enhance the properties, sustainability, and range of applications of these materials. Innovations in material processing, nano-reinforcement, sustainable practices, and biodegradability are expected to drive the development of next-generation PLA hybrid biocomposites that offer superior performance and environmental benefits. As research in this field continues to advance, PLA hybrid biocomposites are likely to play an increasingly important role in various industries, contributing to the global transition toward more sustainable and ecofriendly materials.

5.11 CONCLUSIONS

The exploration of PLA hybrid biocomposites demonstrates their substantial potential for advancing industrial applications while promoting environmental sustainability. These materials excel by combining the strengths of PLA and natural fibers, leading to biocomposites that meet the mechanical performance requirements of various industries and reduce environmental impact. The synergy achieved through hybridization often results in properties superior to those of the individual components, making PLA hybrid biocomposites a compelling choice for sustainable material development.

The use of natural fibers in these biocomposites enhances mechanical properties like tensile strength and flexibility, offering a renewable and biodegradable alternative to synthetic fibers. When paired with PLA, which is known for its biodegradability and sufficient mechanical properties, these fibers help create composites that are both environmentally friendly and mechanically robust. This combination is particularly beneficial in applications requiring lightweight and high-performance materials, such as automotive components and construction materials.

However, several challenges need to be addressed for the successful implementation of PLA hybrid biocomposites. Issues such as moisture absorption, fiber uniformity, and the need for optimized processing techniques must be overcome to ensure consistent quality and long-term durability. Additionally, the limitations of PLA, particularly its thermal stability and degradation behavior, require careful consideration to maintain the composite's performance in its intended application. Future advancements in PLA hybrid biocomposites will likely focus on overcoming these challenges through innovations in material processing, such as additive manufacturing and nano-reinforcement. Research into bio-based additives and alternative natural fibers will further align these materials with sustainable practices. As industries increasingly seek sustainable alternatives, PLA hybrid biocomposites are poised to play a significant role in transitioning to environmentally friendly manufacturing practices, contributing to a more sustainable future.

REFERENCES

Abdal-hay, Abdalla, Faheem A. Sheikh, and Jae Kyoo Lim. 2013. "Air Jet Spinning of Hydroxyapatite/Poly(Lactic Acid) Hybrid Nanocomposite Membrane Mats for Bone Tissue Engineering." *Colloids and Surfaces B: Biointerfaces* 102 (February): 635–643. https://doi.org/10.1016/j.colsurfb.2012.09.017.

Alias, Aisyah Humaira, Mohd Nurazzi Norizan, Fatimah Athiyah Sabaruddin, Muhammad Rizal Muhammad Asyraf, Mohd Nor Faiz Norrahim, Ahmad Rushdan Ilyas, Anton M. Kuzmin, et al. 2021. "Hybridization of MMT/Lignocellulosic Fiber Reinforced Polymer Nanocomposites for Structural Applications: A Review." *Coatings* 11 (11): 1355. https://doi.org/10.3390/coatings11111355.

Alkbir, M.F.M., S.M. Sapuan, A.A. Nuraini, and M.R. Ishak. 2016. "Fibre Properties and Crashworthiness Parameters of Natural Fibre-Reinforced Composite Structure: A Literature Review." *Composite Structures*. https://doi.org/10.1016/j.compstruct.2016.01.098.

Alsubari, S., M.Y.M. Zuhri, S.M. Sapuan, M.R. Ishak, R.A. Ilyas, and M.R.M. Asyraf. 2021. "Potential of Natural Fiber Reinforced Polymer Composites in Sandwich Structures: A Review on Its Mechanical Properties." *Polymers* 13 (3): 423. https://doi.org/10.3390/polym13030423.

Amiandamhen, S.O., M. Meincken, and L. Tyhoda. 2020. "Natural Fibre Modification and Its Influence on Fibre-Matrix Interfacial Properties in Biocomposite Materials." *Fibers and Polymers* 21 (4): 677–689. https://doi.org/10.1007/s12221-020-9362-5.

Arul Jeya Kumar, A., and M. Prakash. 2020. "Mechanical and Morphological Characterization of Basalt/Cissus Quadrangularis Hybrid Fiber Reinforced Polylactic Acid Composites." *Proceedings of the Institution of Mechanical Engineers, Part C: Journal of Mechanical Engineering Science* 234 (14): 2895–2907. https://doi.org/10.1177/0954406220911072.

Asaithambi, B., G. Ganesan, and S. Ananda Kumar. 2014. "Bio-Composites: Development and Mechanical Characterization of Banana/Sisal Fibre Reinforced Poly Lactic Acid (PLA) Hybrid Composites." *Fibers and Polymers* 15 (4): 847–854. https://doi.org/10.1007/s12221-014-0847-y.

Asyraf, M.R.M., M.R. Ishak, M.N.F. Norrahim, N.M. Nurazzi, S.S. Shazleen, R.A. Ilyas, M. Rafidah, and M.R. Razman. 2021. "Recent Advances of Thermal Properties of Sugar Palm Lignocellulosic Fibre Reinforced Polymer Composites." *International Journal of Biological Macromolecules* (November). https://doi.org/10.1016/j.ijbiomac.2021.10.221.

Asyraf, M.R.M., Agusril Syamsir, Nazirul Mubin Zahari, Abu Bakar Mohd Supian, Mohamad Ridzwan Ishak, Salit Mohd Sapuan, Shubham Sharma, et al. 2022. "Product Development of Natural Fibre-Composites for Various Applications: Design for Sustainability." *Polymers* 14 (5): 920. https://doi.org/10.3390/polym14050920.

Atiqah, A., Mohammad Jawaid, S.M. Sapuan, and Mohamad Ridzwan Ishak. 2018. "Effect of Surface Treatment on the Mechanical Properties of Sugar Palm/Glass Fiber-Reinforced Thermoplastic Polyurethane Hybrid Composites." *BioResources* 13 (1): 1174–1188. https://doi.org/10.15376/biores.13.1.1174-1188.

Azlin, M.N.M., S.M. Sapuan, M.Y.M. Zuhri, and E.S. Zainudin. 2022. "Effect of Stacking Sequence and Fiber Content on Mechanical and Morphological Properties of Woven Kenaf/Polyester Fiber Reinforced Polylactic Acid (PLA) Hybrid Laminated Composites." *Journal of Materials Research and Technology* 16: 1190–1201. https://doi.org/10.1016/j.jmrt.2021.12.046.

Baghaei, Behnaz, Mikael Skrifvars, and Lena Berglin. 2013. "Manufacture and Characterisation of Thermoplastic Composites Made from PLA/Hemp Co-Wrapped Hybrid Yarn Prepregs." *Composites Part A: Applied Science and Manufacturing* 50: 93–101. https://doi.org/10.1016/j.compositesa.2013.03.012.

Bajracharya, Rohan M., Dilpreet S. Bajwa, and Sreekala G. Bajwa. 2017. "Mechanical Properties of Polylactic Acid Composites Reinforced with Cotton Gin Waste and Flax Fibers." *Procedia Engineering* 200: 370–376. https://doi.org/10.1016/j.proeng.2017.07.052.

Bandaru, Aswani Kumar, Shivdayal Patel, Yogesh Sachan, Suhail Ahmad, R. Alagirusamy, and Naresh Bhatnagar. 2016. "Mechanical Behavior of Kevlar/Basalt Reinforced Polypropylene Composites." *Composites Part A: Applied Science and Manufacturing* 90: 642–652. https://doi.org/10.1016/j.compositesa.2016.08.031.

Bangar, Sneh Punia, M.M. Harussani, R.A. Ilyas, Adeleke Omodunbi Ashogbon, Arashdeep Singh, Monica Trif, and Seid Mahdi Jafari. 2022. "Surface Modifications of Cellulose Nanocrystals: Processes, Properties, and Applications." *Food Hydrocolloids* 130 (September): 107689. https://doi.org/10.1016/j.foodhyd.2022.107689.

Bledzki, A.K., P. Franciszczak, and A. Meljon. 2015. "High Performance Hybrid PP and PLA Biocomposites Reinforced with Short Man-Made Cellulose Fibres and Softwood Flour." *Composites Part A: Applied Science and Manufacturing* 74 (July): 132–139. https://doi.org/10.1016/j.compositesa.2015.03.029.

Dhand, Vivek, Garima Mittal, Kyong Yop Rhee, Soo-Jin Park, and David Hui. 2015. "A Short Review on Basalt Fiber Reinforced Polymer Composites." *Composites Part B: Engineering* 73 (May): 166–180. https://doi.org/10.1016/j.compositesb.2014.12.011.

Duc, F., P.E. Bourban, C.J.G. Plummer, and J.A.E. Månson. 2014. "Damping of Thermoset and Thermoplastic Flax Fibre Composites." *Composites Part A: Applied Science and Manufacturing* 64: 115–123. https://doi.org/10.1016/j.compositesa.2014.04.016.

Eng, Chern Chiet, Nor Azowa Ibrahim, Norhazlin Zainuddin, Hidayah Ariffin, Wan Md Zin Wan Yunus, and Yoon Yee Then. 2014. "Enhancement of Mechanical and Dynamic Mechanical Properties of Hydrophilic Nanoclay Reinforced Polylactic Acid/Polycaprolactone/Oil Palm Mesocarp Fiber Hybrid Composites." *International Journal of Polymer Science*. https://doi.org/10.1155/2014/715801.

Eselini, Najah, Seha Tirkeş, Alinda Akar, and Umit Tayfun. 2019. "Production and Characterization of Poly (Lactic Acid)-Based Biocomposites Filled with Basalt Fiber and Flax Fiber Hybrid." *Journal of Elastomers & Plastics* 52 (8): 701–716. https://doi.org/10.1177/0095244319884716.

Faruk, Omar, Andrzej K. Bledzki, Hans-Peter Fink, and Mohini Sain. 2012. "Biocomposites Reinforced with Natural Fibers: 2000–2010." *Progress in Polymer Science* 37 (11): 1552–1596. https://doi.org/10.1016/j.progpolymsci.2012.04.003.

Furtado, Samuel C.R., A.L. Araújo, Arlindo Silva, Cristiano Alves, and A.M.R. Ribeiro. 2014. "Natural Fibre-Reinforced Composite Parts for Automotive Applications." *International Journal of Automotive Composites* 1 (1): 18. https://doi.org/10.1504/ijautoc.2014.064112.

Garlotta, Donald. 1978. "A Literature Review of Poly(Lactic Acid)." *Journal of Polymer and the Environment* 9 (2): 63–84.

Holbery, James, and Dan Houston. 2006. "Natural-Fiber-Reinforced Polymer Composites in Automotive Applications." *Jom* 58 (11): 80–86. https://doi.org/10.1007/s11837-006-0234-2.

Ilyas, R.A., S.M. Sapuan, M.M. Harussani, M.Y.A.Y. Hakimi, M.Z.M. Haziq, M.S.N. Atikah, M.R.M. Asyraf, et al. 2021. "Polylactic Acid (PLA) Biocomposite: Processing, Additive Manufacturing and Advanced Applications." *Polymers* 13 (8): 1326. https://doi.org/10.3390/polym13081326.

Ilyas, R.A., S.M. Sapuan, M.R. Ishak, and E.S. Zainudin. 2017. "Effect of Delignification on the Physical, Thermal, Chemical, and Structural Properties of Sugar Palm Fibre." *BioResources* 12 (4): 8734–8754. https://doi.org/10.15376/biores.12.4.8734-8754.

Ilyas, R.A., M.Y.M. Zuhri, H.A. Aisyah, M.R.M. Asyraf, S.A. Hassan, E.S. Zainudin, S.M. Sapuan, et al. 2022. "Natural Fiber-Reinforced Polylactic Acid, Polylactic Acid Blends and Their Composites for Advanced Applications." *Polymers* 14 (1): 202. https://doi.org/10.3390/polym14010202.

Jang, Jun Young, Tae Kyeong Jeong, Hwa Jin Oh, Jae Ryoun Youn, and Young Seok Song. 2012. "Thermal Stability and Flammability of Coconut Fiber Reinforced Poly(Lactic Acid) Composites." *Composites Part B: Engineering* 43 (5): 2434–2438. https://doi.org/10.1016/j.compositesb.2011.11.003.

Kaiser, Mohammad R., Hazleen B. Anuar, Noorasikin B. Samat, and Shamsul B.Abdul Razak. 2013. "Effect of Processing Routes on the Mechanical, Thermal and Morphological Properties of PLA-Based Hybrid Biocomposite." *Iranian Polymer Journal (English Edition)* 22 (2): 123–131. https://doi.org/10.1007/s13726-012-0112-4.

Kamaruddin, Zatil, Ridhwan Jumaidin, Rushdan Ilyas, Mohd Selamat, Roziela Alamjuri, and Fahmi Yusof. 2022. "Influence of Alkali Treatment on the Mechanical, Thermal, Water Absorption, and Biodegradation Properties of Cymbopogan Citratus Fiber-Reinforced, Thermoplastic Cassava Starch – Palm Wax Composites." *Polymers* 14 (14): 2769. https://doi.org/10.3390/polym14142769.

Khomenkova, Rada, and Savkina Larysa. 2019. *Solid State Composites and Hybrid Systems*. 1st ed. Boca Raton, FL, USA: CRC Press (Tailor & Francis Group) LLC. https://doi.org/10.1201/9781351176071.

Lee, Seung Hwan, Siqun Wang, and Yoshikuni Teramoto. 2008. "Isothermal Crystallization Behavior of Hybrid Biocomposite Consisting of Regenerated Cellulose Fiber, Clay, and Poly(Lactic Acid)." *Journal of Applied Polymer Science* 108 (2): 870–875. https://doi.org/10.1002/app.26853.

Mahbubul Islam, Mohammad. 2019. "Varietal Advances of Jute, Kenaf and Mesta Crops in Bangladesh: A Review." *International Journal of Bioorganic Chemistry* 4 (1): 24. https://doi.org/10.11648/j.ijbc.20190401.15.

Manral, Ankit, Furkan Ahmad, and Vijay Chaudhary. 2019. "Static and Dynamic Mechanical Properties of PLA Bio-Composite with Hybrid Reinforcement of Flax and Jute." *Materials Today: Proceedings* (xxxx): 2–5. https://doi.org/10.1016/j.matpr.2019.07.240.

Mazani, N., S.M. Sapuan, M.L. Sanyang, A. Atiqah, and R.A. Ilyas. 2019. "Design and Fabrication of a Shoe Shelf From Kenaf Fiber Reinforced Unsaturated Polyester Composites." In *Lignocellulose for Future Bioeconomy*, edited by Hidayah Ariffin, S.M. Sapuan, and Mohd Ali Hassan, 1st ed., 315–332. Amsterdam, Netherlands: Elsevier. https://doi.org/10.1016/B978-0-12-816354-2.00017-7.

Meng, Linghan, Long Yu, Saud Khalid, Hongsheng Liu, Shaobo Zhang, Qingfei Duan, and Ling Chen. 2019. "Preparation, Microstructure and Performance of Poly (Lactic Acid)-Poly (Butylene Succinate-Co-Butyleneadipate)-Starch Hybrid Composites." *Composites Part B: Engineering* 177. https://doi.org/10.1016/j.compositesb.2019.107384.

Munawar, Rose Farahiyan, Nurul Hayati Jamil, Mohd Khairul Shahril, Skh Muhammad Skh Abdul Rahim, Muhammad Zaimi Zainal Abidin, M.A. Azam, and Kok Tee Lau. 2015. "Development of Green Composite: Pineapple Leaf Fibers (PALF) Reinforced Polylactide (PLA)." *Applied Mechanics and Materials* 761 (May): 520–525. https://doi.org/10.4028/www.scientific.net/AMM.761.520.

Nabilah Haris, Nur Izzah, Mohamad Zaki Hassan, R.A. Ilyas, Mohamed Azlan Suhot, S.M. Sapuan, Rozzeta Dolah, Roslina Mohammad, and M.R.M. Asyraf. 2022. "Dynamic Mechanical Properties of Natural Fiber Reinforced Hybrid Polymer Composites: A Review." *Journal of Materials Research and Technology* (May). https://doi.org/10.1016/j.jmrt.2022.04.155.

Norfarhana, A.S., R.A. Ilyas, and N. Ngadi. 2022. "A Review of Nanocellulose Adsorptive Membrane as Multifunctional Wastewater Treatment." *Carbohydrate Polymers* 291 (September): 119563. https://doi.org/10.1016/j.carbpol.2022.119563.

Norrahim, M.N.F., R.A. Ilyas, N.M. Nurazzi, M.S.A. Rani, M.S.N. Atikah, and S.S. Shazleen. 2021. "Chemical Pretreatment of Lignocellulosic Biomass for the Production of Bioproducts: An Overview." *Applied Science and Engineering Progress* 14 (4): 588–605.

Nyambo, Calistor, Amar K. Mohanty, and Manjusri Misra. 2010. "Polylactide-Based Renewable Green Composites from Agricultural Residues and Their Hybrids." *Biomacromolecules* 11 (6): 1654–1660. https://doi.org/10.1021/bm1003114.

Okubo, Kazuya, Toru Fujii, and Erik T. Thostenson. 2009. "Multi-Scale Hybrid Biocomposite: Processing and Mechanical Characterization of Bamboo Fiber Reinforced PLA with Microfibrillated Cellulose." *Composites Part A: Applied Science and Manufacturing* 40 (4): 469–475. https://doi.org/10.1016/j.compositesa.2009.01.012.

Omran, Abdoulhdi A. Borhana, Abdulrahman A.B.A. Mohammed, S.M. Sapuan, R.A. Ilyas, M.R.M. Asyraf, Seyed Saeid Rahimian Koloor, and Michal Petrů. 2021. "Micro- and Nanocellulose in Polymer Composite Materials: A Review." *Polymers* 13 (2): 231. https://doi.org/10.3390/polym13020231.

Pang, Chaowei, Robert A. Shanks, and Fugen Daver. 2015. "Characterization of Kenaf Fiber Composites Prepared with Tributyl Citrate Plasticized Cellulose Acetate." *Composites Part A: Applied Science and Manufacturing* 70: 52–58. https://doi.org/10.1016/j.compositesa.2014.12.003.

Pappu, Asokan, Kim L. Pickering, and Vijay Kumar Thakur. 2019. "Manufacturing and Characterization of Sustainable Hybrid Composites Using Sisal and Hemp Fibres as Reinforcement of Poly (Lactic Acid) via Injection Moulding." *Industrial Crops and Products* 137 (November 2018): 260–269. https://doi.org/10.1016/j.indcrop.2019.05.040.

Pasquini, Daniel, Eliangela de Morais Teixeira, Antonio Aprigio da Silva Curvelo, Mohamed Naceur Belgacem, and Alain Dufresne. 2008. "Surface Esterification of Cellulose Fibres: Processing and Characterisation of Low-Density Polyethylene/Cellulose Fibres Composites." *Composites Science and Technology* 68 (1): 193–201. https://doi.org/10.1016/j.compscitech.2007.05.009.

Ramesh, P., B. Durga Prasad, and K.L. Narayana. 2020. "Effect of Fiber Hybridization and Montmorillonite Clay on Properties of Treated Kenaf/Aloe Vera Fiber Reinforced PLA Hybrid Nanobiocomposite." *Cellulose* 27 (12): 6977–6993. https://doi.org/10.1007/s10570-020-03268-6.

Raquez, Jean Marie, Youssef Habibi, Marius Murariu, and Philippe Dubois. 2013. "Polylactide (PLA)-Based Nanocomposites." *Progress in Polymer Science* 38 (10–11): 1504–1542. https://doi.org/10.1016/j.progpolymsci.2013.05.014.

Rigolin, T.R., M.C. Takahashi, D.L. Kondo, and S.H.P. Bettini. 2019. "Compatibilizer Acidity in Coir-Reinforced PLA Composites: Matrix Degradation and Composite Properties." *Journal of Polymers and the Environment* 27 (5): 1096–1104. https://doi.org/10.1007/s10924-019-01411-4.

Sanchez-Garcia, Maria D., and Jose M. Lagaron. 2010. "On the Use of Plant Cellulose Nanowhiskers to Enhance the Barrier Properties of Polylactic Acid." *Cellulose* 17 (5): 987–1004. https://doi.org/10.1007/s10570-010-9430-x.

Sanjay, M.R., P. Madhu, Mohammad Jawaid, P. Senthamaraikannan, S. Senthil, and S. Pradeep. 2018. "Characterization and Properties of Natural Fiber Polymer Composites: A Comprehensive Review." *Journal of Cleaner Production* 172 (January): 566–581. https://doi.org/10.1016/j.jclepro.2017.10.101.

Shah, Md, U. Ain, Mohamed T.H. Sultan, Mohammad Jawaid, Francisco Cardona, and Abd R. Abu Talib. 2016. "A Review on the Tensile Properties of Bamboo Fiber Reinforced Polymer Composites." *BioResources*. https://doi.org/10.15376/biores.11.4.Shah.

Siakeng, Ramengmawii, Mohammad Jawaid, Hidayah Ariffin, and S.M. Sapuan. 2018. "Thermal Properties of Coir and Pineapple Leaf Fibre Reinforced Polylactic Acid Hybrid Composites." *IOP Conference Series: Materials Science and Engineering* 368 (1). https://doi.org/10.1088/1757-899X/368/1/012019.

Siakeng, Ramengmawii, Mohammad Jawaid, Hidayah Ariffin, S.M. Sapuan, Mohammad Asim, and Naheed Saba. 2019. "Natural Fiber Reinforced Polylactic Acid Composites: A Review." *Polymer Composites* 40 (2): 446–463. https://doi.org/10.1002/pc.24747.

Siakeng, Ramengmawii, Mohammad Jawaid, Mohammad Asim, and Suchart Siengchin. 2020. "Accelerated Weathering and Soil Burial Effect on Biodegradability, Colour and Textureof Coir/Pineapple Leaf Fibres/PLA Biocomposites." *Polymers* 12 (2): 458. https://doi.org/10.3390/polym12020458.

Södergård, A., M. Stolt, and S. Inkinen. 2022. "Industrial Production of High-Molecular-Weight Poly (Lactic Acid)." In: Rafael A. Auras, Loong-Tak Lim, Susan E. M. Selke, Hideto Tsuji (eds.) *Poly (Lactic Acid) Synthesis, Structures, Properties, Processing, Applications, and End of Life,* 29–44. DOI:10.1002/9781119767480. John Wiley & Sons, Inc.

Sukmawan, Romi, Hitoshi Takagi, and Antonio Norio Nakagaito. 2016. "Strength Evaluation of Cross-Ply Green Composite Laminates Reinforced by Bamboo Fiber." *Composites Part B: Engineering*. https://doi.org/10.1016/j.compositesb.2015.08.072.

Suriani, M.J., R.A. Ilyas, M.Y.M. Zuhri, A. Khalina, M.T.H. Sultan, S.M. Sapuan, C.M. Ruzaidi, et al. 2021. "Critical Review of Natural Fiber Reinforced Hybrid Composites: Processing, Properties, Applications and Cost." *Polymers* 13 (20): 3514. https://doi.org/10.3390/polym13203514.

Suriani, M.J., Fathin Sakinah Mohd Radzi, R.A. Ilyas, Michal Petrů, S.M. Sapuan, and C.M. Ruzaidi. 2021. "Flammability, Tensile, and Morphological Properties of Oil Palm Empty Fruit Bunches Fiber/Pet Yarn-Reinforced Epoxy Fire Retardant Hybrid Polymer Composites." *Polymers* 13 (8): 1282. https://doi.org/10.3390/polym13081282.

Surya, Indra, N.G. Olaiya, Samsul Rizal, Ikramullah Zein, N.A. Sri Aprilia, M. Hasan, Esam Bashir Yahya, K.K. Sadasivuni, and H.P.S. Abdul Khalil. 2020. "Plasticizer Enhancement on the Miscibility and Thermomechanical Properties of Polylactic Acid-Chitin-Starch Composites." *Polymers* 12 (1): 115. https://doi.org/10.3390/polym12010115.

Thiruchitrambalam, M., A. Alavudeen, A. Athijayamani, N. Venkateshwaran, and A. Elaya Perumal. 2009. "Improving Mechanical Properties of Banana/Kenaf Polyester Hybrid Composites Using Sodium Laulryl Sulfate Treatment." *Materials Physics and Mechanics* 8 (2): 165–173.

Trivedi, Alok Kumar, M.K. Gupta, and Harinder Singh. 2023. "PLA Based Biocomposites for Sustainable Products: A Review." *Advanced Industrial and Engineering Polymer Research* 6 (4): 382–395. https://doi.org/10.1016/j.aiepr.2023.02.002.

Veerasimman, Arumugaprabu, Vigneshwaran Shanmugam, Sundarakannan Rajendran, Deepak Joel Johnson, Ajith Subbiah, John Koilpichai, and Uthayakumar Marimuthu. 2022. "Thermal Properties of Natural Fiber Sisal Based Hybrid Composites – A Brief Review." *Journal of Natural Fibers* 19 (12): 4696–4706. https://doi.org/10.1080/15440478.2020.1870619.

Velde, K. Van De, and P. Kiekens. 2002. "Biopolymers: Overview of Several Properties and Consequences on Their Applications." *Polymer Testing* 21 (4): 433–442. https://doi.org/10.1016/S0142-9418(01)00107-6.

6 Nanofiber-Reinforced PLA Bionanocomposites

6.1 INTRODUCTION

PLA is one of the most widely used biopolymers in the production of bionanocomposites due to its excellent properties and renewable nature. PLA is a biodegradable polyester derived from renewable resources such as corn starch or sugarcane, making it an environmentally friendly alternative to traditional petroleum-based polymers. The importance of PLA in bionanocomposites cannot be overstated, as it offers a unique combination of biodegradability, biocompatibility, and favorable mechanical properties. PLA's ability to degrade into non-toxic by-products under composting conditions makes it particularly attractive for applications where environmental impact is a concern. Furthermore, PLA's mechanical properties, including its tensile strength and modulus, are comparable to those of conventional polymers, which makes it a viable option for a wide range of applications.

The reinforcement of PLA with nanofibers further enhances its mechanical and thermal properties, making it suitable for more demanding applications. Nanofibers, with their high aspect ratio and superior mechanical strength, significantly improve the performance of the PLA matrix. Including nanofibers in PLA-based bionanocomposites leads to materials that are not only stronger but also more thermally stable and resistant to environmental degradation. Among the various types of nanofibers used in these composites, cellulose nanofibers, chitin nanofibers, and carbon nanotubes are the most prominent. Each of these nanofibers brings specific benefits to the composite, depending on the intended application.

Cellulose nanofibers, derived from natural sources like wood or agricultural waste, are valued for their high tensile strength, low density, and biodegradability. These properties make cellulose nanofibers an ideal reinforcement material for PLA, especially in applications where sustainability is a key consideration. Chitin nanofibers, extracted from the exoskeletons of crustaceans, offer additional benefits such as antimicrobial properties, which are particularly useful in biomedical and packaging applications.

Meanwhile, carbon nanotubes, although synthetic, are used in bionanocomposites for their exceptional mechanical strength, electrical conductivity, and thermal stability. Integrating them into PLA results in composites with enhanced electrical and thermal properties, which expands the potential applications of these materials. The significance of nanofibers in reinforcing PLA-based bionanocomposites lies in their ability to improve the overall performance of the composite. The nanoscale dimensions of these fibers provide a high surface area to volume ratio, which enhances the interfacial bonding between the nanofibers and the PLA matrix. This strong

DOI: 10.1201/9781032678894-6

interfacial bonding is crucial for achieving the desired mechanical strength and durability of the composite. Additionally, the uniform dispersion of nanofibers within the PLA matrix leads to improved thermal stability and resistance to environmental factors, such as moisture and UV radiation.

This chapter encompasses a detailed exploration of the role of nanofibers in PLA-based bionanocomposites, including the various types of nanofibers used, their properties and characteristics, and the advantages they offer. The chapter also provides an overview of the structure and composition of PLA nanofiber composites, setting the stage for a comprehensive discussion on their applications, challenges, and future research directions. As bionanocomposites continue to gain prominence in the field of materials science, understanding the role of nanofibers in these advanced materials will be essential for driving innovation and expanding their use in sustainable technologies.

6.2 NANOFIBERS IN BIONANOCOMPOSITES

Nanofibers are crucial components in the development of advanced bionanocomposites due to their unique properties and ability to significantly enhance the performance of the composite materials. Various types of nanofibers are employed in bionanocomposites, each contributing specific characteristics that benefit the overall material properties. Cellulose nanofibers, derived from natural sources, offer high tensile strength, low density, and biodegradability, making them an excellent reinforcement for PLA composites. Chitin nanofibers, another naturally occurring nanofiber, provide antimicrobial properties, in addition to mechanical reinforcement, making them suitable for applications in medical and packaging materials. Carbon nanotubes, though synthetic, are widely used in bionanocomposites for their exceptional mechanical strength, electrical conductivity, and thermal stability, greatly enhancing the functionality of the composites.

The properties and characteristics of these nanofibers, such as their aspect ratio, surface chemistry, and interaction with the PLA matrix, play a critical role in determining the overall performance of the bionanocomposite. Nanofibers, owing to their nanoscale dimensions, offer a high surface-area-to-volume ratio, leading to better interfacial bonding with the PLA matrix, which is crucial for the mechanical strength and durability of the composite. Moreover, the uniform dispersion of nanofibers within the PLA matrix can result in significant improvements in thermal stability and resistance to environmental degradation.

Incorporating nanofibers into PLA composites brings several advantages, including enhanced mechanical properties like tensile strength and modulus, improved thermal resistance, and the potential for functionalization to impart additional properties such as antimicrobial activity or electrical conductivity. These enhancements make PLA nanofiber composites highly attractive for a wide range of applications, from packaging and automotive components to biomedical devices and environmental remediation technologies. This chapter delves into the types of nanofibers used, their properties and characteristics, and the specific advantages they confer to PLA-based bionanocomposites.

6.2.1 Nanocellulose/PLA

Adding nanocellulose to PLA significantly enhances the material's properties, primarily due to the large surface area and high aspect ratio of nanocellulose. However, achieving effective dispersion of nanocellulose within the PLA matrix is challenging because of the distinct hydrophilic nature of nanocellulose and the hydrophobic characteristics of PLA. When properly dispersed, nanocellulose can improve the mechanical properties, such as tensile strength and modulus, of PLA composites. It also enhances the thermal stability and reduces the brittleness of PLA, making it more suitable for various applications. However, poor dispersion can lead to aggregation of nanocellulose, which might reduce the expected improvements in the composite's properties.

The key challenge in developing high-performance PLA nanocomposites is enhancing the compatibility between nanocellulose and the PLA polymer matrix to improve the dispersion of nanocellulose particles. Due to the inherent differences in their properties – nanocellulose being hydrophilic and PLA being hydrophobic – achieving uniform dispersion within the PLA matrix is difficult. This poor dispersion often leads to suboptimal performance in the resulting nanocomposites, as the full potential of the nanocellulose reinforcement is not realized (Table 6.1).

Exclusive reported studies on the different processing techniques for producing bast fibers from nanocellulose and fabricating bast fibers nanocellulose-reinforced PLA nanocomposites. *Agave Tequilana Weber* (ATW) waste, rich in cellulose, was utilized to extract cellulose nanocrystals (CNC) as a sustainable reinforcement for PLA nanocomposites (Uribe-Calderón et al. 2023), as shown in Figure 6.1. The surface modification of CNC with poly(2-ethyl hexyl acrylate) (CNCG) improved compatibility with hydrophobic PLA, enhancing both thermal stability (T_d ~370 °C) and tensile properties (2.11 MPa). The SEM images in Figure 6.2 reveal a smooth, brittle fracture surface in pure PLA, while PLA/CNC nanocomposites exhibit rougher surfaces indicative of ductile fracture. Although some CNC agglomerates were present, surface modification of CNC reduced their size and improved dispersion, leading to a smoother fracture surface in grafted-CNC nanocomposites. CNCG also increased the degree of crystallinity (25.1%) and positively influenced the melt rheological behavior, making these nanocomposites ideal for applications like food packaging films through blow-extrusion processing.

6.2.2 Layered Ceramic/PLA

The primary drawback of PLA is its hydrophobic nature, which can be mitigated by incorporating ceramic layered nanofillers with uniform distribution. Blending organic components enhances the bonding between PLA and the exfoliated ceramics (Dash et al. 2008). One of the simplest methods to synthesize layered ceramic PLA nanocomposites involves intercalation using solvents, melt blending, or in situ polymerization. Solvents are crucial in dissolving PLA and facilitating the effective dispersion of ceramic nanofillers within the composite solution (Cumkur, Baouz, and Yilmazer 2015).

TABLE 6.1
Nanocellulose-Reinforced PLA Nanocomposites

Natural Fibers	Chemical Pretreatment	Mechanical/ Physical Pretreatment	Biological Pretreatment	Principal Treatment	Post-Treatment	Fabrication Technique	Reference
Bleached flax yarns	Sodium hydroxide	–	–	Sulfuric acid hydrolysis	Centrifugation, neutralization (by adding aqueous NaOH)	Evaporation casting of PLA in chloroform solvent	(Liu et al. 2010)
Flax fabric	Sodium hydroxide, sodium carbonate; hydrogen peroxide	Grinding	–	Sulfuric acid hydrolysis	Washing, centrifuga-tion, dialysis, ultrasonication	–	(Csiszár and Nagy 2017)
Flax fiber	Elemental chlorine-free (ECF) bleaching sequence treatment	–	–	Hydrochloric acid hydrolysis	Dialysis, sonication, vacuum filtration	–	(Mujtaba et al. 2017)
Hemp fiber	Sodium hydroxide, sodium chlorite, acetic acid	Grinding, high-speed blending	–	Defibrillation (microfluidizer)	–	–	(Pacaphol and Aht-Ong 2016)
Hemp pristine fiber	Toluene, ethanol; sodium chlorite, acetic acid; sodium bisulphate; sodium hydroxide	Carding, chopping, thermal combined sonication treatment	Enzymatic reaction with Pectinase Aspergillus enzyme in deionized water solution	Sulfuric acid hydrolysis	Centrifugation, dialysis, sonication	–	(Luzi et al. 2014)
Hemp yarns	–	Chopping	–	Sodium hydroxide, sodium hypochlorite (oxidation hydrolysis under continuous agitation and sonication)	–	–	(Dai, Fan, and Collins 2013)

Jute felts	Sodium hydroxide, dimethyl sulfoxide	Cutting, shredding	–	Sulfuric acid hydrolysis	Sonication, neutralization (by adding aqueous NaOH), centrifugation	–	(Das et al. 2011)
Jute fiber	Sodium hydroxide; sodium chlorite, sodium sulfite; hydrogen peroxide	Chopping	–	Sulfuric acid hydrolysis	Centrifugation	–	(Orasugh et al. 2018)
Jute fiber	Sodium hydroxide; sodium hypochlorite, sodium sulphate	Chopping, high-energy planetary ball milling	–	Sulfuric acid hydrolysis	Neutralization (by adding aqueous NaOH), discoloring (by adding aqueous NaOCl), vacuum filtration	–	(Jabbar et al. 2017)
Jute fiber	Caustic soda; sodium chlorite	Cutting, steam explosion	–	Oxalic acid hydrolysis; steam explosion	Centrifugation, mechanical agitation, sonication, neutralization (by adding aqueous NaOH)	–	(Thomas et al. 2015)
Jute fiber	Toluene; sodium hydroxide; hydrogen peroxide	Cutting	–	Sulfuric acid hydrolysis; formic acid hydrolysis	Filtration (using nylon membrane), neutralization (by rinsing with distilled water)	–	(Kasyapi, Chaudhary, and Bhowmick 2013)
Jute fiber	Sodium hydroxide, dimethyl sulfoxide	–	–	TEMPO-mediated oxidation	Centrifugation	–	(Cao et al. 2013)
Kenaf core wood	Sodium hydroxide; sodium chlorite, acetic acid	Grinding	–	Sulfuric acid hydrolysis	Centrifugation, dialysis	–	(Chan et al. 2013)

(*Continued*)

TABLE 6.1 (*Continued*)
Nanocellulose-Reinforced PLA Nanocomposites

Natural Fibers	Chemical Pretreatment	Mechanical/ Physical Pretreatment	Biological Pretreatment	Principal Treatment	Post-Treatment	Fabrication Technique	Reference
Kenaf fiber	Sodium hydroxide; sodium chlorite, acetic acid	–	–	Sulfuric acid hydrolysis	Centrifugation, dialysis	–	(Zainuddin et al. 2017)
Kenaf fiber	Sodium hydroxide; sodium chlorite, acetic acid	Grinding; blending	–	Sulfuric acid hydrolysis; hydrochloric acid hydrolysis	Centrifugation, dialysis, homogenization, addition with chloroform	–	(Zaini et al. 2013)
Kenaf stem fiber	Sodium hydroxide	Grinding, blending	–	Homogenization	–	–	(Jonoobi et al. 2011)
Mulberry pulp	–	Cutting, beating, disintegration	–	Sulfuric acid hydrolysis	Centrifugation, ultrasonication, dialysis	–	(Wang, Shankar, and Rhim 2017)
Mulberry pulp	–	Cutting, beating	–	Sulfuric acid hydrolysis	Centrifugation, ultrasonication, dialysis	–	(Reddy and Rhim 2014)
Mulberry pulp	–	Cutting, beating, disintegration	–	Sulfuric acid hydrolysis	Centrifugation, ultrasonication, dialysis	–	(Wang, Shankar, and Rhim 2017)
Mulberry pulp	–	Cutting, beating	–	Sulfuric acid hydrolysis	Centrifugation, ultrasonication, dialysis	–	(Reddy and Rhim 2014)

Pristine hemp fiber	Toluene, ethanol; sodium chlorite, acetic acid; sodium bisulphate; sodium hydroxide	Carding, chopping	–	Sulfuric acid hydrolysis; physical modification with acid phosphate ester of ethoxylatednon-ylphenol surfactant	Centrifugation, dialysis, sonication	Evaporation casting of PLA/PBS blend with functional-ized nanocellulose in chloro-form solvent	(Luzi et al. 2016)
Pristine jute fiber	Sodium hydroxide, dimethyl sulfoxide	Grinding	–	TEMPO-mediated oxidation	Centrifugation, high-speed homogeni-zation, ultrasonication	–	(Lin et al. 2014)
Ramie fiber	Sodium hydroxide	Cutting	–	Sulfuric acid hydrolysis; physical modification with organosilanization	Centrifugation, dialysis, homogenization, filtration, neutraliza-tion (by adding aqueous NaOH)	Melt-extrusion of PLA with functional-ized nanocellulose in absence of solvent	(Raquez et al. 2012)
Ramie fiber	Sodium hydroxide	Cutting	–	Sulfuric acid hydrolysis	Centrifugation, dialysis, homogenization, filtration, addition with chloroform	–	(Alloin et al. 2010)
Refined industrial flax stem fibers	Toluene, ethanol; sodium chlorite, acetic acid; sodium bisulfite	Grinding (ball milling)	–	Sulfuric acid hydrolysis	Centrifugation, ultrasonication, dialysis	–	(Astruc et al. 2017)
Roselle microcrystal-line cellulose	–	–	–	Sulfuric acid hydrolysis	Centrifugation, dialysis, ultrasonication	–	(Kian et al. 2018)

(Continued)

TABLE 6.1 (*Continued*)
Nanocellulose-Reinforced PLA Nanocomposites

Natural Fibers	Chemical Pretreatment	Mechanical/ Physical Pretreatment	Biological Pretreatment	Principal Treatment	Post-Treatment	Fabrication Technique	Reference
Roselle microcrystal-line cellulose	–	–	–	Sulfuric acid hydrolysis	Centrifugation, dialysis, ultrasonication	–	(Kian et al. 2018)
Waste jute bags	Toluene, ethanol; soda cooking (in basic reactor); hydrogen peroxide; hydrochloric acid hydrolysis (in ultrasonicator)	Chopping	–	Homogenization	–	–	(Kian et al. 2018)
Water retted kenaf fiber	Sodium hydroxide, anthraquinone (in rotatory digester); three-stage bleaching	Cutting, screening	–	Mechanical shearing (super mass colloider); grinding	–	–	(Karimi et al. 2014)

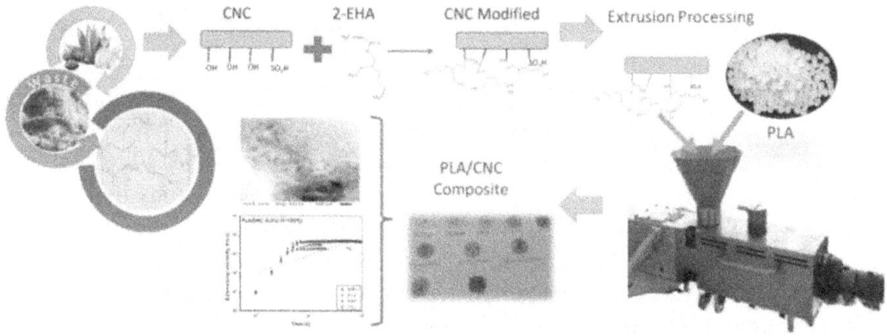

FIGURE 6.1 Cellulose nanocrystal-reinforced PLA nanocomposites.

Source: Uribe-Calderon et al. (2023).

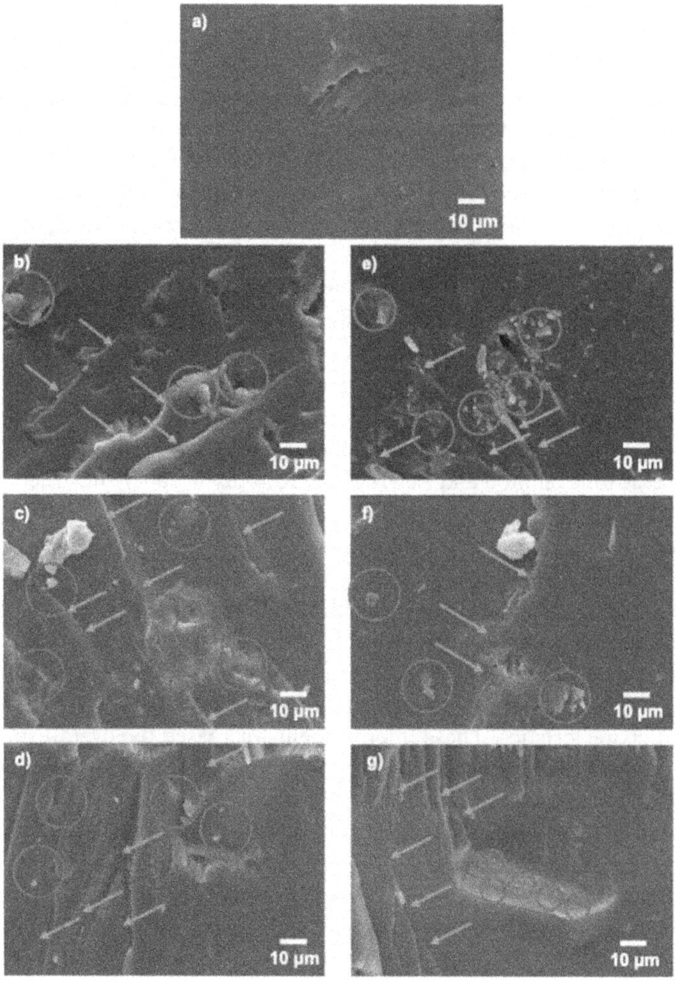

FIGURE 6.2 SEM micrographs of the PLA (a) and PLA nanocomposites with (b) 0.5, (c) 1 and (d) 2 wt.% CNC; and (e) 0.5, (f) 1 and (g) 2 wt.% CNC/2-EHA. The blue arrows indicate ductile zones; the red circles indicate CNC agglomerates.

Source: Uribe-Calderon et al. (2023).

Li et al. (2011) reported that the dispersion of PLLA/PDLA/clay nanocomposites can be improved by increasing shear force during melt blending. They observed a significant increase in ductility, with elongation reaching up to 208% at the break point in tensile tests of PLA/clay layered nanocomposites. Their study highlighted the extensive exfoliation and even distribution of clay platelets, along with a more plasticized interphase surrounding these platelets. In the amorphous PLA structure, this high level of dispersion resulted in multiple shear bonding, caused by the significant exfoliation (Lai et al. 2014; Li and Shimizu 2009).

6.2.3 Layered Double Hydroxide/PLA

LDH exhibits unique properties from those of conventional layered silicates. LDH consists of two layers: One side has positively charged metal–hydroxide sheets, while the other side accommodates intercalated negatively charged anions (Cumkur, Baouz, and Yilmazer 2015). Additionally, water molecules are bonded to the structure. The hydrophobic nature of pure PLAs, a major drawback, is addressed by using organic LDHs for intercalation, which also eliminates the presence of heavy metals. In some cases, pretreated nanofillers are used to facilitate the intercalation of polymeric chains between LDH layers. The pretreatment process involves adding organic or polymeric positive anions, such as alkyl carboxylates, vinylpyridines, and butadiene, between LDH layers.

Katiyar et al. (2010) discussed the synthesis of LDH/PLA nanocomposites through the process of ring-opening polymerization (ROP) of LA, as shown in Fig. 5. This study identified two types of LDH nanofillers: LDH carbonate (LDH-CO3) and laurate-modified LDH (LDH-C12). Exfoliated PLA nanocomposites were obtained by blending with LDH-C12. However, a drawback of this approach is that the molecular weight of PLA is significantly reduced after in situ polymerization (Katiyar et al. 2010; Hoque 2013; Compton and Lewis 2014, 2014; Pyda and Wunderlich 2005).

6.2.4 Glass/PLA

Malinowski et al. (2015) reported that the mechanical properties of PLA composites could be significantly enhanced through the reinforcement of glass nanoparticles. They highlighted that uniformly incorporating glass nanoparticles throughout the PLA matrix improved flexural modulus, modulus of elasticity (E), impact resistance, and thermal characteristics. These improvements increased the overall strength and durability of the PLA composite, making it more suitable for applications where mechanical robustness is critical.

Additionally, incorporating glass nanoparticles improved the impact resistance of PLA composites, making them more resilient to sudden forces or shocks. This enhancement is particularly beneficial in applications where the material may be subjected to dynamic or high-impact conditions. The thermal characteristics of the PLA composite were also significantly improved. The glass nanoparticles increased the material's thermal stability, making it more resistant to degradation at elevated temperatures. This improvement extends the application range of PLA composites, allowing them to be used in environments where higher thermal resistance is required.

Zhang et al. (2015) studied the rheological properties of PLA/ABS/TCS composites, and they established that blending phosphate glass particles with the PLA matrix resulted in a significant increase in the decomposition rate of PLA. The accelerated decomposition of PLA in the presence of phosphate glass particles is particularly advantageous in biomedical applications, where the material's biodegradability is a key consideration. The enhanced decomposition rate facilitates the timely breakdown of the material in biological environments, reducing its environmental impact.

Moreover, incorporating phosphate glass particles also promoted the precipitation of calcium phosphate on the surfaces of the composite (Suryanegara, Nakagaito, and Yano 2009; Škrlová et al. 2019). This is a crucial finding for applications in bone tissue engineering, where the formation of a calcium phosphate layer is essential for promoting bone cell attachment and growth. The presence of this layer mimics the natural bone mineral, providing a favorable environment for bone regeneration.

In addition to these findings, the study also noted the improved rheological properties of the PLA/ABS/TCS composites, which are indicative of the material's flow behavior under stress. The blending of phosphate glass particles improved the viscosity and flow characteristics of the PLA matrix, making it easier to process and mold into desired shapes. This improvement in processability is significant for large-scale manufacturing and production of PLA-based composites.

Malinowski et al. (2015) and Ramanjaneyulu, Venkatachalapathi, and Prasanthi (2022) collectively demonstrated that the reinforcement of PLA with glass nanoparticles and phosphate glass particles offers substantial improvements in mechanical, thermal, and rheological properties. These enhancements expand the potential applications of PLA composites in various industries, including biomedical, automotive, and packaging. The incorporation of glass nanoparticles addresses some of the inherent limitations of PLA, such as its mechanical strength and thermal stability, while also enhancing its biodegradability and processability, making it a more versatile and sustainable material choice for future developments (Suryanegara, Nakagaito, and Yano 2009; Škrlová et al. 2019).

6.2.5 Silica/PLA

Nanosilica, a group of hard fillers, plays a critical role in enhancing the mechanical properties of pure PLA, particularly in terms of adhesion, strength, and scratch resistance. The addition of nano silica fillers to PLA has been widely studied for its potential to address some of the inherent limitations of PLA, such as its relatively low strength and brittleness.

Zhu et al. (2010) explored the effects of oleic acid-modified silicon oxide (SiO_2) nanofillers on the mechanical properties of PLA. They found that while the incorporation of these nanofillers significantly improved the flexural strength of PLA, it also led to a noticeable decrease in tensile strength. This trade-off highlights the complexity of reinforcing PLA with hard fillers, where improvements in one mechanical property may come at the expense of another. The enhanced flexural strength indicates better resistance to bending forces, making the material more suitable for applications where stiffness is crucial. However, the reduced tensile strength made the material more prone to failure under tension, which could limit its use in applications requiring high tensile performance (Pal et al. 2017).

Researchers have investigated the impact of surface-modified silicon nanoparticles on the mechanical properties of PLA composites. For instance, introducing surface-altered silicon nanoparticles into poly(L-lactic acid) (PLLA) increased the toughness and tensile strength of the composite. This enhancement was attributed to the homogeneous dispersion of SiO2 nanoparticles within the PLA matrix, which promotes better stress transfer and reduces the likelihood of stress concentration points that can lead to material failure. The uniform distribution of nanoparticles throughout the PLA matrix helps to mitigate the brittle nature of PLA, making the composite material more ductile and better able to absorb energy before fracturing (Dorgan, Lehermeier, and Mang 2000; Antunes et al. 2020; Wen et al. 2011).

Additionally, the surface modification of SiO_2 nanoparticles is crucial in improving their compatibility with the PLA matrix. Surface-altered nanoparticles tend to have better interaction with the polymer chains, which enhances the overall interfacial adhesion between the filler and the matrix. This improved adhesion contributes to the increased tensile strength and toughness observed in the composites. The role of surface modification is also vital in preventing the agglomeration of nanoparticles, ensuring that they remain evenly distributed within the PLA matrix, which is essential for achieving consistent mechanical properties throughout the material.

Incorporating nanosilica into PLA represents a promising strategy for enhancing the material's mechanical performance, particularly in terms of adhesion, strength, and scratch resistance. However, achieving an optimal balance between different mechanical properties, such as flexural and tensile strength, requires careful consideration of the type and amount of nanofiller used, as well as the surface modification techniques employed. These findings underscore the potential of nano silica-reinforced PLA composites in various applications, from packaging to automotive components, where improved mechanical properties are essential (Dorgan, Lehermeier, and Mang 2000; Antunes et al. 2020; Wen et al. 2011; Pal et al. 2017).

6.2.6 Titanium Dioxide/PLA

Incorporating TiO_2 nanoparticles into the PLA matrix significantly enhances the material's photocatalytic and magnetic properties, particularly in crystalline forms of PLA. This modification has been shown to improve the material's functionality in various advanced applications. According to Liao et al. (2007), the inclusion of TiO_2 nanoparticles not only boosts the photocatalytic activity of PLA but also contributes to the material's magnetic properties, opening up possibilities for its use in areas like environmental remediation and magnetic storage devices.

One of the most notable benefits of TiO_2 incorporation is its antibacterial nature, which significantly improves the performance of PLA-based implants in biomedical applications. The antimicrobial properties of TiO_2 help to reduce the risk of infection, making PLA-TiO_2 composites particularly valuable in medical devices and implants. This enhanced biocompatibility is crucial for long-term implantation, where maintaining a sterile environment is essential for patient safety. Moreover, blending TiO_2 nanoparticles with PLA has been observed to increase the material's photodegradability. This feature is particularly important for applications where environmental sustainability is a concern, as it allows for the breakdown of PLA under UV light, reducing the environmental impact of PLA-based products. Meng et al. (2011) highlighted that this increase in photodegradability is accompanied by a significant improvement in the toughness of the PLA-TiO_2 composite. The toughness enhancement is critical for extending the lifespan and durability of PLA products, making them more suitable for demanding applications where mechanical strength is required (Ikada and Tsuji 2000; Yan et al. 2007; Sharif, Mondal, and Hoque 2019).

Incorporating TiO_2 nanoparticles into the PLA matrix offers a multifaceted improvement in the material's properties, including enhanced photocatalytic activity, antimicrobial behavior, photodegradability, and toughness. These advancements make PLA-TiO_2 composites highly attractive for a wide range of applications, from medical implants to environmentally friendly packaging solutions (Meng et al. 2011; Pantani et al. 2013; Ikada and Tsuji 2000; Yan et al. 2007; Sharif, Mondal, and Hoque 2019).

6.2.7 Zinc Oxide/PLA

Zinc oxide (ZnO) nanoparticles offer notable benefits when incorporated into PLA matrices, primarily due to their strong antibacterial properties and excellent absorption of ultraviolet (UV) radiation. This UV absorption capability makes ZnO a valuable additive in applications where protection from UV degradation is essential, such as in packaging and outdoor materials. However, despite these advantages, the inclusion of ZnO nanoparticles into the PLA matrix presents significant challenges, particularly during melt processing. Murariu et al. (2010) observed that blending ZnO nanoparticles with PLA can lead to dangerous degradation of the polyester matrix at the high temperatures required for melt processing. This degradation severely compromises the thermo-mechanical properties of the PLA, leading to reduced performance in critical applications.

To address these issues, researchers have explored various surface treatment techniques for ZnO nanoparticles to enhance their compatibility with the PLA matrix and to mitigate the negative effects observed during processing. Surface treatments with selective additives, such as fatty amides, stearic acid, and stearates, have proven to be effective in reducing the degradation and improving the overall stability of the PLA-ZnO composites. These treatments create a barrier that prevents direct interaction between the ZnO particles and the PLA matrix at elevated temperatures, thereby preserving the mechanical integrity of the composite.

One particularly effective surface treatment involves the use of triethoxy caprylylsilane. This treatment has been shown to significantly enhance the thermal and mechanical properties of PLA-ZnO nanocomposites, making them more suitable for high-performance applications. The triethoxy caprylylsilane treatment improves the dispersion of ZnO nanoparticles within the PLA matrix, leading to a more uniform distribution and better stress transfer between the PLA and the nanoparticles. As a result, the treated PLA-ZnO composites exhibit improved strength, durability, and resistance to thermal degradation, making them highly desirable for applications where both UV protection and mechanical performance are critical.

While ZnO nanoparticles offer substantial benefits for PLA composites, addressing the challenges of matrix degradation through effective surface treatments is essential for realizing their full potential. By using advanced surface modification techniques, such as triethoxy caprylylsilane treatment, it is possible to enhance the thermal and mechanical properties of PLA-ZnO nanocomposites, expanding their applicability in various demanding environments (Vallet-Regí et al. 1998).

6.2.8 Alumina/PLA

The interconnection via polar coupling and hydrogen bonding significantly enhances the adhesion properties in Alumina-PLA nanocomposites, making this combination particularly advantageous for various applications. Alumina, a ceramic material, is known for its exceptional properties such as bio-inertness, excellent corrosion resistance, and the ability to enhance the mechanical strength of PLA. The integration of alumina into PLA nanocomposites broadens their applicability, especially in the biomedical field. These nanocomposites are increasingly used in dental implants and orthopedic implants due to their ability to support tissue growth and promote healing. Specifically, alumina coatings are recommended for facilitating tissue regeneration in damaged organs and for maxillofacial reconstruction in cases of severe injuries (Zheng et al. 2009; Pinto et al. 2013).

Moreover, Lule et al. (2019) demonstrated the significant improvement in adhesion between the PLA matrix and surface-treated alumina particles. The study revealed that incorporating 40% surface-treated alumina particles into the PLA matrix enhances thermal conductivity by 150% compared to neat PLA. This improvement in thermal properties is crucial for applications requiring materials that can efficiently dissipate heat, such as in electronic packaging and thermal management systems.

Additionally, Tang et al. (2013) observed a notable enhancement in the flame retardant properties of PLA when aluminum hypophosphite and expanded graphite are added. This finding is particularly important for applications where fire resistance is a critical requirement, such as in construction materials and automotive components. The synergistic effects of these additives not only improve the mechanical and thermal properties of PLA but also expand its potential use in more demanding environments, thereby making alumina–PLA nanocomposites a versatile and valuable material in both medical and industrial applications.

6.2.9 Fe_3O_4/PLA

In recent years, polymer-based nanocomposites have garnered significant attention due to their enhanced mechanical, thermal, and functional properties. Among these, incorporating nanoparticles such as Fe_3O_4 into biodegradable polymers like PLA has shown great promise. The unique combination of PLA's biodegradability and Fe_3O_4 nanoparticles' magnetic properties offers a versatile material suitable for a wide range of applications. The interaction between PLA and Fe_3O_4 nanoparticles, particularly through hydrogen bonding, plays a crucial role in enhancing the overall performance of these nanocomposites. This synergy not only improves the physical properties of PLA but also extends its applicability into advanced fields such as biomedicine, electronics, and environmental technology. The following discussion explores the bonding mechanisms, characterization techniques, and the extensive applications of Fe_3O_4-PLA nanocomposites, highlighting their potential impact on various industries.

Zheng et al. (2009) reported the presence of strong bonding interactions between PLA and Fe_3O_4 nanoparticles, primarily facilitated through hydrogen bonding. This strong interaction was confirmed using various characterization techniques, which demonstrated the effective incorporation of Fe_3O_4 nanoparticles into the PLA matrix. These nanoparticles not only enhance the mechanical properties of PLA composites but also impart additional functionalities, making these composites highly versatile.

The incorporation of Fe_3O_4 nanoparticles into PLA broadens the scope of its applications across various fields. Fe_3O_4 nanoparticles are widely recognized for their magnetic properties, which make them invaluable in several domains. In the biomedical field, Fe_3O_4 nanoparticles are utilized for targeted drug delivery, magnetic resonance imaging, and hyperthermia treatment of cancer, owing to their biocompatibility and superparamagnetic behavior. Additionally, these nanoparticles serve as effective catalysts in chemical reactions due to their high surface area and magnetic separability, which allows for easy recovery and reuse (Carus et al. 2010; Tingaut, Zimmermann, and Lopez-Suevos 2010; Madhavan Nampoothiri, Nair, and John 2010).

Beyond biomedical applications, Fe_3O_4 nanoparticles are also employed in pigments, contributing to the production of high-quality, durable inks and coatings. Their magnetic properties are particularly advantageous in the field of data storage, where they are used in the development of magnetic recording devices that require high-density data storage capabilities. Furthermore, Fe_3O_4 nanoparticles are applied in gas sensors, where their sensitivity to changes in the environment allows for the detection of various gases with high accuracy. In the realm of lubrication, these nanoparticles are used to improve the performance of lubricants, reducing wear and friction in mechanical systems (Carus et al. 2010; Tingaut, Zimmermann, and Lopez-Suevos 2010; Madhavan Nampoothiri, Nair, and John 2010).

Carus et al. (2010), Tingaut, Zimmermann, and Lopez-Suevos (2010), and Madhavan Nampoothiri, Nair, and John (2010) further emphasized the multifaceted applications of Fe_3O_4 nanoparticles, highlighting their potential to revolutionize industries ranging from healthcare to electronics. The combination of PLA with Fe_3O_4 nanoparticles not only enhances the properties of the composite material but also opens up new possibilities for its use in advanced technological applications, making it a promising material for future innovations.

6.2.10 PLA/Nanofiller Reinforcement

The integration of nanofillers into polymer matrices like PLA has emerged as a promising strategy to enhance the mechanical properties and broaden the applicability of these composites. By incorporating nanofillers such as cellulose nanofibers into the PLA matrix, significant improvements in tensile strength can be achieved. The performance of PLA composites is closely tied to the quality of dispersion of these nanofillers within the matrix, as uniform dispersion is critical for maximizing the mechanical benefits. This approach not only reinforces the PLA structure but also enables the development of high-performance materials suitable for various advanced applications.

Madhavan Nampoothiri, Nair, and John (2010) and Murariu et al. (2010) demonstrated that the tensile strength of PLA composites is linearly correlated with the inherent strengths of the cellulose nanofibers used. The study highlighted the importance of achieving a homogenous dispersion of individual cellulose nanofillers within the PLA matrix to ensure optimal composite performance. Proper dispersion is essential as it prevents the aggregation of nanofibers, which could otherwise lead to weak points within the material, compromising its overall mechanical integrity. This enhanced strength makes these composites ideal for applications requiring durable and lightweight materials.

In another study, Bodin et al. (2007) explored the purification of toxins from bacterial nanocellulose by treating it with sodium hydroxide. This purification process is crucial for medical applications, as it reduces the water content within the composite fibers, which in turn inhibits the absorption of proteins from the blood. This property is particularly important for developing blood-compatible materials, a key requirement for medical devices such as scaffold materials, medical implants, and artificial blood vessels. The reduced protein absorption minimizes the risk of thrombosis and other complications, thereby enhancing the safety and effectiveness of these medical devices.

The combination of enhanced mechanical properties and improved biocompatibility positions PLA-based composites as a versatile material in the medical field. These composites are increasingly being utilized in the development of scaffolds for tissue engineering, implants, and artificial blood vessels, where their mechanical strength, biocompatibility, and ability to integrate with biological systems are of paramount importance. The ongoing research in optimizing the dispersion of nanofillers and refining purification processes continues to push the boundaries of what PLA composites can achieve, making them a valuable resource in both medical and industrial applications.

6.2.11 PLA/Nanoparticle Reinforcement

Incorporating inorganic nanoparticles into polymer matrices like PLA has gained significant attention for enhancing the material's properties, particularly for applications in environments exposed to heat and microbial activity. One of the critical properties improved by adding nanoparticles is the glass transition temperature (T_g), which determines the material's ability to retain its structural integrity under elevated temperatures. This enhancement is crucial for expanding the use of PLA composites in applications that require thermal stability, such as packaging, automotive parts, and electronics (Govender 2000).

In addition to thermal stability, incorporating nanoparticles into the PLA matrix has been shown to impart antimicrobial properties, making these composites highly effective in combating a range of bacterial and fungal pathogens. Various studies have reported the effectiveness of different nanoparticle-embedded PLA composites against specific microorganisms. For instance, ZnO nanoparticles demonstrated strong antimicrobial activity against *Escherichia coli* and *Staphylococcus aureus*

(Govender 2000), while silanized ZnO nanoparticles showed efficacy against *Listeria monocytogenes* and *Salmonella enterica typhimurium* (Rokbani, Daigle, and Ajji 2018). These properties make ZnO-PLA composites suitable for use in medical devices, food packaging, and other applications where hygiene is paramount.

Further research has expanded the range of antimicrobial PLA composites. ZnO-doped silica nanoparticles have proven effective against *E. coli* (Râpă et al. 2021), and ZnO/Ag composites have shown activity against both *S. aureus* and *Pseudomonas aeruginosa* (Ahmed et al. 2018). Moreover, bimetallic silver-copper nanoparticles, when combined with cinnamon essential oil, exhibited potent antimicrobial effects against *Campylobacter jejuni*, *Listeria monocytogenes*, and *Salmonella typhimurium* (Fonseca et al. 2015). Adding TiO_2 nanoparticles was particularly effective against *E. coli* and *Aspergillus fumigatus* (Asadi and Pirsa 2020), while TiO_2 combined with lycopene pigments showed strong antimicrobial activity against *E. coli* and *S. aureus* (Swaroop and Shukla 2018). Furthermore, MgO nanoparticles were effective against *E. coli*, further broadening the potential applications of PLA composites in antimicrobial settings (Mulla et al. 2021).

The enhanced Tg and antimicrobial properties resulting from incorporating these nanoparticles into PLA composites position them as highly versatile materials. They are increasingly being adopted in areas where both thermal stability and antimicrobial protection are critical, such as in food packaging, health care, and environmental applications. The continued development and refinement of these composites will likely lead to even broader applications, leveraging their unique combination of properties to address challenges in various industries.

6.3 FABRICATION TECHNIQUES FOR BIONANOCOMPOSITES

The fabrication of bionanocomposites, particularly those based on PLA and nanofibers, involves a series of intricate processes designed to ensure optimal dispersion and strong interfacial bonding between the matrix and the reinforcing nanofibers. The choice of fabrication technique plays a crucial role in determining the final properties of the composite, including its mechanical strength, thermal stability, and biodegradability. Several methods are commonly employed in the fabrication of PLA-based bionanocomposites, each with its own set of advantages and challenges.

6.3.1 Solution Casting

Solution casting is a widely used method for fabricating PLA-based bionanocomposites. In this technique, both PLA and nanofibers are dissolved in an appropriate solvent and then cast onto a substrate to form a thin film. As the solvent evaporates, it leaves behind a solid composite film with the nanofibers uniformly distributed within the PLA matrix. This technique is particularly effective for creating thin films and coatings where uniform dispersion is crucial. For instance, one study involved the fabrication of PLA/cellulose nanocrystal composites using chloroform as a solvent. The resulting films showed improved mechanical properties and thermal stability due to the uniform dispersion of cellulose nanocrystals within the PLA matrix (Ramanadha reddy and Venkatachalapathi 2023).

Another example is the fabrication of PLA/chitin nanowhisker composites, where a solution of PLA and chitin nanowhiskers was cast to form films. These composites exhibited enhanced mechanical properties and were found to be promising for packaging applications due to their biodegradability and antimicrobial properties. Solution casting is ideal for laboratory-scale production; however, solvent removal can be challenging and may require additional processing steps to eliminate residual solvents.

6.3.2 Melt Blending

Melt blending is a commonly employed method for fabricating PLA-based bionanocomposites on an industrial scale. In this process, PLA and nanofibers are mixed together in the molten state using an extruder or an internal mixer. The high temperature of the melt ensures that PLA is sufficiently fluid to allow the nanofibers to be evenly dispersed throughout the matrix. For example, a study reported the successful fabrication of PLA/cellulose nanofiber composites via melt blending. The resulting composites exhibited enhanced mechanical properties, making them suitable for automotive applications.

Another instance is the fabrication of PLA/carbon nanotube composites using melt blending. The carbon nanotubes were dispersed within the PLA matrix, resulting in composites with improved electrical conductivity and mechanical strength. This approach is particularly advantageous for large-scale production, as it eliminates the need for solvents and is compatible with standard polymer processing equipment like injection molding machines.

However, one of the challenges associated with melt blending is ensuring the uniform dispersion of nanofibers, particularly those with high aspect ratios like carbon nanotubes. High processing temperatures can also pose a risk of thermal degradation to certain nanofibers, which must be carefully managed to preserve the properties of the composite.

6.3.3 Electrospinning

Electrospinning is a versatile technique that is particularly useful for creating nanofiber-reinforced PLA composites, especially when nanofiber mats or membranes are required. In this method, a high-voltage electric field is applied to a polymer solution or melt, causing it to be drawn into ultrafine fibers that are collected on a grounded substrate. The resulting nanofibrous mats have a high surface area to volume ratio, which is beneficial for applications such as filtration, tissue engineering, and drug delivery.

For example, a study utilized electrospinning to create PLA/cellulose nanofiber mats. These mats exhibited excellent mechanical properties and biocompatibility, making them ideal for use in biomedical applications such as wound dressings. Another study demonstrated the fabrication of PLA/chitin nanofiber composites via electrospinning. The resulting composites were characterized by their high porosity and mechanical strength, which are crucial for applications like filtration. Although electrospinning allows for precise control over fiber diameter and composition, the technique is generally limited to small-scale production and can be time-consuming, which may limit its applicability in industrial-scale manufacturing.

6.3.4 In Situ Polymerization

In situ polymerization is a fabrication technique where the polymerization of the PLA matrix occurs in the presence of nanofibers. This method ensures intimate mixing of the nanofibers with the growing polymer chains, resulting in strong interfacial bonding and uniform dispersion. This technique is particularly beneficial for creating composites with high nanofiber loadings, as it allows better control over the molecular weight of the PLA matrix and the dispersion of the nanofibers.

For instance, in one study, PLA/cellulose nanocrystal composites were fabricated using in situ polymerization. The cellulose nanocrystals were dispersed within the polymerizing PLA matrix, resulting in composites with superior mechanical properties and thermal stability. Another example involved fabricating PLA/carbon nanotube composites via in situ polymerization, where the carbon nanotubes were incorporated into the PLA matrix during polymerization, leading to composites with enhanced electrical and mechanical properties. However, the process requires precise control over polymerization conditions, such as temperature, catalyst concentration, and reaction time, which can complicate the fabrication process.

6.3.5 Layer-by-Layer Assembly

Layer-by-layer (LbL) assembly is a fabrication technique that involves the sequential deposition of PLA and nanofibers onto a substrate, building up the composite structure layer by layer. This method allows for precise control over the thickness and composition of each layer, resulting in composites with tailored properties.

An example of this technique can be found in the fabrication of PLA/cellulose nanofiber composites. By layering PLA and cellulose nanofibers alternately, researchers were able to create composites with gradient properties, such as varying mechanical strength and biodegradability across the layers. This method is particularly useful for producing multifunctional materials, such as membranes with selective permeability or coatings with enhanced durability. Layer-by-layer assembly is particularly advantageous for creating thin films and coatings with gradient or multifunctional properties. However, the technique can be time-consuming, especially when multiple layers are required, and is generally limited to small-scale production.

Each of these fabrication techniques offers unique advantages and challenges, and the choice of method depends on the specific application requirements, the properties of the nanofibers and PLA, and the desired form of the final bionanocomposite. As research in bionanocomposites continues to advance, new fabrication methods and modifications to existing techniques are likely to emerge, further expanding the potential applications of these sustainable materials.

6.4 MECHANICAL PROPERTIES OF NANOFIBER-REINFORCED PLA BIONANOCOMPOSITES

6.4.1 Tensile Strength and Modulus

Nanofiber-reinforced PLA bionanocomposites exhibit significant improvements in tensile strength and modulus compared to pure PLA and conventional composites. Nanofibers such as cellulose nanofibers, chitin nanofibers, and carbon nanotubes

(CNTs) enhances the load transfer between the matrix and the reinforcement due to the high aspect ratio and surface area of nanofibers. This improved load transfer results in composites with higher tensile strength and stiffness.

For instance, PLA reinforced with cellulose nanofibers shows tensile strength increases of up to 50% compared with neat PLA because of the excellent interfacial adhesion between the nanofibers and the PLA matrix, which facilitates efficient stress transfer under tensile loading. The modulus of the composites also increases, which indicates enhanced stiffness, a desirable property for structural applications.

Moreover, the type, dispersion, and concentration of nanofibers play a crucial role in determining the tensile properties. Uniform dispersion of nanofibers within the PLA matrix ensures that stress is evenly distributed across the material, minimizing stress concentrations that can lead to failure. Studies have shown that a 5 wt.% addition of cellulose nanofibers can significantly improve both the tensile strength and modulus of PLA composites, making them comparable to or even superior to some conventional composites.

6.4.2 Flexural Properties

Flexural strength and modulus are critical for applications where materials are subjected to bending forces. Nanofiber-reinforced PLA composites generally show enhanced flexural properties due to the reinforcing effect of the nanofibers. The high modulus of the nanofibers contributes to the overall rigidity of the composite, thereby improving its resistance to bending.

PLA composites reinforced with nanofibers such as chitin and CNTs have demonstrated superior flexural strength compared to neat PLA. For example, the addition of CNTs to PLA has been shown to improve the flexural strength by up to 40%. This improvement is attributed to the strong interaction between the CNTs and the PLA matrix, which resists deformation under bending stress. Additionally, the presence of nanofibers prevents the propagation of cracks during bending, which contributes to the higher flexural modulus observed in these composites.

The flexural properties of nanofiber-reinforced PLA composites are also influenced by the processing techniques used. For example, composites fabricated through melt blending tend to have better flexural properties compared to those produced by solution casting, due to the more uniform distribution of nanofibers in the matrix.

6.4.3 Impact Resistance

Impact resistance is a measure of a material's ability to absorb energy during fracture. Nanofiber-reinforced PLA composites typically exhibit improved impact resistance with neat PLA. Adding nanofibers such as cellulose nanocrystals (CNCs) enhances the toughness of the composite by hindering crack initiation and propagation.

The impact resistance of PLA can be significantly enhanced by incorporating a small percentage of nanofibers. For example, PLA composites reinforced with 3 wt.% CNCs showed a twofold increase in impact strength because of the energy-absorbing capacity of the nanofibers. Moreover, the orientation and dispersion of nanofibers within the PLA matrix play a crucial role in determining the impact properties. Well-dispersed nanofibers with a high aspect ratio provide more surface area for energy absorption, leading to improved impact resistance.

6.4.4 Conventional versus Microfiber Composites

Nanofiber-reinforced PLA composites offer distinct advantages over conventional and microfiber composites in terms of mechanical properties. While conventional composites typically use larger fibers such as glass or carbon fibers, the high surface area and aspect ratio of nanofibers allow for better stress distribution and interfacial bonding, resulting in superior tensile, flexural, and impact properties.

For example, PLA composites reinforced with nanofibers often outperform those reinforced with microfibers in terms of tensile strength and modulus. This is because nanofibers provide more reinforcement at lower loadings, which not only enhances the mechanical properties but also reduces the weight of the composite. Additionally, nanofiber composites exhibit better impact resistance than microfiber composites, as the smaller dimensions of the nanofibers allow for more efficient energy absorption and dissipation during impact. In summary, nanofiber-reinforced PLA bionanocomposites demonstrate superior mechanical properties to those of both neat PLA and conventional composites. These enhancements make them highly suitable for a wide range of applications, from automotive components to packaging materials, where mechanical strength, stiffness, and impact resistance are critical.

6.5 THERMAL PROPERTIES OF NANOFIBER-REINFORCED PLA BIONANOCOMPOSITES

6.5.1 Thermal Stability

The thermal stability of nanofiber-reinforced PLA bionanocomposites is a critical factor that influences their performance in various applications, particularly those involving high-temperature environments. Nanofibers such as cellulose nanofibers, chitin nanofibers, and CNTs significantly enhance the thermal stability of PLA due to their high thermal resistance and the formation of a protective barrier that slows down the degradation of the polymer matrix.

For instance, PLA reinforced with cellulose nanofibers shows improved thermal stability, with higher onset degradation temperature than that with neat PLA. This enhancement is attributed to the high thermal conductivity of the nanofibers, which helps in dissipating heat more efficiently across the composite, thereby delaying the degradation process. Moreover, the strong interfacial adhesion between the nanofibers and the PLA matrix restricts the mobility of polymer chains, reducing the rate of thermal degradation.

The addition of CNTs, known for their exceptional thermal stability, further boosts the thermal resistance of PLA composites. Studies have shown that incorporating CNTs into PLA can raise the thermal degradation temperature by as much as 30–40 °C. This makes CNT-reinforced PLA composites particularly suitable for applications that require materials to withstand elevated temperatures without significant loss of mechanical integrity. Furthermore, the concentration and dispersion of nanofibers within the PLA matrix also play a significant role in determining the thermal stability. Well-dispersed nanofibers create a more uniform barrier to heat transfer, thereby enhancing the overall thermal stability of the composite.

6.5.2 Crystallization

Crystallization is another important thermal property of nanofiber-reinforced PLA bionanocomposites, as it directly affects the material's mechanical and thermal properties. The presence of nanofibers in PLA not only influences the crystallization rate but also alters the crystalline structure, leading to changes in the overall performance of the composite.

Nanofibers act as nucleating agents in the PLA matrix, promoting the formation of crystalline regions, which improves the thermal and mechanical properties of the composite. For example, adding CNCs to PLA significantly accelerates crystallization, leading to a higher degree of crystallinity in the final composite. This increase in crystallinity enhances the rigidity and thermal resistance of the composite, making it more suitable for high-performance applications. The influence of nanofibers on the crystallization behavior of PLA can also lead to the formation of different crystalline forms, such as the α or β phases, which have distinct thermal and mechanical properties. The type of nanofiber used, along with its concentration and dispersion, determines the dominant crystalline phase in the composite. For instance, studies have shown that the presence of chitin nanofibers in PLA can favor the formation of the α-crystalline phase, which is known for its higher thermal stability and mechanical strength.

Additionally, the processing conditions, such as cooling rate and annealing temperature, also play a crucial role in the crystallization behavior of nanofiber reinforced PLA composites. Slow cooling and appropriate annealing can further enhance the crystallinity, leading to improved thermal and mechanical performance.

6.5.3 Influence of Nanofiber Reinforcement

The incorporation of nanofibers into PLA has a profound influence on the overall thermal properties of the composite, including its thermal conductivity, heat deflection temperature (HDT), and glass transition temperature (Tg). Nanofibers, with their high aspect ratio and thermal conductivity, facilitate efficient heat transfer within the composite, thereby improving its thermal management capabilities.

For example, adding CNTs to PLA not only improves the thermal stability but also increases the thermal conductivity of the composite. This makes CNT-reinforced PLA composites ideal for applications where efficient heat dissipation is crucial, such as in electronic packaging or automotive components. The HDT of PLA composites is also enhanced by the presence of nanofibers. Nanofiber reinforcement increases the rigidity of the composite, allowing it to withstand higher temperatures before deforming. This is particularly important for applications that involve prolonged exposure to elevated temperatures. The glass transition temperature (Tg) of PLA composites can be influenced by the type and concentration of nanofibers used. While some nanofibers increase the Tg due to the restricted mobility of polymer chains, others may have a minimal impact depending on their interaction with the PLA matrix. Nanofiber reinforced PLA bionanocomposites exhibit significantly improved thermal properties, including enhanced thermal stability, accelerated crystallization behavior, and better thermal management capabilities. These enhancements make them highly suitable for a wide range of high-performance applications where thermal resistance and stability are critical.

6.6 MORPHOLOGICAL ANALYSIS OF NANOFIBER-REINFORCED PLA BIONANOCOMPOSITES

6.6.1 Microscopy

The morphological analysis of nanofiber-reinforced PLA bionanocomposites is essential for understanding the material's structural integrity, interface interactions, and the effectiveness of nanofiber dispersion. Various microscopy techniques, including scanning electron microscopy (SEM), transmission electron microscopy (TEM), and atomic force microscopy (AFM), are commonly employed to analyze these composites at nano scale. SEM is widely used to examine the surface morphology of PLA bionanocomposites. SEM provides high-resolution images that reveal the distribution and orientation of nanofibers within the PLA matrix. It is particularly useful for identifying surface defects, such as voids or cracks, which can significantly affect the composite's mechanical properties. SEM can also be used to evaluate the fracture surfaces of composites, providing insights into the failure mechanisms. For example, SEM analysis of a fracture surface can reveal whether the failure was due to fiber pull-out, fiber breakage, or matrix cracking.

TEM is another powerful tool used for morphological analysis, especially when investigating the nanofiber–matrix interface and the internal structure of the composite. TEM offers higher resolution than SEM, allowing for the visualization of individual nanofibers within the matrix. This technique is particularly valuable for observing the dispersion of nanofibers, their alignment, and the quality of the interfacial bonding. TEM can also detect the presence of agglomerations or clusters of nanofibers, which can negatively impact the mechanical and thermal properties of the composite. AFM is employed to study the surface topography and mechanical properties at the nanoscale. AFM provides three-dimensional images of the composite surface, enabling the measurement of surface roughness and the identification of nanofiber distribution patterns. It is also used to assess the mechanical properties of the composite at specific points, offering insights into how the presence of nanofibers influences the overall stiffness and elasticity of the material.

6.6.2 Nanofiber–Matrix Interface

The nanofiber–matrix interface plays a critical role in determining the mechanical and thermal properties of nanofiber reinforced PLA bionanocomposites. A strong interface ensures effective load transfer from the matrix to the nanofibers, enhancing the composite's overall strength and stiffness. The nature of the interface is influenced by factors such as the chemical compatibility between the nanofibers and the PLA matrix, the surface functionalization of nanofibers, and the processing techniques used to fabricate the composite.

For instance, cellulose nanofibers often require surface modification to improve their compatibility with the hydrophobic PLA matrix. Surface treatments, such as silanization or grafting with PLA, can enhance interfacial adhesion by promoting covalent bonding or strong intermolecular interactions between the nanofibers and the matrix. Improved interfacial adhesion results in better stress transfer and reduces the likelihood of fiber pull-out during mechanical loading. The quality of the nanofiber-matrix interface can be evaluated using microscopy techniques like TEM, which can reveal the degree of bonding and the presence of any interfacial voids or gaps. A well-bonded interface typically shows no gaps between the nanofibers and the matrix, indicating efficient stress transfer and improved mechanical properties.

6.6.3 Dispersion and Distribution of Nanofibers

The dispersion and distribution of nanofibers within the PLA matrix are crucial factors that significantly affect the performance of the bionanocomposite. Uniform dispersion ensures that the reinforcing effect of the nanofibers is maximized, leading to enhanced mechanical, thermal, and barrier properties. Poor dispersion, on the other hand, can form fiber agglomerates that act as stress concentrators and weaken the composite.

Achieving uniform dispersion of nanofibers in PLA is challenging due to the tendency of nanofibers to agglomerate, driven by strong van der Waals forces and hydrogen bonding between fibers. Several techniques are employed to improve dispersion, including high shear mixing, ultrasonication, and the use of surfactants or compatibilizers. These methods help break down fiber agglomerates and promote a more even distribution of nanofibers within the matrix. SEM and TEM are commonly used to assess the dispersion and distribution of nanofibers in the composite. SEM images can reveal the overall distribution of nanofibers on the composite surface, while TEM provides a more detailed view of the internal structure, allowing for the observation of individual nanofibers within the matrix. AFM can also be used to analyze the surface distribution of nanofibers and assess their impact on the composite's surface properties. In summary, the morphological analysis of nanofiber-reinforced PLA bionanocomposites using techniques like SEM, TEM, and AFM provides valuable insights into the structural characteristics of these materials. Understanding the nanofiber–matrix interface and ensuring uniform dispersion and distribution of nanofibers are key to optimizing the performance of PLA bionanocomposites, making them suitable for a wide range of high-performance applications.

6.7 BARRIER AND FUNCTIONAL PROPERTIES

6.7.1 Gas and Moisture Barrier Properties

Nanofiber-reinforced PLA bionanocomposites exhibit enhanced gas and moisture barrier properties, making them highly suitable for applications in packaging and biomedical fields. The presence of nanofibers such as cellulose, chitin, or carbon nanotubes within the PLA matrix reduces the permeability of gases and moisture through the composite. This improvement is primarily due to the tortuous path that nanofibers create, which hinders the diffusion of gas molecules and moisture.

For instance, cellulose nanofibers, when incorporated into PLA, significantly improve the oxygen barrier properties, which is critical for food packaging applications. The nanofibers increase the crystalline regions within the PLA matrix, which further reduces the free volume available for gas diffusion. Similarly, the incorporation of nanofibers can reduce the moisture absorption rate, enhancing the composite's dimensional stability in humid environments. This moisture resistance is particularly valuable for applications that require prolonged exposure to moisture, such as in medical devices or outdoor equipment.

6.7.2 Antimicrobial and Biodegradation Properties

Nanofiber-reinforced PLA bionanocomposites can be tailored to possess antimicrobial properties, which are highly beneficial for packaging, medical, and hygiene-related applications. By incorporating antimicrobial agents into the nanofibers or functionalizing the nanofibers with antimicrobial compounds, these composites can effectively inhibit the growth of bacteria, fungi, and other pathogens. For example, chitin nanofibers, which inherently possess antimicrobial properties, can be used in PLA composites to prevent microbial growth on surfaces. These properties are particularly useful in food packaging, where they can help extend the shelf life of perishable goods by reducing bacterial contamination. Additionally, silver nanoparticles can be incorporated into nanofibers to impart antimicrobial properties, offering a broad-spectrum defense against pathogens.

The biodegradation properties of PLA composites are also enhanced by the inclusion of natural nanofibers. These fibers, being biodegradable themselves, do not hinder the overall biodegradability of the PLA matrix. In fact, they can accelerate the degradation process by providing additional sites for microbial attack. This is particularly advantageous for applications where environmental sustainability is a priority, as it ensures that the composite materials will break down naturally after their useful life, reducing environmental impact.

6.7.3 Enhanced Functionalities through Nanofiber Incorporation

Incorporating nanofibers into PLA not only enhances the basic mechanical and thermal properties but also introduces a range of additional functionalities that broaden the application scope of these composites. For instance, the inclusion of electrically conductive nanofibers like carbon nanotubes can endow PLA composites with electrical conductivity, making them suitable for use in electronic devices, sensors, and electromagnetic interference shielding.

Moreover, nanofibers can be functionalized with various chemical groups to impart specific functionalities to the PLA composites, such as hydrophobicity, flame retardancy, or UV resistance. These enhanced functionalities make nanofiber-reinforced PLA composites versatile materials for a wide range of applications, from packaging and automotive parts to biomedical devices and electronics.

6.8 FUTURE TRENDS AND RESEARCH DIRECTIONS

The future of nanofiber-reinforced PLA bionanocomposites lies in the continuous innovation of fabrication techniques. Advances in processing methods such as electrospinning, 3D printing, and melt compounding are expected to play a pivotal role in the large-scale production of these composites. These techniques not only improve the dispersion of nanofibers within the PLA matrix but also enable the production of complex composite structures with tailored properties. For instance, electrospinning allows for the precise control of nanofiber alignment and orientation, which can be used to enhance specific properties like tensile strength or electrical conductivity. Additionally, the development of scalable and cost-effective fabrication methods will be crucial for the widespread adoption of these materials in various industries.

The development of sustainable nanofibers is another key area of research in the field of PLA bionanocomposites. Researchers are exploring the use of renewable and biodegradable sources for nanofiber production, such as agricultural waste, algae, and bacterial cellulose. These sustainable nanofibers not only reduce the environmental impact of the composites but also offer unique properties that can be tailored for specific applications. In the future, the integration of biobased nanofibers with PLA could lead to the creation of fully biodegradable and environmentally friendly composites with minimal ecological footprint. This aligns with global efforts to reduce plastic waste and promote sustainable materials in various sectors, including packaging, automotive, and consumer goods.

Ongoing research is focused on further improving the performance of nanofiber reinforced PLA bionanocomposites, particularly in areas such as mechanical strength, thermal stability, and barrier properties. By optimizing the type, size, and concentration of nanofibers, as well as enhancing the interfacial bonding between the nanofibers and the PLA matrix, researchers aim to develop composites with superior performance characteristics. In addition to performance improvements, there is a growing emphasis on reducing the environmental impact of these composites. This includes exploring methods to enhance the biodegradability of the composites, reduce energy consumption during production, and utilize sustainable resources for nanofiber and PLA production. The development of life cycle assessment models specific to these composites will also play a critical role in evaluating and minimizing their environmental impact.

6.9 CONCLUSIONS

Nanofiber reinforced PLA bionanocomposites represent a significant advancement in the field of sustainable materials, offering a unique combination of enhanced mechanical, thermal, and barrier properties with the added benefits of biodegradability and environmental sustainability. The incorporation of nanofibers into PLA not only improves the structural integrity and performance of the composites but also introduces a range of functional properties that broaden their application potential. The use of advanced microscopy techniques has provided valuable insights into the morphology and interface characteristics of these composites, highlighting the importance of uniform nanofiber dispersion and strong interfacial bonding in achieving optimal performance. The development of antimicrobial properties and enhanced biodegradation through nanofiber incorporation further underscores the versatility and environmental benefits of these materials. Future trends in the field are expected to focus on innovations in fabrication techniques, advances in sustainable nanofiber production, and continuous improvements in composite performance and environmental impact. The ongoing research and development efforts will likely lead to the widespread adoption of nanofiber reinforced PLA bionanocomposites in various industries, contributing to a more sustainable and environmentally friendly future. The findings of this study have important implications for both industry and research, providing a foundation for the development of next-generation bionanocomposites with tailored properties for specific applications. As the field continues to evolve, the potential for further advancements in material performance and sustainability remains high, offering exciting opportunities for innovation and growth.

REFERENCES

Ahmed, Jasim, Yasir Ali Arfat, Anibal Bher, Mehrajfatema Mulla, Harsha Jacob, and Rafael Auras. 2018. "Active Chicken Meat Packaging Based on Polylactide Films and Bimetallic Ag – Cu Nanoparticles and Essential Oil." *Journal of Food Science* 83 (5): 1299–1310. https://doi.org/10.1111/1750-3841.14121.

Alloin, Fannie, Alessandra D'Aprea, Nadia El Kissi, Alain Dufresne, and Frédéric Bossard. 2010. "Nanocomposite Polymer Electrolyte Based on Whisker or Microfibrils Polyoxyethylene Nanocomposites." *Electrochimica Acta* 55 (18): 5186–5194. https://doi.org/10.1016/j.electacta.2010.04.034.

Antunes, Ana, Anton Popelka, Omar Aljarod, Mohammad K. Hassan, and Adriaan S. Luyt. 2020. "Effects of Rutile – TiO2 Nanoparticles on Accelerated Weathering Degradation of Poly(Lactic Acid)." *Polymers* 12 (5): 1096. https://doi.org/10.3390/polym12051096.

Asadi, Sima, and Sajad Pirsa. 2020. "Production of Biodegradable Film Based on Polylactic Acid, Modified with Lycopene Pigment and TiO2 and Studying Its Physicochemical Properties." *Journal of Polymers and the Environment* 28 (2): 433–444. https://doi.org/10.1007/s10924-019-01618-5.

Astruc, Jérémy, Malladi Nagalakshmaiah, Gaétan Laroche, Michel Grandbois, Saïd Elkoun, and Mathieu Robert. 2017. "Isolation of Cellulose-II Nanospheres from Flax Stems and Their Physical and Morphological Properties." *Carbohydrate Polymers* 178: 352–359. https://doi.org/10.1016/j.carbpol.2017.08.138.

Bodin, Aase, Henrik Bäckdahl, Helen Fink, Lena Gustafsson, Bo Risberg, and Paul Gatenholm. 2007. "Influence of Cultivation Conditions on Mechanical and Morphological Properties of Bacterial Cellulose Tubes." *Biotechnology and Bioengineering* 97 (2): 425–434. https://doi.org/10.1002/bit.21314.

Cao, Xinwang, Bin Ding, Jianyong Yu, and Salem S. Al-Deyab. 2013. "In Situ Growth of Silver Nanoparticles on TEMPO-Oxidized Jute Fibers by Microwave Heating." *Carbohydrate Polymers* 92 (1): 571–576. https://doi.org/10.1016/j.carbpol.2012.08.091.

Carus, Michael, D. Carrez, Harald Kaeb, J. Ravenstijn, and Joachim Venus. 2010. "Level Playing Field for Bio-Based Chemistry and Materials." *Policy Paper on Bio-Based Economy in the EU.*

Chan, Chi Hoong, Chin Hua Chia, Sarani Zakaria, Ishak Ahmad, and Alain Dufresne. 2013. "Production and Characterisation of Cellulose and Nano-Crystalline Cellulose from Kenaf Core Wood." *BioResources* 8 (1): 785–794. http://dx.doi.org/10.15376/biores.8.1.785-794.

Compton, Brett G., and Jennifer A. Lewis. 2014. "3D-Printing of Lightweight Cellular Composites." *Advanced Materials* 26 (34): 5930–5935. https://doi.org/10.1002/adma.201401804.

Csiszár, Emilia, and Sebestyén Nagy. 2017. "A Comparative Study on Cellulose Nanocrystals Extracted from Bleached Cotton and Flax and Used for Casting Films with Glycerol and Sorbitol Plasticisers." *Carbohydrate Polymers* 174: 740–749. https://doi.org/10.1016/j.carbpol.2017.06.103.

Cumkur, Eda Acik, Touffik Baouz, and Ulku Yilmazer. 2015. "Poly(Lactic Acid)–Layered Silicate Nanocomposites: The Effects of Modifier and Compatibilizer on the Morphology and Mechanical Properties." *Journal of Applied Polymer Science* 132 (38). https://doi.org/10.1002/app.42553.

Dai, Dasong, Mizi Fan, and Philip Collins. 2013. "Fabrication of Nanocelluloses from Hemp Fibers and Their Application for the Reinforcement of Hemp Fibers." *Industrial Crops and Products* 44: 192–199. https://doi.org/10.1016/j.indcrop.2012.11.010.

Das, Kunal, Dipa Ray, N.R. Bandyopadhyay, Saswata Sahoo, Amar K. Mohanty, and Manjusri Misra. 2011. "Physico-Mechanical Properties of the Jute Micro/Nanofibril Reinforced Starch/Polyvinyl Alcohol Biocomposite Films." *Composites Part B: Engineering* 42 (3): 376–381. https://doi.org/10.1016/j.compositesb.2010.12.017.

Dash, Sukalyan, Soumyashri Mishra, Sabita Patel, and Bijay K. Mishra. 2008. "Organically Modified Silica: Synthesis and Applications Due to Its Surface Interaction with Organic Molecules." *Advances in Colloid and Interface Science* 140 (2): 77–94. https://doi.org/10.1016/j.cis.2007.12.006.

Dorgan, John R., Hans Lehermeier, and Michael Mang. 2000. "Thermal and Rheological Properties of Commercial-Grade Poly(Lactic Acid)s." *Journal of Polymers and the Environment* 8 (1): 1–9. https://doi.org/10.1023/A:1010185910301.

Fonseca, Carmen, Almudena Ochoa, Maria Teresa Ulloa, Eduardo Alvarez, Daniel Canales, and Paula A. Zapata. 2015. "Poly(Lactic Acid)/TiO2 Nanocomposites as Alternative Biocidal and Antifungal Materials." *Materials Science and Engineering: C* 57 (December): 314–320. https://doi.org/10.1016/j.msec.2015.07.069.

Govender, T. 2000. "Defining the Drug Incorporation Properties of PLA – PEG Nanoparticles." *International Journal of Pharmaceutics* 199 (1): 95–110. https://doi.org/10.1016/S0378-5173(00)00375-6.

Hoque, Md. Enamul. 2013. "Processing and Characterization of Cockle Shell Calcium Carbonate (CaCO3) Bioceramic for Potential Application in Bone Tissue Engineering." *Journal of Material Science & Engineering* 02 (04). https://doi.org/10.4172/2169-0022.1000132.

Ikada, Yoshito, and Hideto Tsuji. 2000. "Biodegradable Polyesters for Medical and Ecological Applications." *Macromolecular Rapid Communications* 21 (3): 117–132. https://doi.org/10.1002/(SICI)1521-3927(20000201)21:3<117::AID-MARC117>3.0.CO;2-X.

Jabbar, Abdul, Jiří Militký, Jakub Wiener, Bandu Madhukar Kale, Usman Ali, and Samson Rwawiire. 2017. "Nanocellulose Coated Woven Jute/Green Epoxy Composites: Characterization of Mechanical and Dynamic Mechanical Behavior." *Composite Structures* 161: 340–349. https://doi.org/10.1016/j.compstruct.2016.11.062.

Jonoobi, Mehdi, Jalaludin Harun, Paridah Md. Tahir, Alireza Shakeri, Syeed SaifulAzry, and Majid Davoodi Makinejad. 2011. "Physicochemical Characterization of Pulp and Nanofibers from Kenaf Stem." *Materials Letters* 65 (7): 1098–1100. https://doi.org/10.1016/j.matlet.2010.08.054.

Karimi, Samaneh, Paridah Md. Tahir, Ali Karimi, Alain Dufresne, and Ali Abdulkhani. 2014. "Kenaf Bast Cellulosic Fibers Hierarchy: A Comprehensive Approach from Micro to Nano." *Carbohydrate Polymers* 101: 878–885. https://doi.org/10.1016/j.carbpol.2013.09.106.

Kasyapi, Nibedita, Vidhi Chaudhary, and Anil K. Bhowmick. 2013. "Bionanowhiskers from Jute: Preparation and Characterization." *Carbohydrate Polymers* 92 (2): 1116–1123. https://doi.org/10.1016/j.carbpol.2012.10.021.

Katiyar, Vimal, Nathalie Gerds, Christian Bender Koch, Jens Risbo, Hans Christian B. Hansen, and David Plackett. 2010. "Poly L-Lactide-Layered Double Hydroxide Nanocomposites via in Situ Polymerization of L-Lactide." *Polymer Degradation and Stability* 95 (12): 2563–2573. https://doi.org/10.1016/j.polymdegradstab.2010.07.031.

Kian, Lau Kia, Mohammad Jawaid, Hidayah Ariffin, and Zoheb Karim. 2018. "Isolation and Characterization of Nanocrystalline Cellulose from Roselle-Derived Microcrystalline Cellulose." *International Journal of Biological Macromolecules* 114: 54–63. https://doi.org/10.1016/j.ijbiomac.2018.03.065.

Lai, Sun-Mou, Sheng-Huang Wu, Gwo-Geng Lin, and Trong-Ming Don. 2014. "Unusual Mechanical Properties of Melt-Blended Poly(Lactic Acid) (PLA)/Clay Nanocomposites." *European Polymer Journal* 52 (March): 193–206. https://doi.org/10.1016/j.eurpolymj.2013.12.012.

Li, Yonghui, Caihong Chen, Jun Li, and Xiuzhi Susan Sun. 2011. "Synthesis and Characterization of Bionanocomposites of Poly(Lactic Acid) and TiO2 Nanowires by in Situ Polymerization." *Polymer* 52 (11): 2367–2375. https://doi.org/10.1016/j.polymer.2011.03.050.

Li, Yongjin, and Hiroshi Shimizu. 2009. "Improvement in Toughness of Poly(L-Lactide) (PLLA) through Reactive Blending with Acrylonitrile – Butadiene – Styrene Copolymer (ABS): Morphology and Properties." *European Polymer Journal* 45 (3): 738–746. https://doi.org/10.1016/j.eurpolymj.2008.12.010.

Liao, Ruogu, Bin Yang, Wei Yu, and Chixing Zhou. 2007. "Isothermal Cold Crystallization Kinetics of Polylactide/Nucleating Agents." *Journal of Applied Polymer Science* 104 (1): 310–317. https://doi.org/10.1002/app.25733.

Lin, Jinyou, Liangbo Yu, Feng Tian, Nie Zhao, Xiuhong Li, Fenggang Bian, and Jie Wang. 2014. "Cellulose Nanofibrils Aerogels Generated from Jute Fibers." *Carbohydrate Polymers* 109: 35–43. https://doi.org/10.1016/j.carbpol.2014.03.045.

Liu, D.Y., X.W. Yuan, D. Bhattacharyya, and A.J. Easteal. 2010. "Characterisation of Solution Cast Cellulose Nanofibre – Reinforced Poly(Lactic Acid)." *Express Polymer Letters* 4 (1): 26–31. https://doi.org/10.3144/expresspolymlett.2010.5.

Lule, Zelalem, and Jooheon Kim. 2019. "Thermally Conductive and Highly Rigid Polylactic Acid (PLA) Hybrid Composite Filled with Surface Treated Alumina/Nano-Sized Aluminum Nitride." *Composites Part A: Applied Science and Manufacturing* 124 (September): 105506. https://doi.org/10.1016/j.compositesa.2019.105506.

Luzi, F., Elena Fortunati, A. Jiménez, Debora Puglia, Daniela Pezzolla, Giovanni Gigliotti, Jose M. Kenny, Amparo Chiralt, and L. Torre. 2016. "Production and Characterization of PLA_PBS Biodegradable Blends Reinforced with Cellulose Nanocrystals Extracted from Hemp Fibres." *Industrial Crops and Products* 93: 276–289. https://doi.org/10.1016/j.indcrop.2016.01.045.

Luzi, F., Elena Fortunati, D. Puglia, M. Lavorgna, C. Santulli, J.M. Kenny, and L. Torre. 2014. "Optimized Extraction of Cellulose Nanocrystals from Pristine and Carded Hemp Fibres." *Industrial Crops and Products* 56 (May): 175–186. https://doi.org/10.1016/j.indcrop.2014.03.006.

Madhavan Nampoothiri, K., Nimisha Rajendran Nair, and Rojan Pappy John. 2010. "An Overview of the Recent Developments in Polylactide (PLA) Research." *Bioresource Technology* 101 (22): 8493–8501. https://doi.org/10.1016/j.biortech.2010.05.092.

Malinowski, Rafał, Katarzyna Janczak, Piotr Rytlewski, Aneta Raszkowska-Kaczor, Krzysztof Moraczewski, and Tomasz Żuk. 2015. "Influence of Glass Microspheres on Selected Properties of Polylactide Composites." *Composites Part B: Engineering* 76 (July): 13–19. https://doi.org/10.1016/j.compositesb.2015.02.013.

Meng, Bing, Jian Tao, Jingjing Deng, Zhihua Wu, and Mingbo Yang. 2011. "Toughening of Polylactide with Higher Loading of Nano-Titania Particles Coated by Poly(ε-Caprolactone)." *Materials Letters* 65 (4): 729–732. https://doi.org/10.1016/j.matlet.2010.11.029.

Mujtaba, Muhammad, Asier M. Salaberria, María A. Andres, Murat Kaya, Ayse Gunyakti, and Jalel Labidi. 2017. "Utilization of Flax (Linum Usitatissimum) Cellulose Nanocrystals as Reinforcing Material for Chitosan Films." *International Journal of Biological Macromolecules* 104 (November): 944–952. https://doi.org/10.1016/j.ijbiomac.2017.06.127.

Mulla, Mehrajfatema Zafar, Md. Ramim Tanver Rahman, Begonya Marcos, Brijesh Tiwari, and Shivani Pathania. 2021. "Poly Lactic Acid (PLA) Nanocomposites: Effect of Inorganic Nanoparticles Reinforcement on Its Performance and Food Packaging Applications." *Molecules* 26 (7): 1967. https://doi.org/10.3390/molecules26071967.

Murariu, Marius, Anne Laure Dechief, Leila Bonnaud, Yoann Paint, Antoine Gallos, Gaëlle Fontaine, Serge Bourbigot, and Philippe Dubois. 2010. "The Production and Properties of Polylactide Composites Filled with Expanded Graphite." *Polymer Degradation and Stability* 95 (5): 889–900. https://doi.org/10.1016/j.polymdegradstab.2009.12.019.

Orasugh, Jonathan Tersur, Nayan Ranjan Saha, Dipak Rana, Gunjan Sarkar, Md. Masud Rahaman Mollick, Atiskumar Chattoapadhyay, Bhairab Chandra Mitra, Dibyendu Mondal, Swapan Kumar Ghosh, and Dipankar Chattopadhyay. 2018. "Jute Cellulose Nano-Fibrils/Hydroxypropylmethylcellulose Nanocomposite: A Novel Material with Potential for Application in Packaging and Transdermal Drug Delivery System." *Industrial Crops and Products* 112: 633–643. https://doi.org/10.1016/j.indcrop.2017.12.069.

Pacaphol, Kamonwan, and Duangdao Aht-Ong. 2016. "Preparation of Hemp Nanofibers from Agricultural Waste by Mechanical Defibrillation in Water." 142: 1283–1295. https://doi.org/10.1016/j.jclepro.2016.09.008.

Pal, Nidhi, Poornima Dubey, P. Gopinath, and Kaushik Pal. 2017. "Combined Effect of Cellulose Nanocrystal and Reduced Graphene Oxide into Poly-Lactic Acid Matrix Nanocomposite as a Scaffold and Its Anti-Bacterial Activity." *International Journal of Biological Macromolecules* 95 (February): 94–105. https://doi.org/10.1016/j.ijbiomac.2016.11.041.

Pantani, Roberto, Giuliana Gorrasi, Giovanni Vigliotta, Marius Murariu, and Philippe Dubois. 2013. "PLA-ZnO Nanocomposite Films: Water Vapor Barrier Properties and Specific End-Use Characteristics." *European Polymer Journal* 49 (11): 3471–3482. https://doi.org/10.1016/j.eurpolymj.2013.08.005.

Pinto, Artur M., Joana Cabral, David A. Pacheco Tanaka, Adélio M. Mendes, and Fernão D. Magalhães. 2013. "Effect of Incorporation of Graphene Oxide and Graphene Nanoplatelets on Mechanical and Gas Permeability Properties of Poly(Lactic Acid) Films." *Polymer International* 62 (1): 33–40. https://doi.org/10.1002/pi.4290.

Pyda, M., and B. Wunderlich. 2005. "Reversing and Nonreversing Heat Capacity of Poly(Lactic Acid) in the Glass Transition Region by TMDSC." *Macromolecules* 38 (25): 10472–10479. https://doi.org/10.1021/ma051611k.

Ramanadha reddy, S., and Dr. N. Venkatachalapathi. 2023. "A Review on Characteristic Variation in PLA Material with a Combination of Various Nano Composites." *Materials Today: Proceedings* (May). https://doi.org/10.1016/j.matpr.2023.04.616.

Ramanjaneyulu, B., N. Venkatachalapathi, and G. Prasanthi. 2022. "Melt Rheology Properties of PLA/ABS/TCS Polymer Blends." *International Journal of Ambient Energy* 43 (1): 2536–2539. https://doi.org/10.1080/01430750.2020.1745885.

Râpă, Maria, Maria Stefan, Paula Adriana Popa, Dana Toloman, Cristian Leostean, Gheorghe Borodi, Dan Cristian Vodnar, et al. 2021. "Electrospun Nanosystems Based on PHBV and ZnO for Ecological Food Packaging." *Polymers* 13 (13): 2123. https://doi.org/10.3390/polym13132123.

Raquez, J.-M., Y. Murena, A.-L. Goffin, Y. Habibi, B. Ruelle, F. DeBuyl, and P. Dubois. 2012. "Surface-Modification of Cellulose Nanowhiskers and Their Use as Nanoreinforcers into Polylactide: A Sustainably-Integrated Approach." *Composites Science and Technology* 72 (5): 544–549. https://doi.org/10.1016/j.compscitech.2011.11.017.

Reddy, Jeevan Prasad, and Jong-Whan Rhim. 2014. "Characterization of Bionanocomposite Films Prepared with Agar and Paper-Mulberry Pulp Nanocellulose." *Carbohydrate Polymers* 110 (22): 480–488. https://doi.org/10.1016/j.carbpol.2014.04.056.

Rokbani, Hajer, France Daigle, and Abdellah Ajji. 2018. "Combined Effect of Ultrasound Stimulations and Autoclaving on the Enhancement of Antibacterial Activity of ZnO and SiO2/ZnO Nanoparticles." *Nanomaterials* 8 (3): 129. https://doi.org/10.3390/nano8030129.

Sharif, Ahmed, Sudipta Mondal, and Md. Enamul Hoque. 2019. "Polylactic Acid (PLA)-Based Nanocomposites: Processing and Properties." In *Bio-Based Polymers and Nanocomposites*, 233–254. Cham: Springer International Publishing. https://doi.org/10.1007/978-3-030-05825-8_11.

Škrlová, Kateřina, Kateřina Malachová, Alexandra Muñoz-Bonilla, Dagmar Měřinská, Zuzana Rybková, Marta Fernández-García, and Daniela Plachá. 2019. "Biocompatible Polymer Materials with Antimicrobial Properties for Preparation of Stents." *Nanomaterials* 9 (11): 1548. https://doi.org/10.3390/nano9111548.

Suryanegara, Lisman, Antonio Norio Nakagaito, and Hiroyuki Yano. 2009. "The Effect of Crystallization of PLA on the Thermal and Mechanical Properties of Microfibrillated Cellulose-Reinforced PLA Composites." *Composites Science and Technology* 69 (7–8): 1187–1192. https://doi.org/10.1016/j.compscitech.2009.02.022.

Swaroop, Chetan, and Mukul Shukla. 2018. "Nano-Magnesium Oxide Reinforced Polylactic Acid Biofilms for Food Packaging Applications." *International Journal of Biological Macromolecules* 113 (July): 729–736. https://doi.org/10.1016/j.ijbiomac.2018.02.156.

Tang, Gang, Rui Zhang, Xin Wang, Bibo Wang, Lei Song, Yuan Hu, and Xinglong Gong. 2013. "Enhancement of Flame Retardant Performance of Bio-Based Polylactic Acid Composites with the Incorporation of Aluminum Hypophosphite and Expanded Graphite." *Journal of Macromolecular Science, Part A* 50 (2): 255–269. https://doi.org/10.1080/10601325.2013.742835.

Thomas, Martin George, Eldho Abraham, P. Jyotishkumar, Hanna J. Maria, Laly A. Pothen, and Sabu Thomas. 2015. "Nanocelluloses from Jute Fibers and Their Nanocomposites with Natural Rubber: Preparation and Characterization." *International Journal of Biological Macromolecules* 81: 768–777. https://doi.org/10.1016/j.ijbiomac.2015.08.053.

Tingaut, Philippe, Tanja Zimmermann, and Francisco Lopez-Suevos. 2010. "Synthesis and Characterization of Bionanocomposites with Tunable Properties from Poly(Lactic Acid) and Acetylated Microfibrillated Cellulose." *Biomacromolecules* 11 (2): 454–464. https://doi.org/10.1021/bm901186u.

Uribe-Calderón, Jorge, Denis Rodrigue, Valerian Hirschberg, Victor Vladimir Fernández-Escamilla, and Soledad Cecilia Pech-Cohuo. 2023. "Influence of Surface-Modified Cellulose Nanocrystal on the Rheological, Thermal and Mechanical Properties of PLA Nanocomposites." *Polymer Bulletin* 80 (9): 10193–10213. https://doi.org/10.1007/s00289-022-04556-w.

Vallet-Regí, M., S. Granado, D. Arcos, M. Gordo, M.V. Cabanas, C.V. Ragel, A.J. Salinas, A.L. Doadrio, and J. San Román. 1998. "Preparation, Characterization, Andin Vitro Release of Ibuprofen from Al2O3/PLA/PMMA Composites." *Journal of Biomedical Materials Research* 39 (3): 423–428. https://doi.org/10.1002/(SICI)1097-4636(19980305)39:3<423::AID-JBM11>3.0.CO;2-B.

Wang, Longfeng, Shiv Shankar, and Jong-Whan Rhim. 2017. "Properties of Alginate-Based Films Reinforced with Cellulose Fibers and Cellulose Nanowhiskers Isolated from Mulberry Pulp." *Food Hydrocolloids* 63: 201–208. https://doi.org/10.1016/j.foodhyd.2016.08.041.

Wen, Xin, Kunyu Zhang, Yan Wang, Lijing Han, Changyu Han, Huiliang Zhang, Shan Chen, and Lisong Dong. 2011. "Study of the Thermal Stabilization Mechanism of Biodegradable Poly(L-Lactide)/Silica Nanocomposites." *Polymer International* 60 (2): 202–210. https://doi.org/10.1002/pi.2927.

Yan, Shifeng, Jingbo Yin, Yan Yang, Zhengzhan Dai, Jia Ma, and Xuesi Chen. 2007. "Surface-Grafted Silica Linked with L-Lactic Acid Oligomer: A Novel Nanofiller to Improve the Performance of Biodegradable Poly(L-Lactide)." *Polymer* 48 (6): 1688–1694. https://doi.org/10.1016/j.polymer.2007.01.037.

Zaini, Lukmanul Hakim, Mehdi Jonoobi, Paridah Md. Tahir, and Samaneh Karimi. 2013. "Isolation and Characterization of Cellulose Whiskers from Kenaf (Hibiscus Cannabinus L.) Bast Fibers." *Biomaterials and Nanobiotechnology* 4: 37–44. https://doi.org/10.4236/jbnb.2013.41006.

Zainuddin, Norhidayu, Ishak Ahmad, Hanieh Kargarzadeh, and Suria Ramli. 2017. "Hydrophobic Kenaf Nanocrystalline Cellulose for the Binding of Curcumin." *Carbohydrate Polymers* 163: 261–269. https://doi.org/10.1016/j.carbpol.2017.01.036.

Zhang, Haichen, Jintao Huang, Li Yang, Rongyuan Chen, Wei Zou, Xiangkun Lin, and Jinping Qu. 2015. "Preparation, Characterization and Properties of PLA/TiO2 Nanocomposites Based on a Novel Vane Extruder." *RSC Advances* 5 (6): 4639–4647. https://doi.org/10.1039/C4RA14538K.

Zheng, Xiaotong, Shaobing Zhou, Yu Xiao, Xiongjun Yu, Xiaohong Li, and Peizhuo Wu. 2009. "Shape Memory Effect of Poly(D,L-Lactide)/Fe3O4 Nanocomposites by Inductive Heating of Magnetite Particles." *Colloids and Surfaces B: Biointerfaces* 71 (1): 67–72. https://doi.org/10.1016/j.colsurfb.2009.01.009.

Zhu, Aiping, Huaxin Diao, Qianping Rong, and Auyun Cai. 2010. "Preparation and Properties of Polylactide – Silica Nanocomposites." *Journal of Applied Polymer Science* 116 (5): 2866–2873. https://doi.org/10.1002/app.31786.

7 Mechanical Properties of Natural Fiber PLA Composites

7.1 INTRODUCTION

Natural fiber polymer composites are a unique class of materials that combine a polymer matrix with natural fibers for reinforcement. One such polymer that has gained significant attention is polylactic acid (PLA), a biodegradable and renewable thermoplastic polyester derived from organic sources such as corn starch or sugar cane (Lee et al., 2021). The combination of PLA and natural fibers results in composites that are not only biodegradable and recyclable but also exhibit high mechanical strength, low toxicity, good barrier properties, and excellent processing characteristics. These PLA-based composites have found extensive use in various sectors including packaging, medicine, and agriculture (Ilyas et al., 2022).

Natural fiber composites (NFCs) are a subset of biocomposites that consist of natural fibers embedded within a biodegradable matrix like PLA. NFCs have garnered considerable interest due to their environmental benefits such as biodegradability, recyclability, renewability, and low carbon footprint. Additionally, they offer advantages such as low cost, low density, high specific strength and stiffness, good thermal insulation, and easy processing. NFCs have been widely used in various applications ranging from packaging to automotive, construction, medical, and consumer products. However, NFCs also face some challenges that limit their performance and durability. These include poor interfacial adhesion between the fiber and the matrix, high moisture absorption of the fiber, low fire resistance, low impact strength, and low resistance to weathering and biological degradation (Gholampour & Ozbakkaloglu, 2020). Therefore, many researchers have investigated the effects of different factors on the mechanical properties of NFCs.

The research gap in the field of natural fiber and PLA composites is quite extensive. The major focus of recent research on natural fiber composites includes fiber surface treatment, fiber agglomeration and dispersion, interfacial transcrystallinity, impact strength, foaming, inflaming retardance, biodegradable resin matrix, and nanofiber reinforcement (Periasamy et al., 2024; Sun, 2018; Wang et al., 2016). However, there are still challenges in improving the mechanical properties of these composites. The type of fiber used plays an important role in fiber/matrix adhesion and thereby affects the mechanical performance of the biocomposites. There are also limitations in the use of PLA in many applications due to its properties. The combined technique of hybridization and annealing has been suggested as a strategy

DOI: 10.1201/9781032678894-7

for improving the properties of 100% biocomposites and helping overcome some limitations of the use of PLA (Yaisun & Trongsatitkul, 2023).

This chapter aims to review the recent literature on the mechanical behavior of NFCs based on PLA matrix and various natural fibers. The focus will be on analyzing the tensile, flexural, and impact properties of these composites while considering different parameters that influence these properties. The chapter will also identify current trends and gaps in knowledge while suggesting directions for future research.

7.2 MECHANICAL BEHAVIOR OF PLA

PLA is a thermoplastic aliphatic polyester that is biodegradable and generated from renewable resources such as sugar, corn, beet, and cassava (Satsum et al., 2022). It has received a lot of attention in a variety of applications, including medical devices, packaging, and 3D printing (Bakar et al., 2022; DeStefano et al., 2020). PLA is well-known for having high durability and strength compared with synthetic polymers.

7.2.1 PLA as a Structural Polymer

PLA is a potential structural polymer with high stiffness and strength that can be used in various applications. It has a high mechanical strength, with flexural strength of up to 140 MPa and a Young's modulus of approximately 2.5 GPa (Oksiuta et al., 2020). However, it has numerous drawbacks, including brittleness, flexibility, long degradation periods, and expensive production costs (Satsum et al., 2022). Researchers have investigated numerous modifications and composites with other materials to increase their mechanical qualities.

7.2.2 Mechanical Performance of PLA

PLA has excellent mechanical strength, but its brittleness and severe degradation periods limit its applicability. To improve its mechanical behavior, researchers have been experimenting with various modifications and composites. Researchers, for example, investigated the mechanical and thermal properties of PLA composites that had been changed using Mg, Fe, and polyethylene (PE) additives (Oksiuta et al., 2020). The results showed that adding Mg and Fe increased PLA's mechanical qualities, whereas adding PE improved its thermal properties.

Another study looked at how temperature affected the physical qualities of PLA structures (Bakar et al., 2022). Temperature and environmental factors, such as humidity and air pressure, were found to impact the physical properties of PLA. Understanding these aspects can aid in the optimization of PLA performance in a variety of applications.

7.2.3 Factors Affecting Mechanical Behavior

Several factors can influence the mechanical behavior of PLA. For instance, molecular weight (Mw) has a significant effect on qualities including degradation, mechanical strength, and solubility. High Mw PLA is more suited for medical applications because it is completely absorbed into the body (Farah et al., 2016). Additionally, PLA's mechanical characteristics are influenced by its crystallinity. Pectin, for example, improves the tensile and impact strength of PLA composites by stimulating polymer recrystallization and the formation of microspherical crystal formations (Satsum et al., 2022). Finally, temperature and environmental variables like humidity and air pressure can affect the physical qualities of PLA. Understanding these aspects can aid in optimizing PLA performance in various applications (Bakar et al., 2022).

7.3 MATERIAL TESTING AND CHARACTERIZATION

7.3.1 Testing Standards and Procedures

PLA composites can be tested using a variety of standards and methodologies. These standards are useful for assessing and determining the physical, shear, tensile, flexural, and compressive properties of different composite materials. Some of the relevant PLA composite testing standards and processes include ASTM and International Standards Organization (ISO) standards, both standard testing methods for mechanical characteristics and failure of polymers and polymer composites manufactured by additive manufacturing procedures (Forster, 2015). ASTM also provides a variety of composite testing standards such as ASTM D3039, ASTM D3410, and ASTM D7264, which provide tensile, compression, flexural, and shear testing standards for composite material. Finally, investigators can rely on research findings, such as those published by the National Centre for Biotechnology Information (NCBI) on the production and evaluation of PLA matrix composites reinforced with regenerated cellulose fibers for fused deposition modeling, demonstrating the importance of testing procedures in the additive manufacturing process (Gauss et al., 2022).

7.3.2 Tensile Testing

Tensile testing is an established method to assess the mechanical properties of PLA composites. Tensile and compression testing were performed on PLA specimens manufactured using fused deposition modeling (FDM) in one study. The researchers discovered that the Young's modulus, proportional limit, and maximum strength of the PLA specimens allow them to confirm the distinct behavior of FDM-printed PLA material in tensile and compressive states (Sabik et al., 2022). Other researcher has investigated the tensile behavior of 3D-printed PLA composites reinforced with natural fibers, carbon fibers, and other materials (Agaliotis et al., 2022; Maqsood & Rimašauskas, 2021; Grabowik et al., 2017). This research provided insights into the performance and prospective applications of PLA composites by reporting on their tensile strength, Young's modulus, and other mechanical properties as shown in Table 7.1.

7.3.3 Flexural Testing

Flexural testing procedures such as ASTM D790-17 can be used to measure the flexural strength and stiffness of these composites (John & Naidu, 2004).

TABLE 7.1
PLA Tensile Test Data

S. No.	Peak Load [N]	Peak Stress [Mpa]	Strain at Peak Stress [%]	Modulus [Mpa]
1	1461.407	36.5	4.9	1081.730
2	1724.771	43.1	5.0	1228.285
3	1537.018	38.4	5.3	1086.438
4	1518.917	38.0	5.0	1100.184
5	1530.403	38.3	5.3	1074.699

Source: Grabowik et al. (2017).

TABLE 7.2
Physical Characterization of PLA/Natural Fibers Composites

Composites	Fiber (%)	Physical characterization
PLA/jute	40	Moisture absorption, ageing behavior in hygrothermal environment
PLA/hemp	30	Influence of accelerated ageing on physico-mechanical properties of composite
PLA/hemp	30	Water absorption, Influence of hygrothermal ageing on physico-mechanical properties
PLA/KDC	0–60	Density, water absorption
PLA/bamboo fabric	51	Density, moisture absorption
PLA/ramie/flax/cotton	0–50	Effect of water content of natural fibers on PLA degradation

Source: Siakeng et al. (2019).

7.3.3.1 Effect of Fiber Content and Treatment

Factors such as fiber content and surface treatment influence the flexural characteristics of natural fiber-reinforced PLA composites. According to studies, the flexural strengths of these composites rise with fiber content, reaching a maximum at a particular percentage (Ilyas et al., 2022). Furthermore, fiber treatment, such as alkali treatment, can improve composite flexural strength (Singh et al., 2021).

7.3.3.2 Flexural Strength Micromechanics Models

The flexural strength of natural fiber-reinforced PLA composites is predicted using micromechanics models. These models provide some insight into the relationship between composite microstructure and flexural characteristics as presented in Table 7.2.

7.3.4 Impact Testing

There is limited data on the impact testing of natural fiber-reinforced PLA composites. However, in one study, Kalu et al. (2023) examined the impact test analysis of natural fiber-reinforced polymer composites for industrial/household applications and discovered that increasing the fiber content increased the impact strength of the composites, while adding coupling agents improved the interface bonding between the fibers and the matrix, resulting in higher impact strength. Specifically, bamboo fibers were alkali-treated, which contributed to their rigidity. KH560 solution was applied to crushed cellulose fibers. KH560-modified composites demonstrated improved impact toughness and ductility. Maleic anhydride grafting also improved composite stiffness and ductility.

Other researchers examined the impact strength of natural fiber-reinforced PLA composites and learned that fiber content, fiber type, and matrix properties influenced the impact strength of the composites. In that study, the authors also emphasized the importance of future research into the impact behavior of these composites. Table 7.3 demonstrates the average chemical composition of popular natural fibers, whereas Table 7.4 indicates natural fiber mechanical and physical qualities.

While there is minimal information on the impact testing of natural fiber-reinforced PLA composites, the available research indicates that the impact strength of these composites is controlled by a variety of parameters such as fiber content, fiber type, and matrix properties. More research is needed to fully comprehend the impact behavior of these composites and their potential uses in a variety of industries.

TABLE 7.3
Average Chemical Composition of Natural Fibers

Type of fibre	Cellulose	Hemicelluloses	Lignin	Pectin	Wax	Ash	Moisture	Others
Abaca	56–64	25–29	11–14	-	-	-	-	-
Jute	64.4	112	0.2	11.8	0.5	0.5–2.1	10	-
Sisal	65.8	12	0.8	9.9	1.2	0.3	10	-
Kenaf	44.4	-	20.1	-	-	4.6	-	-
Coconut	37–43	24–28	26–28	-	-	-	-	7
Bamboo	78.83	-	10.15	-	-	-	-	-

Source: Sinha et al. (2017).

TABLE 7.4
Mechanical and Physical Properties of Natural Fibers

Type of fibre	Diameter (μm)	Density (g / cm^3)	Tensile strength (MPa)	Young/s Modulus (GPa)
Abaca	250–300	1.5	717	18.6
Jute	250–2500	1.3–1.49	393–800	13–26.5
Sisal	205–230	1.41	350–370	12.8
Kenaf	83.5	1.2	282.60	7.13
Coconut	396.98	1.2	140–225	3–5
Bamboo	-	1.2–1.5	500–575	27–40

Source: Sinha et al. (2017).

7.3.5 Hardness Testing

The mechanical evaluation of natural fiber-reinforced PLA composites includes hardness testing. Several studies have been conducted to assess the hardness of PLA composites using various methodologies such as Rockwell hardness testing and Brinell hardness testing.

The Rockwell hardness test was used to analyze the materials in a study on mechanical characterization of PLA composites created using FDM (Vinyas et al., 2019). Researchers found that the Rockwell hardness of pure PLA was 88 RHN and that adding 15% pine needle fibers lowered the hardness to 32 RHN (Ramachandran, 2018). Furthermore, Brinell hardness was 298 BHN for recycled PLA reinforced with kenaf fiber, PES, and sand.

The hardness of PLA composites varies when blended with different natural fibers, emphasizing the need for hardness testing in describing these materials' mechanical properties. Furthermore, the hardness test is one of the mechanical tests used to evaluate PLA-based biodegradable composites, demonstrating its importance in understanding the material's behavior and performance. As a result, hardness testing is critical in the thorough evaluation of natural fiber-reinforced PLA composites, providing significant insights into their mechanical properties and applicability.

7.4 MECHANICAL PROPERTIES OF PLA COMPOSITES

PLA composites are combined structures between PLA and other materials to improve the mechanical properties to the desired specifications for usage. There are many distinct mechanical properties to observe such as tensile strength, flexural strength, compressive strength, shear strength, and impact resistance. All these features will be specified further.

7.4.1 Tensile Strength

The tensile strength of the PLA composites depends on the type and percentage of reinforcement material used in the composite. Pure PLA has a tensile strength range between 40 to 60 MPa. This range of tensile strength is useful for low-stress applications but adding reinforcement material will significantly improve the tensile strength of PLA composite. For example, Ko et al. in 2021 stated that the tensile strength of PLA composites containing poly (glycolic acid) fiber and hydroxyapatite particles was found to be higher than that of PLA composites containing only hydroxyapatite particles. The tensile strength of PLA composites can also be improved by adding unidirectionally aligned poly (glycolic acid) fibers. The findings proved that the mechanical properties of PLA can be improved by adding and experimenting with new materials.

Adding carbon to PLA filament is a ground-breaking strategy for improving its structural stability. Through the deliberate combination of PLA with carbon-based materials, scientists have attempted to utilize the distinct mechanical characteristics that are present in carbon compounds. In addition to reinforcing the filament, this thoughtful combination adds several beneficial properties that go beyond those found in traditional PLA formulations. Renowned for its exceptional tensile strength and resilience, carbon serves as a strong ally to reinforce the overall strength of PLA filament. The resulting composite material exhibits a significant increase in tensile strength, which makes it especially desirable for uses where mechanical performance is critical. With this increased tensile strength, the filament will be more resilient and dependable in situations when it is subjected to various levels of stress. Hence, adding carbon to PLA filament can enhance its tensile strength (Atakok et al., 2022).

PLA shows better performance with higher mechanical strength compared to ABS (acrylonitrile butadiene styrene). The flexural strength of PLA is generally greater than its compressive and tensile strength (Gao et al., 2022). This unique feature gives PLA a significant edge in applications where bending force resistance is essential. PLA's adaptability in real-world applications is demonstrated by its ability to handle bending stresses more effectively than both compressive and tensile forces. This gives engineers and designers a material that performs well in a variety of mechanical circumstances.

7.4.2 Flexural Strength

Like tensile strength, the flexural strength of PLA composites is determined depending on various factors, including the type and amount of reinforcement, manufacturing processes, and testing conditions. Flexural strength refers to a material's ability to withstand bending or deformation under applied loads. Typically, the flexural strength of pure PLA ranges from 60 to 70 MPa (Oksiuta et al., 2020). This baseline understanding provides a benchmark against which the impact of various additives and manufacturing modifications can be assessed.

Thangavel et al. (2023) provide an interesting example of how reinforcing materials can potentially increase flexural strength. They concentrated on a composite that had a remarkable flexural strength of 156.68 MPa and was made up of fly ash with wt.1%, tamarind kernel with wt.2%, and PLA. The significantly higher values for the PLA composite than for the baseline highlight the critical function of reinforcement materials in enhancing the material's resistance to bending forces. This remarkable improvement is the result of a synergistic interaction between the constituents, their concentrations, and the combination of additives.

Additionally, Hassan et al. (2017) determined that the flexural strength of kenaf fiber mat-reinforced PLA composite increased from 28.9 MPa to 38.8 MPa when the fiber content increased from 10 wt.% to 40 wt.%. The result highlights the positive relationship between the concentration of reinforcing fibers and the resulting flexural strength, suggesting that a composite material with a higher percentage of kenaf fibers is more durable and resilient.

7.4.3 Compressive Strength

Compressive strength is the mechanical property of PLA composites that measure the material's ability to withstand forces applied along the material's length (axial load). It helps the material to resist the axial load to keep its size or volume. Pure PLA typically displays compressive strength in the range of 40 MPa to 70 MPa, influenced by factors like molecular weight and processing parameters (Qin et al., 2022).

Similarly, carbon fiber-reinforced PLA composites can achieve compressive strengths of 80 MPa to 200 MPa, influenced by fiber type and alignment (Nunna et al., 2023). PLA's compatibility with various reinforcements, such as natural fibers and nanoparticles, further diversifies compressive strength outcomes. The manufacturing process crucially impacts composite properties, with factors like reinforcement dispersion and orientation within the matrix.

In another example, investigators reported that the compressive strength of henequen fiber/PLA composite increased from 32.4 MPa to 44.8 MPa when the fiber content increased from 0 wt.% to 30 wt.% (Agaliotis et al., 2022). This significant improvement indicates that henequen fibers contribute to the material's ability to resist compressive forces, making it more robust and better suited for applications where compressive strength is crucial.

7.4.4 Shear Strength

The shear strength in PLA composites is a mechanical property that characterizes the material's ability to resist forces acting parallel to its surface. This property is particularly important in applications where materials are subjected to sliding or shearing forces, such as in joints, connections, or components that experience lateral loads. The shear strength of pure PLA can vary based on factors like molecular weight and processing conditions. Generally, PLA exhibits shear strength of 33 MPa (Anderson, 2017). However, the introduction of reinforcing materials, such as glass or carbon fibers, can significantly enhance the shear strength of PLA composites. The specific shear strength values depend on the types, fillers, and orientation of the reinforcement.

For instance, Landes and Letcher (2020) found that their composite material was brittle, and interestingly, there were no significant differences observed in the ultimate shear strength among various fiber orientations. The 0° orientation still had a slight advantage over the other orientations at 29.13 (±0.97) MPa. Since the material is brittle in shear, its usage in applications where ductility and deformation capacity under shear are crucial for preserving structural integrity. The interlaminar shear strength of a 316 stainless steel fiber-reinforced 3D-printed PLA composites (PLA-SSF) with a volume fraction of 30% was found to be 28.5 MPa (±2.0) (Clarke et al., 2023). This research shows the effectiveness of stainless-steel reinforcement and indicates a significant improvement above the basic compressive strength of standard PLA.

7.4.5 Impact Resistance

A material's ability to withstand sudden or dynamic loads without fracture or failure is known as the impact resistance of the material. This property is particularly important in applications where the material may be subjected to impacts, such as in packaging, automotive components, or consumer goods.

The 2–3 kJ/m^2 Charpy impact resistance of neat PLA makes it brittle; plasticization is one of the numerous approaches to improving its impact resistance. PLA is also sensitive to moisture during processing and crystallizes slowly. Moreover, PLA

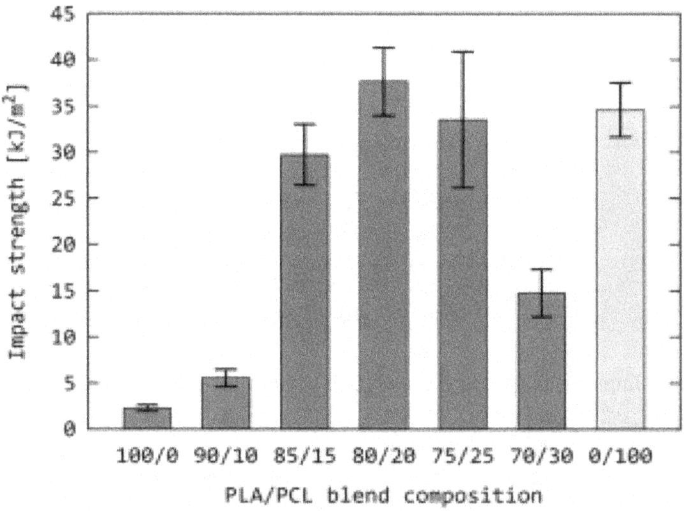

FIGURE 7.1 Charpy notched impact strength of PLA/PCL.

Source: Ostafinska et al. (2015).

becomes even more brittle due to its quite fast physical aging (Ferdinánd et al., 2023), although Ostafinska et al. (2015) obtained a much higher 35 kJ/m^2 at the same elastomer weight fraction. Toughness of the prepared PLA/PCL blends achieved local maximum at the composition 80/20, which was 16 times higher than that of pure PLA (Figure 7.1). This synergistic effect was distinctly evident, resulting from a careful optimization of the blend preparation and, in part, from the serendipitous combination of favorable factors, including the phase morphology of the blend and the crystallinities of both components.

Wang et al. (2019) performed the study of glass fiber reinforced PLA composite and the glass fiber (GF) increases the impact strength significantly. Notably, as the GF content increased from 0 wt.% to 20 wt.%, the impact strength increased from 30.9 to 102.8 J/m, with an increment of more than twofold as depicted in Figure 7.2. The following two reasons may be responsible for the impact strength to have more noticeable enhancement. Initially, the fibers disperse uniformly and separately throughout the PLA matrix, minimizing the possibility of fiber aggregation leading to fracture initiation and propagation. Second, because of the functionalization of the fiber surfaces, there is good adhesion between the GF and the PLA matrix. This prevents crack formation by reducing stress concentration around the GF and increasing energy consumption when the GF is removed from the PLA matrix.

PLA composites exhibit a range of mechanical properties that make them versatile for various applications. The reinforcement materials are helpful to the PLA composites to widen the variety of mechanical properties for certain applications. Table 7.5 shows data on the types of PLA that have been through mechanical properties testing and experiments.

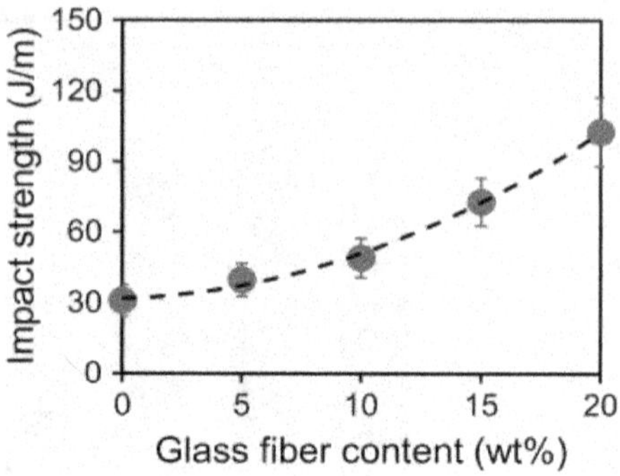

FIGURE 7.2 Impact strength graph of the PLA/m-GF composites.

Source: Wang et al. (2019).

TABLE 7.5
PLA Mechanical Properties

PLA composite	UTS (MPa)	Flexural Strength (MPa)	Elastic Modulus (Gpa)	Compressive Strength (MPa)	Impact Energy (J/M)	Shear Strength (MPa)	Reference
Unreinforced PLA	50.1	78.32	3.47	-	-	-	(Oksiuta et al., 2020)
Polyethylene (5%)-PLA	42.7	-	3.39	-	-	-	(Oksiuta et al., 2020)
Magnesium (5%)-PLA	49.1	-	3.57	-	-	-	(Oksiuta et al., 2020)
Iron (5%)-PLA	52.8	-	3.53	-	-	-	(Oksiuta et al., 2020)
Carbon fiber-PLA	43.75	76.33	-	-	-	-	(Maqsood & Rimašauskas, 2021)
PLA (70%)-tapioca starch (30%)	9.7	-	-	-	-	-	(Yusoff et al., 2021)
Bamboo-PLA (-45/45)	32.50	56.76	263	48.55	21.01	25.11	(Landes & Letcher, 2020)
Fly ash (1%)-tamarind kernel (2%)-PLA	76.24	156.68	-	-	-	-	(Thangavel et al., 2023)
PLA-Graphene	63.2	83.4	3.59	-	-	-	(Caminero et al., 2019)

7.5 FACTORS INFLUENCING MECHANICAL PROPERTIES

7.5.1 Type and Content of Fillers

Various fillers can be incorporated into PLA composites to enhance their mechanical properties, contributing to their versatility and suitability for diverse applications. In a study assessing the impact of wood flour (WF), adding WF improved the tensile strength and flexural modulus of PLA composites (Thongsang et al., 2019). This signifies the potential for natural fiber fillers to positively influence mechanical performance, offering advantages in specific applications where increased strength and stiffness are desired.

Another filler, talc (TC), was examined for its impact on PLA composites, and incorporating TC enhanced elongation and flexural strength, showing that talc improved ductility and strength in PLA composites, expanding the range of mechanical properties that can be tailored for specific applications. Calcium carbonate (CaCO3) represents another filler explored for its impact on PLA composite mechanical properties. The authors found that adding (CaCO3) improved both tensile strength and Young's modulus in the composites (Sit et al., 2023). This demonstrates the effectiveness of inorganic fillers like (CaCO3) in augmenting the structural and tensile properties of PLA composites.

Microballoons were also investigated as fillers for PLA composites, with a study revealing enhancements in impact strength and flexural modulus upon their addition. This suggests that lightweight microballoons can play a role in improving impact resistance while maintaining or even enhancing stiffness in PLA composites. Silicon dioxide (SiO_2) was studied for its impact on PLA composite mechanical properties, demonstrating improvements in tensile strength and Young's modulus. The addition of SiO_2 contributes to reinforcing the composite structure, enhancing its ability to withstand tensile loads and maintain structural integrity.

7.5.2 Processing Conditions

The processing conditions of PLA composites play a crucial role in determining their final mechanical properties, and these conditions were linked to the type and quantity of added materials. The processing of natural fiber composites is influenced by several key parameters, including moisture content, fiber type, fiber volume fraction, and processing temperature. Effective control of moisture in both the fibers and the matrix is important, and necessary modifications must be undertaken if moisture is present to ensure optimal processing. The length, aspect ratio, and chemical composition of the fibers significantly impact the overall processing, sustainability, and performance of the composites (Ho et al., 2012).

Increasing the fiber volume fraction of the composite enhances the stiffness and strength of the composite. However, it also leads to elevated water uptake and decreased deformability. Thus, achieving the right balance in fiber volume fraction is essential for obtaining the desired mechanical properties and performance (Ciprian et al., 2015).

It's important to note that the maximum temperature during processing should not exceed 200 °C within 20 minutes to prevent degradation of most natural fibers (Hart & Summerscales, 2017). Temperatures beyond this limit can induce degradation, shrinkage, and result in suboptimal performance for the natural fiber composite. This is due to changes in the chemical, physical, and mechanical properties of the natural fibers, including depolymerization, oxidation, hydrolysis, decarboxylation, dehydration, and recrystallization.

7.5.3 Temperature and Environmental Effects

The temperature characteristics of PLA can affect its processing and performance. PLA has a glass transition temperature of 60–65 °C, where it transitions from a glassy to a rubbery state, offering flexibility. The melting temperature ranges from 150–160 °C, making it suitable for processing at relatively low temperature (Webb et al., 2012). However, prolonged exposure to temperatures above 200 °C can lead to thermal degradation, reducing its molecular weight and mechanical properties (Iglesias-Montes et al., 2023). Controlling processing temperatures during manufacturing is essential to avoid degradation and achieve desired material properties.

PLA's environmental behavior is a significant aspect of its appeal. As a biodegradable polymer, PLA undergoes microbial degradation under specific conditions, breaking down into carbon dioxide and water (Zeenat et al., 2021). Composting, especially in industrial composting facilities with temperatures above 50 °C, accelerates this process. However, home composting takes longer. PLA is also susceptible to hydrolytic degradation in the presence of moisture, impacting its mechanical properties over time. UV degradation can occur when exposed to sunlight, and UV stabilizers may be used to counteract this effect, ensuring durability in outdoor applications. Understanding these environmental factors is important for optimizing the usage and disposal of PLA in an ecofriendly manner.

7.6 FRACTURE MECHANICS

7.6.1 Stress Concentrations

The study of stress concentrations is important, especially in the context of polymer composites like PLA. Stress concentrations are distinct places within a material where the stress level is significantly higher than in the surrounding areas. These areas are frequently connected with geometric defects such as holes, notches, or sharp edges. Bikiaris et al. (2023) describes the importance of these concentrations.

Understanding stress concentrations is important because of their impact on the mechanical characteristics of materials. These stress concentrations in polymer composites, such as PLA, can have a significant impact on the material's strength and failure mechanisms. This insight is essential to the study of fracture mechanics because it contributes to a more thorough understanding of how materials perform under different stress circumstances. Thus, stress concentration analysis is critical for predicting material behavior and improving the design and durability of polymer-based composites.

Within the domain of PLA composites, the consequences of stress concentrations are dependent upon numerous influencing elements, all of which contribute to the material's overall mechanical behavior. The fiber-matrix interface is the most important of these elements. Table 7.6 shows the shear strength values for different matrices and varieties of fibers. The way the PLA matrix and reinforcing fibers interact determines how stress is distributed throughout the composite. At this point, irregularities or insufficient bonding are likely to cause stress concentrations to rise, which could hasten the crack formation process (Scarponi & Pizzinelli, 2009). This phenomenon emphasizes the importance of achieving a consistent and strong interface in composite design.

Utilizing the single fiber fragmentation test (SFFT) and single fiber tensile (SFT) testing, Joffe et al. (2003) carried out a comprehensive investigation of flax fibers. Both unmodified and modified vinylester and polyester matrices were used in their experiments. Interfacial shear strength (ISS) ranged from 18 to 42 MPa, with treated vinyl ester matrix treatments showing the best outcomes.

The way the fibers are arranged inside the composite structure is important. This factor has a significant impact on how stress is distributed and transmitted. A key

TABLE 7.6
Interfacial Shear Strengths for Different Flax Types and Matrices

Flax Fiber Type	Matrix	Modifying Factor	ISS (MPa)	Test Method
Green	PP	-	6.33	SFFT
		Acetylation	11.61	
		Stearic acid treatment	9.46	
		-	7.21	Micro-debond
		MA-PP treated	7.20	
Dew retted	PP	-	12.75	SFFT
		Acetylation	13.05	
		Stearic acid treatment	13.36	
		Transcrystaline layer	23.05	
		-	17.3	
	HDPE	-	18.0	Pull-out
	LDPE	-	5.6	
	MA-PP	-	17.8	
Duralin™	PP	-	7.45	Micro-debond
		Hot-cleaned	6.63	
		MA-PP treated	7.17	
	HDPE	-	16.2	Pull-out
	LDPE	-	7.1	

Source: Joffe et al. (2003).

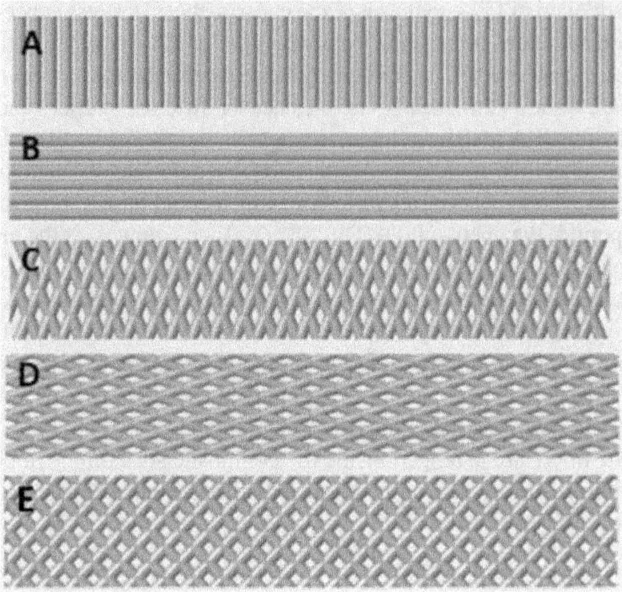

FIGURE 7.3 Specimen with different fibre orientation (angle measured from horizontal) (a) 0° specimen (b) 90° specimen (c) 22.5°/–22.5° specimen (d) 67.5°/–67.5° specimen (e) 45°/–45° specimen.

Source: Sharan Gupta et al. (2021).

element in reducing stress concentrations is the strategic alignment of fibers with respect to expected loading circumstances (Sharan Gupta et al., 2021). Maximizing mechanical performance and durability, fiber orientation takes into consideration while designing PLA composites as illustrated in Figure 7.3.

Lastly, it is important to pay attention to the voids and faults that exist within the PLA composite. The positions and intensities of stress concentrations are largely determined by these structural defects (Mehdikhani et al., 2019). These defects act as spots where stress builds up, making the material more vulnerable to breaking under relatively light pressures. Precise manufacturing procedures are required to reduce flaws and improve the integrity and load-bearing capability of PLA composites.

7.6.2 Crack Propagation

Due to the distinct mechanical properties that natural fibers like hemp, flax, and jute offer, their incorporation into PLA composites has attracted a lot of attention. The type, size, and orientation of these natural fibers are important factors that influence how cracks propagate within the composite material. For example, compared with more flexible fibers like jute, the tensile strength and stiffness of flax fibers can result in a different stress distribution pattern, which can affect the crack formation dynamics in the composite (Sanivada et al., 2020).

The function of interfacial bonding between the PLA matrix and natural fibers in natural fiber-reinforced PLA composites has been highlighted in several studies, including those by Ilyas et al. (2022), Jia et al. (2014), Mohammed et al. (2022), Oksman et al. (2003), and Vargas et al. (2023). Effective stress transmission and by extension, the composite's overall mechanical performance depends heavily on this bonding. Effective stress transfer from the PLA matrix to the fibers is made possible by strong adhesion at the interface, which slows the progression of cracks and increases the load-bearing capacity of the composite. Inter-diffusion, electrostatic adhesion, chemical bonding, and mechanical interlocking are some of the mechanisms that contribute to this crucial interfacial connection. Maximizing the mechanical properties of these composites requires optimizing these bonding mechanisms because it has a direct impact on their usefulness and longevity under different circumstances.

Ilyas et al. (2022) and Oksman et al. (2003) highlight the critical importance of microstructural properties in natural fiber-reinforced PLA composites, especially about crack initiation and propagation. Important variables are the natural fibers' volume percentage and distribution inside the PLA matrix. A constructed microstructure featuring a uniform distribution of fibers makes a substantial contribution to the uniform distribution of stress, which in turn stops the propagation of cracks and increases the toughness of the composite. However, there are difficulties since the hydrophobic PLA matrix and the hydrophilic natural fibers have intrinsic compatibility problems. According to Ilyas et al. (2022), this incompatibility frequently causes nonuniform fiber dispersion, which compromises the composite's mechanical properties. Furthermore, as the same authors observed, fiber surface roughness plays a function in promoting mechanical interaction with the matrix.

Mohammed et al. (2023) and Brebu (2020) demonstrated that the integrity of natural fiber-reinforced PLA composites is significantly threatened by environmental conditions. Important elements including UV exposure, temperature changes, and moisture absorption can cause the PLA matrix and natural fibers to deteriorate. The mechanical characteristics of the composite are drastically changed by this deterioration process, making it less resistant to crack propagation. For example, extended exposure to moisture can significantly deteriorate the link between the fibers and matrix, compromising the structural stability of the composite. Similarly, exposure to UV light negatively affects the fibers, causing them to break down and accelerating the formation of cracks in the material.

7.6.3 Fracture Toughness

In fracture mechanics, fracture toughness is a mechanical characteristic, especially when studying composites made of natural fibers reinforced with PLA. These characteristic measures a material's resistance to fracture propagation and is a measure of how well it can tolerate stress before failing. One important aspect affecting fracture toughness in PLA composites is the incorporation of natural fibers. Various fibers with various mechanical properties such as jute, hemp, and flax contribute in diverse ways to this attribute. The composite's total fracture toughness is largely determined by how these fibers interact with the PLA matrix, which is defined by properties like strength, stiffness, and bonding efficiency (Chittimenu et al., 2021).

Natural fiber-reinforced PLA composites' fracture toughness varies depending on several important variables, mainly the fibers' orientation and length inside the PLA matrix. The mechanical coupling and stress distribution inside the composite are greatly influenced by these physical characteristics of the fibers, as well as its stiffness and ductility. The total response of the composite to applied stresses is determined by the mechanical characteristics of the matrix. Thirdly, another important consideration is the volume percentage of fibers in the composite. This factor establishes how much reinforcement the fibers offer (Madhusudhana et al., 2021).

These composites' fracture toughness is also influenced by environmental conditions. Their mechanical properties can be changed by variables like temperature, moisture content, and UV exposure. Experimental tests to measure fracture toughness, including the single-edge notched bending and compact tension tests, are commonly used in this field of study. To comprehend and forecast how these materials would behave under stress, sophisticated computational models are being employed (Jasim & Mahmood, 2023).

7.7 APPLICATIONS OF PLA COMPOSITES

In the automotive industry, PLA has become a key component, particularly for the manufacture of interior parts. PLA is a more ecologically sustainable substitute for conventional plastics in the production of door panels, seat covers, and dashboard components. High-visibility interior car components benefit from its visual appeal, tactile feel, and outstanding UV resistance.

PLA's use in the food packaging industry is commonly used. Shao et al. (2022) state that PLA is valued for its remarkable mechanical and physical qualities, cell compatibility, and biodegradability. The high volatility of natural antibacterial agents must be countered by the addition of antibacterial agents, which calls for certain processing techniques for the best possible usage in food packaging. The adaptable qualities of PLA are also advantageous to the biomedical industry.

According to Singhvi et al. (2019), PLA is a favored material in tissue engineering, drug delivery systems, and medical implants because of its biocompatibility, biodegradability, mechanical strength, and processability. These qualities can be strengthened even more by using different processing and reinforcement strategies. Finally, there has been an increase in the usage of PLA-based composites in consumer goods. PLA composites have advantages, according to Ilyas et al. (2021), including lighter weight, lower heat conductivity, and better mechanical qualities. Wu et al. (2023) provide additional evidence about their appropriateness for ecofriendly uses, such as the production of athletic gear.

7.8 CHALLENGES AND FUTURE DIRECTIONS

There are many challenges in the way of the development of natural fiber PLA composites, which are for the aviation, defense, construction, and automotive industries. These include the fibers' weak interfacial adhesion, greater water absorption, and inherent incompatibility with the polymer matrix. Pornwannachai et al. (2021) claim that academics have investigated several tactics to lessen these problems. Notably,

Amjad et al. (2022) showed that adding nanofillers and treating the surface of natural fibers improves their mechanical qualities and decreases their absorption of water. As such, these composites are finding more and more use in semi-structural and structural applications.

Additionally, as stated by Pornwannachai et al. (2021) and Amjad et al. (2022), strengthening fiber-matrix adhesion and improving the overall performance of composites depend on optimizing nanofiller concentrations and developing efficient surface treatment techniques. Furthermore, as Ding et al. (2018) point out, there is a growing trend toward the reinforcement of PLA matrices with renewable fibers including flax, cotton, and bamboo. He et al. (2022) highlight the possibility of enhancing the electrical, thermal, and mechanical properties of composites by functionalizing natural fibers and adding nanomaterials such as graphene oxide. Furthermore, Wang et al. (2022) identified two new developments in this sector, the scalability of manufacturing techniques and 3D printing parameters for these composites.

7.9 CONCLUSIONS

In conclusion, research on natural fiber-reinforced PLA composites shows that a variety of factors, including interfacial bonding, material optimization techniques, and environmental circumstances, interact intricately to affect the mechanical properties of these materials. The research emphasizes how these composites are susceptible to environmental elements that can compromise the integrity of the PLA matrix and natural fibers, such as moisture, temperature changes, and UV exposure.

The strength of the bond between the PLA matrix and the natural fibers is essential to improving mechanical performance because it prevents cracks from spreading and increases load-bearing capacity. The mechanical stability is further complicated by challenges resulting from the natural hydrophilic fibers and hydrophobic PLA matrix's intrinsic incompatibility, which can cause non-uniform fiber dispersion. However, there are interesting directions for future research due to the possibility for improvement using renewable fibers, enhanced surface treatment processes, and optimized nanofiller concentrations.

The functionalization of natural fibers and the addition of nanoparticles offer chances to improve the mechanical, electrical, and thermal characteristics of these composites in addition to addressing issues with interfacial adhesion and environmental degradation. Thus, this sector must see ongoing research and progress. To improve our knowledge of the use of natural fiber PLA composites in manufacturing and engineering, more research in these areas is needed. This will also greatly advance the field of composite materials science and engineering.

REFERENCES

Agaliotis, E. M., Ake-Concha, B. D., May-Pat, A., Morales-Arias, J. P., Bernal, C., Valadez-Gonzalez, A., Herrera-Franco, P. J., Proust, G., Koh-Dzul, J. F., Carrillo, J. G., & Flores-Johnson, E. A. (2022). Tensile behavior of 3D printed polylactic acid (PLA) based composites reinforced with natural fiber. *Polymers*, *14*(19), 3976. https://doi.org/10.3390/polym14193976

Amjad, A., Awais, H., Anjang Ab Rahman, A., & Abidin, M. S. Z. (2022). Effect of nanofillers on mechanical and water absorption properties of alkaline treated flax/PLA fibre reinforced epoxy hybrid nanocomposites. *Advanced Composite Materials*, *31*(4), 351–369. https://doi.org/10.1080/09243046.2021.1993563

Anderson, I. (2017). Mechanical properties of specimens 3D printed with virgin and recycled polylactic acid. *3D Printing and Additive Manufacturing*, *4*(2), 110–115. https://doi.org/10.1089/3dp.2016.0054

Atakok, G., Kam, M., & Koc, H. B. (2022). Tensile, three-point bending and impact strength of 3D printed parts using PLA and recycled PLA filaments: A statistical investigation. *Journal of Materials Research and Technology*, *18*, 1542–1554. https://doi.org/10.1016/j.jmrt.2022.03.013

Bakar, A. A. B. A., Zainuddin, M. Z. B., Adam, A. N. B., Noor, I. S. B. M., Tamchek, N. B., Alauddin, M. S. B., & Ghazali, M. I. B. M. (2022). The study of mechanical properties of poly(lactic) acid PLA-based 3D printed filament under temperature and environmental conditions. *Materials Today: Proceedings*, *67*, 652–658. https://doi.org/10.1016/j.matpr.2022.06.198

Bikiaris, N. D., Koumentakou, I., Samiotaki, C., Meimaroglou, D., Varytimidou, D., Karatza, A., Kalantzis, Z., Roussou, M., Bikiaris, R. D., & Papageorgiou, G. Z. (2023). Recent advances in the investigation of poly(lactic acid) (PLA) nanocomposites: Incorporation of various nanofillers and their properties and applications. *Polymers*, *15*(5), 1196. https://doi.org/10.3390/polym15051196

Brebu, M. (2020). Environmental degradation of plastic composites with natural fillers – a review. *Polymers*, *12*(1), 166. https://doi.org/10.3390/polym12010166

Caminero, M., Chacón, J., García-Plaza, E., Núñez, P., Reverte, J., & Becar, J. (2019). Additive manufacturing of PLA-based composites using fused filament fabrication: Effect of graphene nanoplatelet reinforcement on mechanical properties, dimensional accuracy and texture. *Polymers*, *11*(5), 799. https://doi.org/10.3390/polym11050799

Chittimenu, H., Pasupureddy, M., Muthukumar, C., Krishnasamy, S., Thiagamani, S. M. K., & Siengchin, S. (2021). Fracture toughness of the natural fiber-reinforced composites: A review. In *Mechanical and Dynamic Properties of Biocomposites* (pp. 293–304). Wiley. https://doi.org/10.1002/9783527822331.ch16

Ciprian, L., Radu, P., & Ioana, E. (2015). The effects of fibre volume fraction on a glass-epoxy composite material. *INCAS Bulletin*, *7*(3), 113–119. https://doi.org/10.13111/2066-8201.2015.7.3.10

Clarke, A. J., Dickson, A., & Dowling, D. P. (2023). Fabrication and performance of continuous 316 stainless steel fibre-reinforced 3D-printed PLA composites. *Polymers*, *16*(1), 63. https://doi.org/10.3390/polym16010063

DeStefano, V., Khan, S., & Tabada, A. (2020). Applications of PLA in modern medicine. *Engineered Regeneration*, *1*, 76–87. https://doi.org/10.1016/j.engreg.2020.08.002

Ding, W. D., Pervaiz, M., & Sain, M. (2018). *Cellulose-Enabled Polylactic Acid (PLA) Nanocomposites: Recent Developments and Emerging Trends* (pp. 183–216). Springer. https://doi.org/10.1007/978-3-319-66417-0_7

Farah, S., Anderson, D. G., & Langer, R. (2016). Physical and mechanical properties of PLA, and their functions in widespread applications – a comprehensive review. *Advanced Drug Delivery Reviews*, *107*, 367–392. https://doi.org/10.1016/j.addr.2016.06.012

Ferdinánd, M., Várdai, R., Móczó, J., & Pukánszky, B. (2023). A novel approach to the impact modification of PLA. *Engineering Fracture Mechanics*, *277*, 108950. https://doi.org/10.1016/j.engfracmech.2022.108950

Forster, A. M. (2015). *Materials Testing Standards for Additive Manufacturing of Polymer Materials: State of the Art and Standards Applicability*. National Institute of Standards and Technology. https://doi.org/10.6028/NIST.IR.8059

Gao, G., Xu, F., Xu, J., & Liu, Z. (2022). Study of material color influences on mechanical characteristics of fused deposition modeling parts. *Materials*, *15*(19), 7039. https://doi.org/10.3390/ma15197039

Gauss, C., Pickering, K. L., Tshuma, J., & McDonald-Wharry, J. (2022). Production and assessment of poly(lactic acid) matrix composites reinforced with regenerated cellulose fibres for fused deposition modelling. *Polymers*, *14*(19), 3991. https://doi.org/10.3390/polym14193991

Gholampour, A., & Ozbakkaloglu, T. (2020). A review of natural fiber composites: Properties, modification and processing techniques, characterization, applications. *Journal of Materials Science*, *55*(3), 829–892. Springer. https://doi.org/10.1007/s10853-019-03990-y

Grabowik, C., Kalinowski, K., Ćwikła, G., Paprocka, I., & Kogut, P. (2017). Tensile tests of specimens made of selected group of the filament materials manufactured with FDM method. *MATEC Web of Conferences*, *112*, 04017. https://doi.org/10.1051/matecconf/201711204017

Hart, A., & Summerscales, J. (2017). Effect of time at temperature for natural fibres. *Procedia Engineering*, *200*, 269–275. https://doi.org/10.1016/j.proeng.2017.07.038

Hassan, F., Zulkifli, R., Ghazali, M. J., & Azhari, C. H. (2017). Flexural properties of kenaf fibre mat reinforced PLA composites. *Journal of Applied Environmental and Biological Sciences*, *7*(1), 30–35.

He, Y., Wu, S., Yuen, A. C. Y., Huang, F., Boyer, C., Wang, C. H., & Zhang, J. (2022). Scalable manufacturing process and multifunctional performance of cotton fibre-reinforced poly(lactic acid) (pla) bio-composites coated by graphene oxide. *Polymers*, *14*(19), 3946. https://doi.org/10.3390/polym14193946

Ho, M., Wang, H., Lee, J.-H., Ho, C., Lau, K., Leng, J., & Hui, D. (2012). Critical factors on manufacturing processes of natural fibre composites. *Composites Part B: Engineering*, *43*(8), 3549–3562. https://doi.org/10.1016/j.compositesb.2011.10.001

Iglesias-Montes, M. L., D'Amico, D. A., Malbos, L. B., Seoane, I. T., Cyras, V. P., & Manfredi, L. B. (2023). Thermal degradation kinetics of completely biodegradable and biobased PLA/PHB blends. *Thermochimica Acta*, *725*, 179530. https://doi.org/10.1016/j.tca.2023.179530

Ilyas, R. A., Sapuan, S. M., Harussani, M. M., Hakimi, M. Y. A. Y., Haziq, M. Z. M., Atikah, M. S. N., Asyraf, M. R. M., Ishak, M. R., Razman, M. R., Nurazzi, N. M., Norrrahim, M. N. F., Abral, H., & Asrofi, M. (2021). Polylactic acid (PLA) biocomposite: Processing, additive manufacturing and advanced applications. *Polymers*, *13*(8), 1326. https://doi.org/10.3390/polym13081326

Ilyas, R. A., Zuhri, M. Y. M., Aisyah, H. A., Asyraf, M. R. M., Hassan, S. A., Zainudin, E. S., Sapuan, S. M., Sharma, S., Bangar, S. P., Jumaidin, R., Nawab, Y., Faudzi, A. A. M., Abral, H., Asrofi, M., Syafri, E., & Sari, N. H. (2022). Natural fiber-reinforced polylactic acid, polylactic acid blends and their composites for advanced applications. *Polymers*, *14*(1). MDPI. https://doi.org/10.3390/polym14010202

Jasim, A. M., & Mahmood, I. A. (2023). Study the tensile and fracture toughness properties of hybrid composite consist of blend matrix and natural fibers reinforcement. *AIP Conference Proceedings*, *2787*. https://doi.org/10.1063/5.0148108

Jia, W., Gong, R. H., & Hogg, P. J. (2014). Poly (lactic acid) fibre reinforced biodegradable composites. *Composites Part B: Engineering*, *62*, 104–112. https://doi.org/10.1016/j.compositesb.2014.02.024

Joffe, R., Andersons, J., & Wallström, L. (2003). Strength and adhesion characteristics of elementary flax fibres with different surface treatments. *Composites Part A: Applied Science and Manufacturing*, *34*(7), 603–612. https://doi.org/10.1016/S1359-835X(03)00099-X

John, K., & Naidu, S. V. (2004). Effect of fiber content and fiber treatment on flexural properties of sisal fiber/glass fiber hybrid composites. *Journal of Reinforced Plastics and Composites*, *23*(15), 1601–1605. https://doi.org/10.1177/0731684404039799

Kalu, E. I., Skrifvas, M., Khalili, P., Akesson, D., & Ikezue, E. N. (2023). Impact test analysis of natural fibre reinforced polymer composites fabricated with bast fibres for industrial/household applications. *International Journal of Materials Engineering and Technology*, *22*(2), 123–140. https://doi.org/10.17654/0975044423008

Ko, H.-S., Lee, S., Lee, D., & Jho, J. Y. (2021). Mechanical properties and bioactivity of poly(-lactic acid) composites containing poly(glycolic acid) fiber and hydroxyapatite particles. *Nanomaterials*, *11*(1), 249. https://doi.org/10.3390/nano11010249

Landes, S., & Letcher, T. (2020). Mechanical strength of bamboo filled PLA composite material in fused filament fabrication. *Journal of Composites Science*, *4*(4), 159. https://doi.org/10.3390/jcs4040159

Lee, C. H., Padzil, F. N. B. M., Lee, S. H., Ainun, Z. M. A., & Abdullah, L. C. (2021). Potential for natural fiber reinforcement in PLA polymer filaments for fused deposition modeling (FDM) additive manufacturing: A review. *Polymers*, *13*(9). MDPI AG. https://doi.org/10.3390/polym13091407

Madhusudhana, H. K., Kumar, M. P., Patil, A. Y., Keshavamurthy, R., Khan, T. M. Y., Badruddin, I. A., & Kamangar, S. (2021). Analysis of the effect of parameters on fracture toughness of hemp fiber reinforced hybrid composites using the ANOVA method. *Polymers*, *13*(17), 3013. https://doi.org/10.3390/polym13173013

Maqsood, N., & Rimašauskas, M. (2021). Characterization of carbon fiber reinforced PLA composites manufactured by fused deposition modeling. *Composites Part C: Open Access*, *4*, 100112. https://doi.org/10.1016/j.jcomc.2021.100112

Mehdikhani, M., Gorbatikh, L., Verpoest, I., & Lomov, S. V. (2019). Voids in fiber-reinforced polymer composites: A review on their formation, characteristics, and effects on mechanical performance. *Journal of Composite Materials*, *53*(12), 1579–1669. https://doi.org/10.1177/0021998318772152

Mohammed, M., Jawad, A. J. M., Mohammed, A. M., Oleiwi, J. K., Adam, T., Osman, A. F., Dahham, O. S., Betar, B. O., Gopinath, S. C. B., & Jaafar, M. (2023). Challenges and advancement in water absorption of natural fiber-reinforced polymer composites. *Polymer Testing*, *124*, 108083. https://doi.org/10.1016/j.polymertesting.2023.108083

Mohammed, M., Rasidi, M., Mohammed, A., Rahman, R., Osman, A., Adam, T., Betar, B., & Dahham, O. (2022). Interfacial bonding mechanisms of natural fibre-matrix composites: An overview. *BioResources*, *17*(4). https://doi.org/10.15376/biores.17.4.Mohammed

Nunna, S., Ravindran, A. R., Mroszczok, J., Creighton, C., & Varley, R. J. (2023). A review of the structural factors which control compression in carbon fibres and their composites. *Composite Structures*, *303*, 116293. https://doi.org/10.1016/j.compstruct.2022.116293

Oksiuta, Z., Jalbrzykowski, M., Mystkowska, J., Romanczuk, E., & Osiecki, T. (2020). Mechanical and thermal properties of polylactide (PLA) composites modified with Mg, Fe, and polyethylene (PE) additives. *Polymers*, *12*(12), 2939. https://doi.org/10.3390/polym12122939

Oksman, K., Skrifvars, M., & Selin, J.-F. (2003). Natural fibres as reinforcement in polylactic acid (PLA) composites. *Composites Science and Technology*, *63*(9), 1317–1324. https://doi.org/10.1016/S0266-3538(03)00103-9

Ostafinska, A., Fortelny, I., Nevoralova, M., Hodan, J., Kredatusova, J., & Slouf, M. (2015). Synergistic effects in mechanical properties of PLA/PCL blends with optimized composition, processing, and morphology. *RSC Advances*, *5*(120), 98971–98982. https://doi.org/10.1039/C5RA21178F

Periasamy, K., Chakravarthy, K. S., Md, J. S., & Madhu, S. (2024). A detailed evaluation of mechanical properties in newly developed cellulosic fiber: Cissus vitiginea L as a reinforcement for polymer composite. *Biomass Conversion and Biorefinery*, *14*(1), 1237–1250. https://doi.org/10.1007/s13399-023-04229-2

Pornwannachai, W., Horrocks, A. R., & Kandola, B. K. (2021). Surface modification of commingled flax/PP and flax/PLA fibres by silane or atmospheric argon plasma exposure to improve fibre – matrix adhesion in composites. *Fibers*, *10*(1), 2. https://doi.org/10.3390/fib10010002

Qin, D., Sang, L., Zhang, Z., Lai, S., & Zhao, Y. (2022). Compression performance and deformation behavior of 3D-printed PLA-based lattice structures. *Polymers*, *14*(5), 1062. https://doi.org/10.3390/polym14051062

Ramachandran, M. (2018). Review on characterization of various poly lactic acid based biodegradable composites. *Juniper Online Journal Material Science*, *4*(1). https://doi.org/10.19080/JOJMS.2018.04.555626

Sabik, A., Rucka, M., Andrzejewska, A., & Wojtczak, E. (2022). Tensile failure study of 3D printed PLA using DIC technique and FEM analysis. *Mechanics of Materials*, *175*, 104506. https://doi.org/10.1016/j.mechmat.2022.104506

Sanivada, U. K., Mármol, G., Brito, F. P., & Fangueiro, R. (2020). PLA composites reinforced with flax and jute fibers – a review of recent trends, processing parameters and mechanical properties. *Polymers*, *12*(10), 2373. https://doi.org/10.3390/polym12102373

Satsum, A., Busayaporn, W., Rungswang, W., Soontaranon, S., Thumanu, K., & Wanapu, C. (2022). Structural and mechanical properties of biodegradable poly(lactic acid) and pectin composites: Using bionucleating agent to improve crystallization behavior. *Polymer Journal*, *54*(7), 921–930. https://doi.org/10.1038/s41428-022-00637-9

Scarponi, C., & Pizzinelli, C. S. (2009). Interface and mechanical properties of natural fibres reinforced composites: A review. *International Journal of Materials and Product Technology*, *36*(1–4), 278. https://doi.org/10.1504/IJMPT.2009.027837

Shao, L., Xi, Y., & Weng, Y. (2022). Recent advances in PLA-based antibacterial food packaging and its applications. *Molecules*, *27*(18), 5953. https://doi.org/10.3390/molecules27185953

Sharan Gupta, U., Dharkar, A., Dhamarikar, M., Choudhary, A., Wasnik, D., Chouhan, P., Tiwari, S., & Namdeo, R. (2021). Study on the effects of fiber orientation on the mechanical properties of natural fiber reinforced epoxy composite by finite element method. *Materials Today: Proceedings*, *45*, 7885–7893. https://doi.org/10.1016/j.matpr.2020.12.614

Siakeng, R., Jawaid, M., Ariffin, H., Sapuan, S. M., Asim, M., & Saba, N. (2019). Natural fiber reinforced polylactic acid composites: A review. *Polymer Composites*, *40*(2), 446–463. https://doi.org/10.1002/pc.24747

Singh, J. I. P., Singh, S., Dhawan, V., & Gulati, P. (2021). Flax fiber reinforced polylactic acid composites for non-structural engineering applications: Effect of molding temperature and fiber volume fraction on its mechanical properties. *Polymers and Polymer Composites*, *29*(Suppl 9), S780–S789. https://doi.org/10.1177/09673911211025159

Singhvi, M. S., Zinjarde, S. S., & Gokhale, D. V. (2019). Polylactic acid: Synthesis and biomedical applications. *Journal of Applied Microbiology*, *127*(6), 1612–1626. https://doi.org/10.1111/jam.14290

Sinha, A. K., Narang, H. K., & Bhattacharya, S. (2017). Mechanical properties of natural fibre polymer composites. *Journal of Polymer Engineering*, *37*(9), 879–895. https://doi.org/10.1515/polyeng-2016-0362

Sit, M., Dashatan, S., Zhang, Z., Dhakal, H. N., Khalfallah, M., Gamer, N., & Ling, J. (2023). Inorganic fillers and their effects on the properties of flax/PLA composites after UV degradation. *Polymers*, *15*(15), 3221. https://doi.org/10.3390/polym15153221

Sun, Z. (2018). Progress in the research and applications of natural fiber-reinforced polymer matrix composites. *Science and Engineering of Composite Materials*, *25*(5), 835–846. https://doi.org/10.1515/secm-2016-0072

Thangavel, P., Vignesh, S., Sureshkumar, R., Gowrishankar, A., & Girimurugan, R. (2023). Study on mechanical properties of polylactic acid matrix added with fly ash and tamarind kernel powder micro fillers. *NanoWorld Journal*, *9*(Suppl 1). https://doi.org/10.17756/nwj.2023-s1-121

Thongsang, S., Sontikaew, S., Kachapol, K., Kaewkate, T., & Aunchanlung, W. (2019). Comparison of filler types in polylactic acid composites for 3d printing applications. *Matter: International Journal of Science and Technology*, *5*(3), 98–109. https://doi.org/10.20319/mijst.2019.53.98109

Vargas, J., Akrami, R., Marín-Alzate, N., Mesa, S. M., Idarraga, G., Isaza, C. A., Yang, L., & Meza, J. M. (2023). Determination of adhesion between modified PLA matrix and natural fibers using SEM and micro-Raman spectroscopy. *Microscopy and Microanalysis*, *29*(Suppl 1), 1250–1252. https://doi.org/10.1093/micmic/ozad067.641

Vinyas, M., Athul, S. J., Harursampath, D., & Nguyen Thoi, T. (2019). Mechanical characterization of the poly lactic acid (PLA) composites prepared through the fused deposition modelling process. *Materials Research Express*, *6*(10), 105359. https://doi.org/10.1088/2053-1591/ab3ff3

Wang, G., Zhang, D., Wan, G., Li, B., & Zhao, G. (2019). Glass fiber reinforced PLA composite with enhanced mechanical properties, thermal behavior, and foaming ability. *Polymer*, *181*, 121803. https://doi.org/10.1016/j.polymer.2019.121803

Wang, H., Qin, Q.-H., & Xiao, Y. (2016). Special n-sided Voronoi fiber/matrix elements for clustering thermal effect in natural-hemp-fiber-filled cement composites. *International Journal of Heat and Mass Transfer*, *92*, 228–235. https://doi.org/10.1016/j.ijheatmasstransfer.2015.08.093

Wang, Y., Sultana, J., Rahman, M. M., Ahmed, A., Azam, A., Mushtaq, R. T., & Rehman, M. (2022). A sustainable and biodegradable building block: Review on mechanical properties of bamboo fibre reinforced PLA polymer composites and their emerging applications. *Fibers and Polymers*, *23*(12), 3317–3342. https://doi.org/10.1007/s12221-022-4871-z

Webb, H., Arnott, J., Crawford, R., & Ivanova, E. (2012). Plastic degradation and its environmental implications with special reference to poly(ethylene terephthalate). *Polymers*, *5*(1), 1–18. https://doi.org/10.3390/polym5010001

Wu, Y., Gao, X., Wu, J., Zhou, T., Nguyen, T. T., & Wang, Y. (2023). Biodegradable polylactic acid and its composites: Characteristics, processing, and sustainable applications in sports. *Polymers*, *15*(14), 3096. https://doi.org/10.3390/polym15143096

Yaisun, S., & Trongsatitkul, T. (2023). PLA-based hybrid biocomposites: Effects of fiber type, fiber content, and annealing on thermal and mechanical properties. *Polymers*, *15*(20), 4106. https://doi.org/10.3390/polym15204106

Yusoff, N. H., Pal, K., Narayanan, T., & de Souza, F. G. (2021). Recent trends on bioplastics synthesis and characterizations: Polylactic acid (PLA) incorporated with tapioca starch for packaging applications. *Journal of Molecular Structure*, *1232*, 129954. https://doi.org/10.1016/j.molstruc.2021.129954

Zeenat, Elahi, A., Bukhari, D. A., Shamim, S., & Rehman, A. (2021). Plastics degradation by microbes: A sustainable approach. *Journal of King Saud University – Science*, *33*(6), 101538. https://doi.org/10.1016/j.jksus.2021.101538

8 Physical Properties of PLA Composites

8.1 INTRODUCTION

Amid growing environmental awareness, industries are forced to find alternatives of an environmental, ecologically friendly and a sustainable material to be used for the product. The idea that is currently widely being used in industries is utilizing natural fibers with polymers of renewable material as matrix. One of the most remarkable research and innovations is the use of PLA as it provides various benefits.

PLA is a biodegradable and environmentally friendly thermoplastic as it is extracted and produced from a natural source. PLA is even more adaptable because it makes a great basis material for composites. These PLA composites are produced by combining PLA with different materials to create materials with a range of improved physical properties that can be used in a variety of applications (Ilyas et al., 2022). PLA is biodegradable and environmentally friendly because it is derived from natural sources such as cornstarch or sugarcane (Xu et al., 2022). Over time, PLA has been used for a wide range of applications in various industries. Its physical property has become the topic of main research and development to explore the application that it can provide. PLA is versatile and competitive in the current growing industries due to the advantages and properties it provides.

The physical properties of PLA can be enhanced by strategically incorporating additives into the polymer (Ebrahimi & Ramezani Dana, 2022). PLA composites provide new opportunities for applications in the automotive, aviation (Feng et al., 2017) and electronics industries (Mattana et al., 2015), among others, thanks to its enhanced strength, heat resistance, and electrical conductivity (Murariu & Dubois, 2016). Natural fibers are fibers that are extracted from natural resources such as animal, vegetable or mineral (Sathish et al., 2021) and have been proven to be renewable, sustainable, and biodegradable as well as only minimally contributing to any sort of pollution. In addition to being more environmentally friendly, the fibers are tougher and more durable than synthetic fibers (Gholampour & Ozbakkaloglu, 2020). It is possible to grow and collect natural fibers without the use of hazardous chemicals or fossil fuels.

There are a variety of natural fibers used in industries such as flax, which is used to make linen, and jute, a strong and coarse fiber that is used to make industrial products (Mia et al., 2017). Bamboo is also a strong and versatile fiber that is most common in composite production (Yu et al., 2014). Natural fibers are used as reinforcement in PLA composite. This is because incorporating natural fibers in the development of composite can influence certain properties such as dimensional stability, thermal property, impact resistance, physical properties, etc. Firstly, natural fibers such as jute or flax are known for their high strength and stiffness. PLA

DOI: 10.1201/9781032678894-8

composite's tensile strength and stiffness can be enhanced significantly by reinforcing these natural fibers in the composite (Sujaritjun et al., 2013). This can make the composite more durable, strong, and more rigid (Lv, Gu, et al., 2018).

However, the properties of the composite are highly influenced by several variables such as the amount of natural fiber used, the type of natural fiber used and the manufacturing method (Ho et al., 2012). The proportion or ratio of natural fiber to PLA depends on the required functionality and application. To produce a high-quality, durable PLA composite, it is necessary to have proper processing and conduct testing to verify if the material meets the necessary requirement. The remarkable property of PLA composites causes it to be widely used in various applications depending on the requirement. The properties of PLA composites can be tailored according to the requirement of specific application by reinforcing the PLA polymer with a natural fiber. Natural fibers such as jute, hemp, flax or kenaf are frequently used to reinforce PLA polymer to improve its mechanical properties and reduce environmental effect. Manipulating the properties of PLA using various natural fibers allows the industry to manufacture a versatile material that can be used to meet a wide range of applications. This flexibility in material design is an important element for industries to consider that provides diverse properties while maintaining its effectiveness and environmentally friendly (Sood & Dwivedi, 2018). Specific requirements applied to different applications. For example, various characteristics may be required for consumer goods, building materials, and automotive components. It is possible to tailor the composite to precisely match the requirements of the desired use by choosing the right natural fiber and modifying its composition (Kamarudin et al., 2022).

8.2 PLA COMPOSITES

8.2.1 Natural Fibers and PLA

In this chapter, we explain synthesis, production, and environmental sustainability of natural fiber and PLA. Natural fibers are fibers formed from the bodies of plants or animals through geological processes. Natural fibers can be found as continuous strings, weaved fibers, long fibers, mats, and other types of shape and other forms depending on their uses (Khan et al., 2018). Natural fiber can be utilized as a component of composite materials. One of the most common natural fibers composites is natural fiber-reinforced composite. These composites are utilized in a variety of industrial and engineering applications (Khalid et al., 2021).

Raw materials for PLA are renewable plant-based materials formed through fermentation process. These renewable plant materials can be easily obtained from agricultural and food waste (Ruz-Cruz et al., 2022). PLA forms lightweight, biodegradation, biodegradable, thermoplastic polymers. PLA is a material with distinct properties and with the excellent ability to modify its structure, which makes it a more common polymer in various industries. PLA also has higher mechanical properties than other polymers and has good optical and processing abilities (Oksiuta et al., 2020). Natural fibers such as cellulosic fibers can be used as reinforcement to increase the strength of the PLA (Ruz-Cruz et al., 2022).

8.2.2 Synthesis and Production

There are many ongoing experiments and research to improve the properties of PLA to increase its range of usages, specifically its thermal, mechanical, physical, chemical, and oxygen barrier properties. To improve these properties, some aspects need to be considered in PLA production method and synthesis (Figure 8.1).

One of the PLA syntheses that researchers are trying to create is PLA–polystyrene (PS) blends. The problem with this type of synthesis is that the blend is not miscible, exhibits significant phase separation, and causes defects in the PLA. The cause for the lack of miscibility is the high interfacial energy between PLA and PS and absence of favorable secondary interactions. To overcome this problem few production methods can be used. The most suitable method is using PLA-co-PS copolymers as PLA additives or as PLA/PS blend compatibles. This method improves PS chain miscibility within the PLA matrix. The synthesis of PLA-co-PS copolymers form a bottlebrush or star-shaped macrostructures through a multistage process (Gazzotti et al., 2022).

Co-continuous polymer blends method also other methods to improve the properties of PLA. This method enhances miscible PLA and PS mixes since the -OH and -COOH chain groups can be utilized for better mechanical and thermal properties. Co-continuous polymer blends are used to create porous structures with open-channel holes that are coupled and can create polymer material with a hierarchical porous structure. When considering different types of polymers for co-continuous polymer blends, they are mostly immiscible due to molecular weight. This molecular factor can be overcome using binary immiscible blends. Where low volume percentage of either component leads the minor phase to scatter as droplets or threads inside the large phase. Phase inversion occurs when the volume percentage of the minor phase

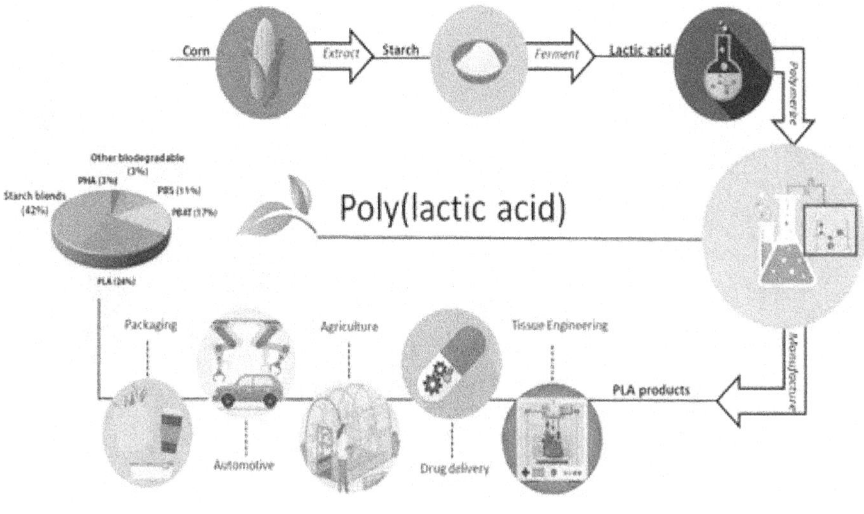

FIGURE 8.1 The Production Flow of Polylactic Acid PLA.

Source: European Bioplastics (2019).

rises, resulting in a continuous morphology characterized by two interpenetrating polymer phase networks, with stability determined by parameters such as viscosity and interfacial tension (Lv, Gu, et al., 2018).

8.2.3 Environmental Sustainability

Polymers are among the widely used materials in human life. From household products to advanced aircraft, all use polymers in their manufacturing because of polymers' excellent mechanical properties and durability (Fahim et al., 2019). Currently, the manufacturing industries are trying to use more environmentally friendly polymers than petroleum-based polymers. The reason for pursuing environmentally friendly polymers is the negative effects of petroleum-based polymers on the environment. Every year, nearly 9 billion metric tons of polymer are produced around the world, and only a small portion is recycled or reused; the rest (78%–79%) ends up in sea, landfills, and other environmental sources. The condition not only affects the environment but also affects human health (Balla et al., 2021). To replace the petroleum-based polymers, manufacturing industries are currently developing many different biopolymers derived from natural sources and either chemically synthesized from biological material or completely biosynthesized by live organisms (Figure 8.2) (Brigham, 2018). PLA minimizes reliance on fossil resources and greenhouse gas emissions (Fahim et al., 2019).

Usages of PLA in product manufacturing plays a major role in reducing waste. For example, current wall insulation products are made from nonrenewable materials,

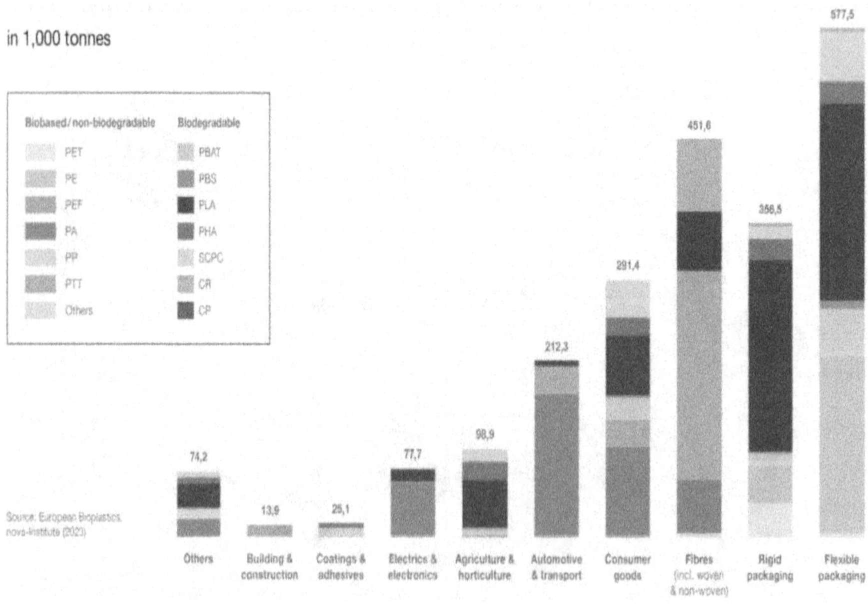

FIGURE 8.2 Global production and capacities of bioplastics.

Source: Brigham (2018).

and studies are ongoing to develop wall insulation made from PLA materials. This replacement is more energy efficient than conventional materials and reduces gas emissions in the building (Xu et al., 2022).

8.3 PHYSICAL PROPERTIES OF PLA COMPOSITES

8.3.1 Density

Density plays an important role in the physical properties of PLA composites. The density of a composite is a measure of how compact the mass is. In the engineering field, the density is used for calculating the mechanical properties of a material. For example, density functional theory (DFT) is used for calculating a material's mechanical characteristics and can be used for any materials in allowed size and shape. The main objective of DFT computations is to provide approximate solution to the many-body Schrödinger equation that later can be converted into mechanical property values. The method shows that density influences the mechanical and physical properties of a material or a composite (Kiely et al., 2021). Reinforcing PLA aims to increase its mechanical and physical properties. The most common reinforcement used in PLA is natural fiber to increase strength and stiffness (Sanivada et al., 2020). When a high-density reinforcement is added into the composite, it will cause the composite density to increase (Annigeri & Veeresh Kumar, 2018).

8.3.2 Transparency

Transparency in PLA composites is a very important physical property. PLA is one major candidate for replacing petroleum polymers in the packaging sector. Packing needs transparent materials so the inner product can be seen clearly, and many researchers are currently researching how to increase the transparency of PLA products. Using core-shell nanoparticles is one of the methods to increase the transparency of PLA composite; seed emulsion polymerization was used to toughen PLA without significantly reducing transparency. The methyl methacrylate-butyl acrylate copolymer had rubbery cores that resisted impact, and the glassy shells provided rigidity and compatibility to the polymer matrix. The nanoparticles increased the transparency to 75% (550 nm) and increased the PLA tensile strength as well, which is also important in the packaging sector (Chen et al., 2018).

Another application of PLA composite is air filters. Particulate matter (PM) pollution is harmful to humans and the environment. To overcome this pollution, the market is currently using PLA nanofiber filters to coat window screens. A type of air filter that can remove PM efficiency and has good transparency with low air resistance. Transparency in this type of air filter is important because the view of the window should not be blocked or blurred. Using electrostatic induction-assisted solution blowing to create fabricate polylactide stereocomplex produced PLA nanofiber filters with up to 83% transparency with 99.5% PM2.5 removal. This method also improved mechanical properties by 140% (Lv, Zhao, et al., 2018).

In addition to its ecofriendly characteristics, the manufacturing industry is using PLA because of its toughness and brittleness, but improving toughness reduces transparency. To overcome this problem, researchers explored a range of imidazolium-functionalized polyether-based ionomers derived from renewable epichlorohydrin elastomers and different comprehensive imidazoles via simple quaternization. This manufacturing method increased the toughness of PLA without compromising its transparency of PLA; transparency exceeded 80%, and strain-to-failure rate exceeded 230% (Wang et al., 2021).

8.3.3 Melting and Glass Transition Temperatures

The melting (T_{m-}) and glass transition (T_g) temperatures of composites are critical properties (Chen et al., 2013). T_{m-} is the temperature at which the PLA composite transitions from a solid to liquid state at which the rigid arrangement of particles converts to a free-moving particle (Kaiser et al., 2013).

The T_m and T_g of natural fiber reinforced PLA composites are higher than those of PLA alone (Sun et al., 2017). This is because fiber reinforcement restricts the mobility of PLA polymer chain making it more difficult for the bond in between the polymer to break to melt from a solid state. The type and quantity of fiber reinforcement also impact the melting point of PLA composites made of natural fiber (Xu et al., 2014).

There are several factors that affect T_m and T_g of a natural fiber PLA composite including the type and content of fiber reinforcement. Fiber reinforcement generally raises the T_g of PLA composites because it constrains PLA polymer chains' mobility, which makes it more difficult for the polymers to transition from a glassy to a rubbery state (Zaleha Mustafa et al., 2022). Processing conditions such as temperature and pressure also affect the T_g of PLA composites, and T_g is also influenced by the interaction between the PLA matrix and the fiber reinforcement. A high T_g depends on the two phases possessing good interfacial adhesion (Simmons et al., 2019).

8.3.4 Solubility

Solubility is the ability of a composite to completely dissolve in a particular solvent to form a homogenous solution. Natural fibers are often hydrophilic, tending to dissolve in water. PLA composites are not highly resistant to certain chemicals. In general, PLA composites are soluble in dioxane, acetonitrile, chloroform, methylene chloride, 1,1,2-trichloroethane and dichloroacetic acid. However, reinforcement with natural fiber does not necessarily improve the chemical resistance but influences the solubility of PLA composites (Farah et al., 2016). The biodegradability of a composite is also influenced by the solubility of the composite. Both PLA and natural fibers are biodegradable materials and will biodegrade under suitable circumstances. The rate of biodegradability depends on the type and composition of natural fiber used (Jiang et al., 2019).

PLA degrades mainly by hydrolysis process, after several months of exposure to moisture. PLA degrades in two phases. First, molecular weight decreases as a result of random non-enzymatic chain scission of the ester groups. In the second stage, the Mw reduces until microorganisms naturally metabolize the lactic acid and low Mw oligomers to produce water and carbon dioxide. The degradation rate of a PLA polymer is mainly affected by the reactivity and accessibility such as particle shape and size, temperature, moisture, crystallinity, lactic acid concentration. The degradable property of PLA composite is one of the reasons for its use in many surgical and medical fields (Siakeng et al., 2019). The solubility property natural fiber PLA composite is affected by several factors. The type and composition of fiber reinforcement affects the solubility of PLA. Natural fiber reinforcement often reduces the mobility of PLA polymer chains, making them harder to break. In order for a polymer to completely dissolve in a solvent, the bond or chain in between the polymer has to be broken, and adding natural fiber reinforcement reduces the solubility. Fiber treatment can enhance the interfacial adhesion between the PLA and natural fiber reinforcement which reduces the mobility of the polymer chain making it difficult to dissolve (Sudamrao Getme & Patel, 2020). The solubility of natural fiber PLA composites can be altered to meet the requirements of a specific application.

8.4 CHARACTERIZING PHYSICAL PROPERTIES

8.4.1 Scanning Electron Microscopy

Scanning electron microscopy (SEM) is an important tool that can be used for characterizing the physical properties of PLA composites. SEM is used to evaluate the morphology, dispersion, and interfacial interactions between the PLA matrix and the reinforcement. This provides information regarding the mechanical, physical and thermal properties of PLA composites (Mohammed & Abdullah, 2018). SEM images showed that adding abaca and synthetic cellulose fibers increased the material's tensile strength and impact resistance. Additionally, the mechanical characteristics were improved by adding carbon and nylon glass fibers (Bledzki et al., 2009). SEM is the best analysis method to obtain a precise analysis of a solid surface; fractured surfaces during tensile tests were observed during the SEM analysis (Figure 8.3) (Zakaria et al., 2013). The fractured surface of PLA could be categorized as ductile fracture. Fractures producing fibrils less than 1 μm long are categorized as brittle fracture (Jeevitha & Amarnath, 2013).

8.4.2 Fourier Transform Infrared Spectroscopy

SEM indicates the surface morphology of the PLA (Mohammed & Abdullah, 2018). However, Fourier transform infrared (FTIR) spectroscopy analyzes the chemical composition, functional groups and the interactions between PLA and the reinforcement. This technique has been used to identify the presence of functional groups such as methyl, carbonyl, and hydroxyl groups in PLA films (Siriprom et al., 2018).

Figure 8.4 shows the FTIR images of pristine PLA, PEG, PLA/PEG, and PLA/PEG nanocomposites. PLA has unique stretching frequencies at 1746, 2995, 2946, and 1080 cm−1 for C=O, $-CH_3$ asymmetric, –CH3symmetric, and C–O, respectively. The bending frequencies of –CH3 asymmetric and –CH3 symmetric were 1452 and 1361 cm−1, respectively (Chieng et al., 2013).

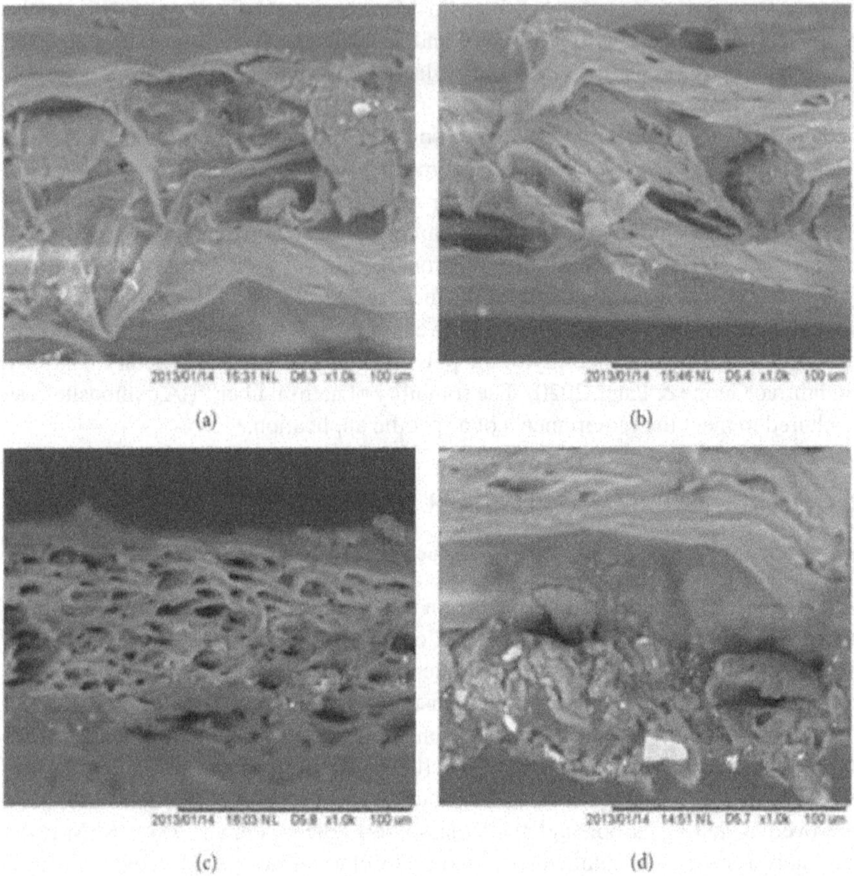

FIGURE 8.3 Typical SEM of (a) PLA, (b) PLA/CS (10% CS), (c) PLA/ENR, and (d) PLA/CS/ENR (10% CS).

Source: Zakaria et al. (2013).

8.4.3 Differential Scanning Calorimetry

Differential scanning calorimetry (DSC) is a widely used technique for determining the thermal characteristics of PLA composites. DSC monitors the flow of heat into or out of a sample as a function of temperature or time, revealing important information regarding phase transitions, thermal stability, and heat capacity (Schick, 2009). Using DSC, few physical properties can be determined such as melting temperature (Tm), crystallization temperature (Tc), and glass transition temperature (Tg). DSC can be used to physically age completely amorphous PLA. To perform this physical age, the PLA sample needs to be heated up to a specific temperature and held at that temperature for a few minutes, then cooled down using a specific cooling rate. Then the physical property of the sample such as glass transition can be observed. The test needs to be performed with a different cooling rate and sample temperature size to obtain accurate results (Monnier et al., 2017).

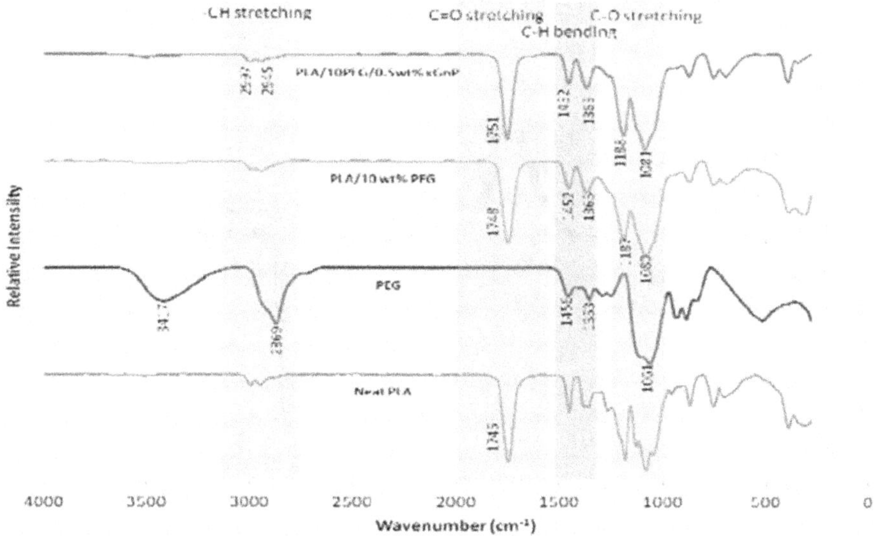

FIGURE 8.4 FTIR spectroscopy of pristine PLA, PEG, PLA/PEG, and PLA/PEG nanocomposites.

Source: Chieng et al. (2013).

PLA composites are widely used in 3D printing due to their physical, thermal, chemical properties and mechanical properties. DSC testing can be used to study thermal properties, morphological qualities, and heat transport behavior of 3D-printed PLA parts. In one study, the authors tested 3D-printed parts with different contents of milled carbon fiber (MCF, wt.%) and PLA (wt.%) for their specific heat capacity (Cp), Tg, Tc, the enthalpy of cold crystallization (DHc), and Tm. Figure 8.5 shows representative DSC thermograms of the investigated samples from that study. The researchers determined that adding MCF to the PLA matrix improved the conductive and physical properties of 3D-printed PLA parts (Barbés et al., 2013).

In another study, DSC was used to study the thermal characterization of biodegradable materials made from plasticized PLA/chitosan and rosemary extract. This study was focused on testing the thermal behavior and the biocomposition properties of the specimen. According to the study, rosemary ethanolic extract provides effective protection against thermo-oxidative degradation to newly created plasticized PLA-based biocomposites with good thermal characteristics, making them appropriate for both medical and food packaging applications (Vasile et al., 2020).

8.4.4 Physical Aging and Environmental Testing

The purpose of physical aging and environmental testing on a composite material, such as PLA composite, is to determine how the material responds to various external variables over time. Aspects that are considered in physical aging and environmental testing for PLA composite are UV exposure, humidity, moisture, thermal aging, chemical resistance, abrasion and wear. All these aspects are tested in the long

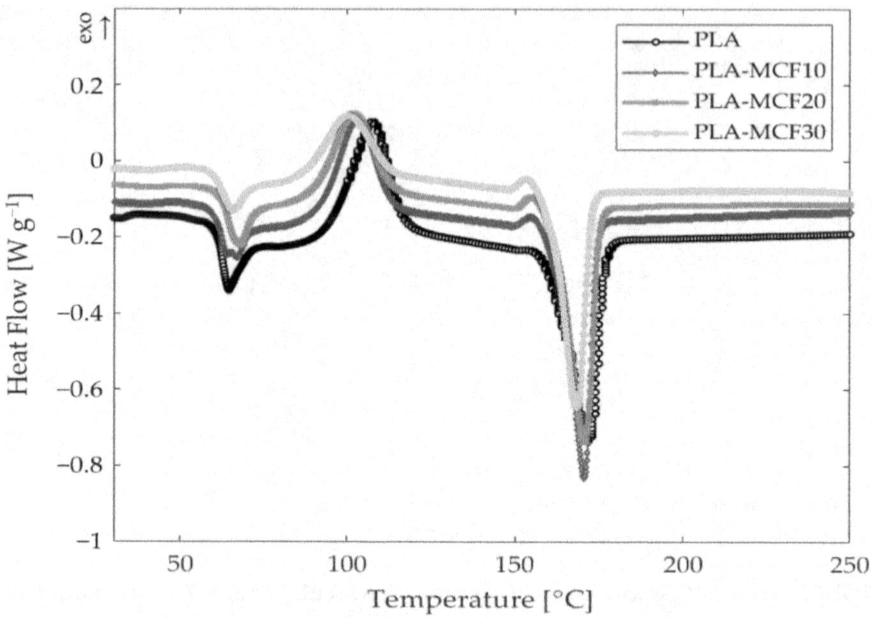

FIGURE 8.5 DSC curves for PLA (black) and MCF-reinforced grades at 10 wt.% (blue), 20 wt.% (red), and 30 wt.% (green) MCF content.

Source: Barbés et al. (2013).

term to study the performance of PLA after specific lifespan (Giv et al., 2021). In a study to understand the ductile-brittle transition behavior of PLA composite over a specific time, three different tests are conducted on the physical aging at 30 °C of PLA/o-MMT samples. These three conducted tests are essential work of fracture test, enthalpy relaxation analysis and small punch tests. After one week of storage at 30 °C, all tested samples displayed comparable thermal, mechanical, and physical properties (Cailloux et al., 2014).

8.5 FACTORS INFLUENCING PHYSICAL PROPERTIES

8.5.1 Fillers

Glass, quartz or silica, natural fiber are common fillers into PLA to improve its mechanical physical and thermal properties (Andrzejewski et al., 2019). Nanofiller is one of many fillers that are commonly added to PLA. Nanofillers must be mixed into the polymer matrix in a certain proportion and different nanofiller have different functions and improve different properties of PLA (Rostami et al., 2018). Nanocellulose fillers used in advanced material such as artificial blood vessels and other artificial tissue products due to mechanical characteristics and biocompatibility, ultrafine fiber network, and high porosity are all features of this material.

Layered silicate, also known as nanoclay, fillers commonly used for biodegradable PLA because of mechanical and obstructive qualities, as well as reduced burnable points, are characteristics of each native polymer. Additionally, carbon nanotubes and graphene filler are used for storage of electrochemical energy, conversion, and hydrogen storage. This filler combined with PLA enhances thermal and mechanical qualities while also providing extra functions like as flame resistance and barrier properties. One of the main constraints of using PLA in the packaging sector is its poor barrier property, but adding fiber filler into the PLA enhanced the barrier property (35%) and tensile strength (13%) over that of pure PLA. Researchers discovered that adding varied quantities of $CaCO_3$ improved the tensile characteristics of PLA material to varying degrees (Guo et al., 2021).

Fused deposition modeling (FDM) is a type of 3D printing that uses PLA as printing materials with different fillers to make products with different properties as needed. Carbon fiber in PLA enhances the 25% flexural stiffness and other mechanical properties such as formability and durability. FDM also uses microcrystalline cellulose as filler (Figure 8.6) because of its high mechanical strength and stiffness combined with low weight, renewability, and biodegradability. Core-shell rubber filler increased a PLA composite's impact strength by 18%; adding glycidyl methacrylate gave the highest durability of all the fillers tested (Arockiam et al., 2022).

8.5.2 Processing Conditions

The processing conditions are the methods and parameters that need to be considered before, during and after forming a PLA composite. The processing conditions play an important role in the properties and quality of PLA composites. The most common parameters that need to be monitored are the pressure, temperature, and base material quality. The cheapest ways of making PLA are extrusion and injection molding even though these processes are expensive compared to conventional

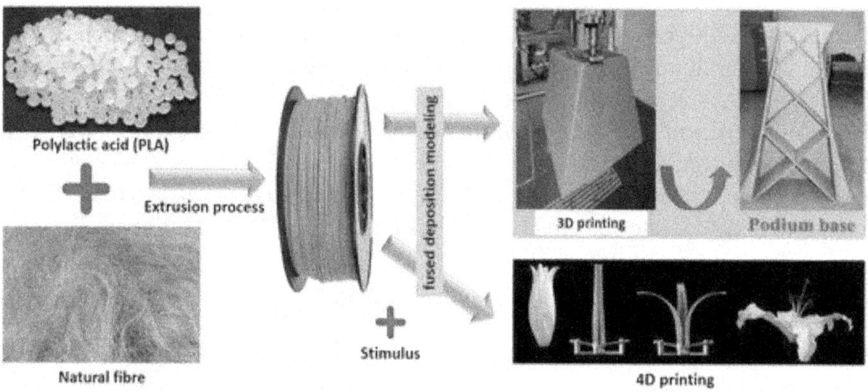

FIGURE 8.6 3-D printing inputs and outputs made from PLA.

Source: Ilyas et al. (2021).

polymer processes (Balla et al., 2021). Both extrusion and injection molding are carried out using specific machines, so the machine parameters also affect the properties of PLA such as the temperature of the machine and mold, the design of the plasticizing system, and the speed of the rotating screw that pushes the materials into mold. Changes in temperature affect the crystallization structure of PLA, which affects its physical properties, surface finish, and appearance of PLA composite products (Tabi et al., 2010)

The PLA needs to dry completely before the processing starts because water particles can cause the cleavage of hydrolytic bonds and degrading of the material by interaction between the polymer chains. This condition can lower T_g and give a negative effect to the final PLA product's physical properties (Ogunsona et al., 2017). Another effect of water present in PLA before processing is degradation. Degradation can cause a random main chain disruption process and causes a reduction in molar mass. This condition can lead to cracks on the surface of the PLA composite and create rough surfaces on the end products (Celestine et al., 2020).

8.5.3 Aging and Environmental Factors

Aging is a natural process that occurs over time and can cause changes in the properties of the PLA composites. The durability of PLA is restricted by various chemical ageing mechanisms, including hydrolysis and heat degradation (Parvatareddy et al., 1995). The functionality of a composite should be tested in real time and current environment. Therefore, accelerated aging techniques are required to prevent delays in product development. Measured variables in accelerated weathering include exposure duration, UV irradiation exposure within a certain wavelength range, and moisture exposure as a number of cycles or time; changing these changes the properties of PLA composites. Extensive exposure to UV rays can damage the structure of the material which might deteriorate the surface of the composite (Boubakri et al., 2010).

Temperature, humidity, and exposure to chemicals also affect the physical properties of materials (Moudood et al., 2019). High humidity causes materials to absorb moisture, which can lead to swelling and changes in their strength and stiffness (Alawsi et al., 2009); Harris & Lee, 2010). This is especially crucial for applications like outdoor furniture and packaging where PLA is subjected to various weather conditions (Varsavas & Kaynak, 2018). Processing temperature can affect the overall thermal properties of the PLA composite such as T_m and T_g. At high temperature, PLA can become ductile and become easier to be deformed, while at low temperature, it becomes more brittle (Barkhad et al., 2020). Exposure to chemicals also causes the composite to deteriorate and affect the material properties (Csizmadia et al., 2013).

8.6 APPLICATIONS OF PLA COMPOSITES

PLA is used for various applications such as automotive, packaging, aviation and biomedical fields as PLA is a biodegradable polyester extracted from renewable resources (Figure 8.7). PLA has good mechanical properties such as high tensile strength, good flexural strength, and high Young's modulus (Hamad et al., 2015). Currently, the primary goal of many research and development projects is to develop

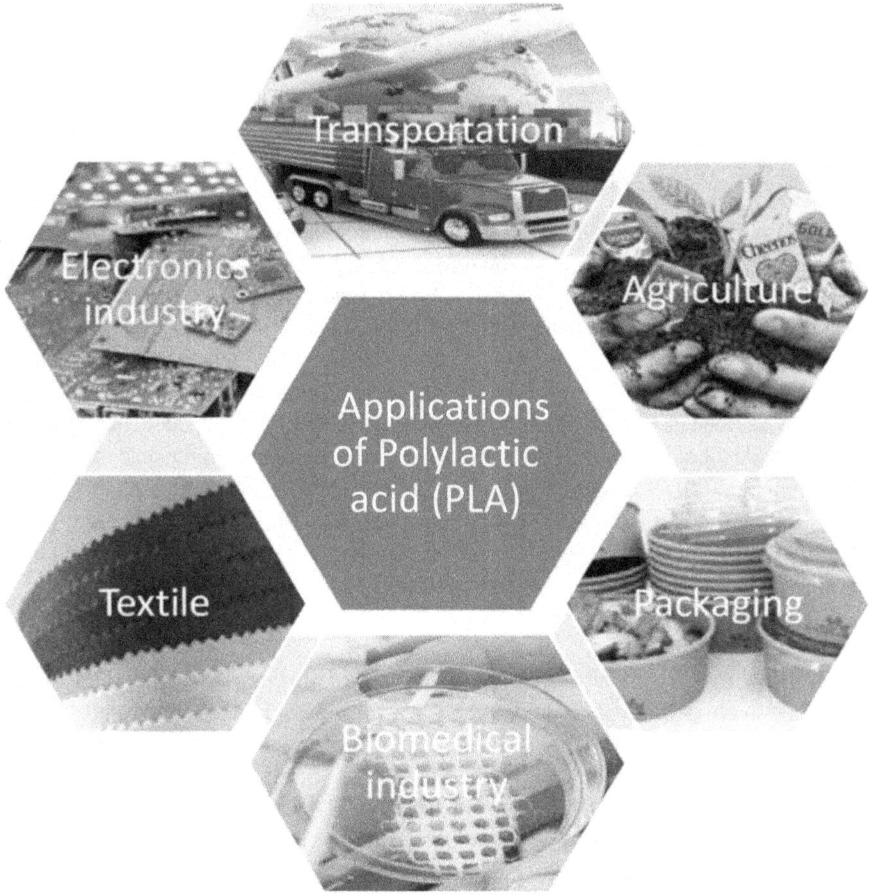

FIGURE 8.7 Applications of PLA composites.

Source: Rezvani Ghomi et al. (2021).

through various methods new classes of biopolymers with enhanced properties making it suitable for various engineering sectors and in applications that require enhanced property. PLA is a great alternative to petrochemical-based synthetic polymers (PET, PS, PE, etc.) (Murariu & Dubois, 2016).

8.6.1 Automotive Industry

In the automotive Industry, PLA composites are attracted by many industries due to their light weight, high strength, and biodegradability (Arjmandi et al., 2017). PLA offers various advantages over petrochemical-based synthetic polymers including light weight, enhanced mechanical and thermal properties, and biodegradability. However, in order to utilize the PLA to its utmost efficiency, the heat resistance, mechanical properties, and durability must be improved (Bouzouita et al., 2017). The mechanical properties of the PLA-based compositions are improved by the addition of specially designed additives, which makes them suitable for use in the automotive industry. In addition, by optimizing injection molding and adding reinforcements, PLA composites can further enhance their performance in automotive interior applications (Notta-Cuvier et al., 2014).

PLA in the automotive industry is used to fabricate various components, particularly as both interior components and mechanical components. Due to its high bio-content content, PLA is well-known for reducing the carbon footprint and offers various advantages, such as impact and UV resistance, high gloss, and dimensional stability. All these characteristics make PLA the ideal replacement for most of the conventional thermoplastics (Balla et al., 2021). Natural biofibers are the most environmentally friendly method to enhance the thermal and mechanical properties of PLA. The Mercedes-Benz E-class makes substantial use of plant fiber, including in headliners, dashboards, door panels, package trays, and some interior components (Kabirian et al., 2018). However, although PLA is widely used in this industry, there are several challenges such as cost, thermal resistance, impact resistance, and compatibility with current manufacturing processes (Tripathi et al., 2021).

8.6.2 Packaging

One of the primary applications of PLA is the packaging industry, particularly food packaging, such as bottles, containers, and fresh food packaging. The trend toward ecofriendly packaging and the increasing number of environmental alarms around the world are driving companies to incorporate the product into many packaging solutions (Raquez et al., 2013). PLA has been widely used in food packaging as it provides high biocompatibility and biodegradability, great physical properties such as high strength, processability, and nontoxicity. For packaging applications, the mechanical properties of PLA films are comparable with those of PET. The disadvantage of PLA films is their lower T_{g-}, especially in applications with high temperature resistance (Muller et al., 2017).

In the past decade, PLA has garnered increasing interest as a material for food packaging due to its ease of extraction from renewable sources, its carbon dioxide-neutral production process, recyclability, composability, and the ability to customize its physical and mechanical properties through polymer architecture (Siracusa et al., 2008). Although PLA provides various benefits in the packaging industry, limitations such as reduced flexibility, suboptimal crystallization characteristics, and inadequate barrier properties hinder the broader utilization of PLA. Hence, to enhance its performance, PLA is frequently blended with suitable additives or other polymers possessing superior gas barrier properties, such as furoates (Fredi et al., 2021).

8.6.3 Aviation

PLA composites are showing great potential for a variety of applications in the aviation sector. The utilization of product lifecycle management has optimized project management efficiency in the development and design of aviation products (Georgievna, 2021). The favorable characteristics of PLA, including its biodegradability and environmentally friendly nature, make it well-suited for a range of industrial applications, including aviation (Domenek & Ducruet, 2016). Utilizing PLA composites in 3D printing, a technology with prospective applications in aviation manufacturing, can be employed for the production of numerous components (Tümer & Erbil, 2021).

PLA composites, especially those incorporating natural fibers such as flax and jute, are attracting interest in the aviation sector because of their lightweight, strong, and bio-derived characteristics (Sanivada et al., 2020). Enhancements are needed in the dispersion of silica nanoparticles within the PLA matrix, as there is anticipation of discovering novel applications in diverse sectors like aviation, energy, and the chemical industry in the near future. The mechanical properties of PLA composites can be optimized by adjusting the processing parameters, such as fiber volume fraction and molding temperature (Kaseem et al., 2021).

PLA composites in aviation improve fuel efficiency and decrease the weight of aircraft, with their elevated strength, making them well-suited for structural elements and impact absorption systems (Parveez et al., 2022). The potential of PLA composites in the development of sustainable aircraft faces obstacles due to challenges such as cost, fire resistance, durability, and compatibility with manufacturing processes. Notwithstanding these challenges, PLA composites are anticipated to assume a noteworthy role in the advancement of sustainable aircraft as efforts are made to address and overcome these issues (Tripathi et al., 2021).

8.6.4 Biomedical

PLA's diverse properties allows it to be used widely in biomedical applications (Figure 8.8). The rate of degradation is frequently deemed a crucial factor in the selection process for biomedical applications (Farah et al., 2016) such as surgical sutures, plates, and screws for craniomaxillofacial bone fixation (DeJong et al., 2004). In the latest

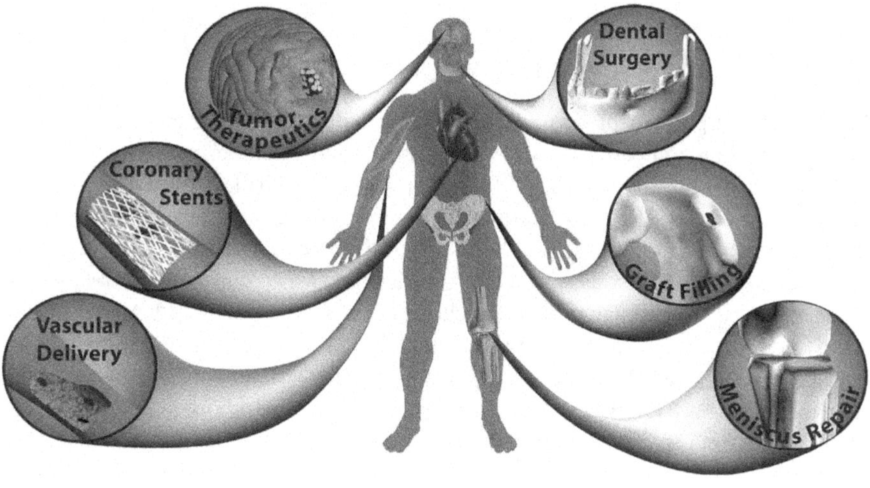

FIGURE 8.8 Applications of PLA biomedical parts in the human body.

Source: Tyler et al. (2016).

progress within tissue engineering and regenerative medicine, there has been a concentrated effort to advance 3D structured scaffolds and hydrogels. These components play a pivotal role in directing tissue development and offering essential mechanical support during the implantation process (Pina et al., 2019). PLA composites, specifically, exhibit potential in the fields of tissue engineering and drug delivery. They demonstrate the capability to offer a supportive framework for cell growth and the controlled discharge of bioactive molecules (Guo et al., 2021).

PLA finds application in diverse biomedical uses, including its utilization in surgical sutures as well as in the manufacturing of plates and screws for craniomaxillofacial bone fixation (Farah et al., 2016). Incorporating an amphiphilic block copolymer into PLA/Mg composites has been discovered to improve their interfacial adhesion and bioactivity, making them appropriate for use in bone implants (Ben Abdeljawad et al., 2021). The potential use of PLA composites in orthopedic and dental applications has been emphasized, specifically addressing their manufacturing processes, inherent properties, and the incorporation of reinforcing fibers and fillers to augment their characteristics (Murariu & Dubois, 2016).

PLA and its copolymers are attracting attention within the realm of cardiovascular applications, thanks to their biocompatibility, mechanical characteristics, and biodegradability. These attributes render them well-suited for applications in cardiovascular devices and tissue engineering (Hadasha & Bezuidenhout, 2017). For instance, PLA composites have found utilization in bioresorbable stents and patches designed to repair compromised heart tissue, fostering regeneration, and enhancing cardiac function (Hadasha & Bezuidenhout, 2017; Toong et al., 2020). Nonetheless, it has certain limitations such as slow degradation, low thermal resistance and poor toughness (Raquez et al., 2013). In response to these constraints, diverse approaches have been investigated, encompassing copolymerization, blending, and the incorporation of nanofillers. Copolymerization has proven effective in improving the characteristics of PLA, rendering it more apt for biomedical applications (Puthumana et al., 2020).

8.7 CHALLENGES AND FUTURE DIRECTIONS

Despite being a popular biodegradable polymer with outstanding mechanical characteristics and biocompatibility, PLA composites face a number of challenges. One of them is poor thermal stability. PLA has a low T_g and Tm, limiting its usage in high-temperature applications. Reinforcing elements such as carbon nanotubes in PLA/PEO blends can increase the thermal stability and morphology; however, their effect in thermal stability is dependent on PLA and CNT concentrations (Zare & Rhee, 2019). Next, brittleness is a major challenge when considering PLA composites in the manufacturing industry, a disadvantage in applications where toughness and impact resistance are critical. Reinforcements help improve toughness, although balancing strength and ductility can be difficult. Adding silane-modified bamboo cellulose nanowhiskers into PLA composites can increase elongation at break, lowering brittleness and enhancing their potential usage in green plastics (Qian et al., 2018).

Other than that, processing certain types of PLA can be challenging. PLA melt processing might be difficult because of its rather restricted processing window.

To minimize thermal deterioration or insufficient dispersion of reinforcing materials, production variables such as temperature and shear rate must be carefully managed. For example, processing stereocomplex-type polylactide is difficult because of its low heat resistance and poor thermal stability, although new improvements such as low-temperature sintering offer promise for better characteristics (Bai et al., 2017). Researchers and manufacturers are working hard to overcome these issues by developing new processing processes, novel reinforcing materials, and gaining a better knowledge of the structure-property connections in PLA composites. These advancements are critical for broadening the application range and increasing the overall performance of PLA-based materials.

8.8 CONCLUSIONS

In conclusion, research on the mechanical properties of PLA, particularly in the context of natural fiber-reinforced composites, has revealed an intriguing narrative. The interaction of natural fibers with PLA in composite materials results in considerable physical property improvements, expanding their potential across sectors. This chapter on natural fiber and PLA has provided the groundwork for understanding their synergistic connection, giving material scientists and engineers critical insights. PLA's renewable and biodegradable nature corresponds with the rising need for ecofriendly alternatives in the manufacturing sector, emphasizing environmental sustainability. Characterization approaches have been helpful in uncovering the structural complexities of PLA composites, providing a path for specific applications. Despite progress, obstacles remain, necessitating continued research to solve concerns like thermal stability and brittleness, guiding the trajectory of PLA's future uses and paving the way for a more sustainable industrial paradigm.

REFERENCES

Alawsi, G., Aldajah, S., & Rahmaan, S. A. (2009). Impact of humidity on the durability of E-glass/polymer composites. *Materials & Design*, *30*(7), 2506–2512. https://doi.org/10.1016/j.matdes.2008.10.002

Andrzejewski, J., Szostak, M., Barczewski, M., & Łuczak, P. (2019). Cork-wood hybrid filler system for polypropylene and poly(lactic acid) based injection molded composites. Structure evaluation and mechanical performance. *Composites Part B: Engineering*, *163*, 655–668. https://doi.org/10.1016/j.compositesb.2018.12.109

Annigeri, U. K., & Veeresh Kumar, G. B. (2018). Effect of reinforcement on density, hardness and wear behavior of aluminum metal matrix composites: A review. *Materials Today: Proceedings*, *5*(5), 11233–11237. https://doi.org/10.1016/j.matpr.2018.02.002

Arjmandi, R., Hassan, A., & Zakaria, Z. (2017). *Polylactic Acid Green Nanocomposites for Automotive Applications* (pp. 193–208). Springer. https://doi.org/10.1007/978-3-319-49382-4_9

Bai, H., Deng, S., Bai, D., Zhang, Q., & Fu, Q. (2017). Recent advances in processing of stereocomplex-type polylactide. *Macromolecular Rapid Communications*, *38*(23), 1700454.

Balla, E., Daniilidis, V., Karlioti, G., Kalamas, T., Stefanidou, M., Bikiaris, N. D., Vlachopoulos, A., Koumentakou, I., & Bikiaris, D. N. (2021). Poly(lactic Acid): A versatile biobased polymer for the future with multifunctional properties – from monomer synthesis, polymerization techniques and molecular weight increase to PLA applications. *Polymers*, *13*(11), 1822. https://doi.org/10.3390/polym13111822

Barbés, B., Páramo, R., Blanco, E., Pastoriza-Gallego, M. J., Piñeiro, M. M., Legido, J. L., & Casanova, C. (2013). Thermal conductivity and specific heat capacity measurements of Al_2O_3 nanofluids. *Journal of Thermal Analysis and Calorimetry*, *111*(2), 1615–1625. https://doi.org/10.1007/s10973-012-2534-9

Barkhad, M. S., Abu-Jdayil, B., Mourad, A. H. I., & Iqbal, M. Z. (2020). Thermal insulation and mechanical properties of polylactic acid (PLA) at different processing conditions. *Polymers*, *12*(9), 2091. https://doi.org/10.3390/polym12092091

Ben Abdeljawad, M., Carette, X., Mincheva, R., Odent, J., & Raquez, J.-M. (2021). Interfacial compatibilization of PLA and Mg in composites for bioresorbable bone implants. *IOP Conference Series: Materials Science and Engineering*, *1056*, 012004. https://doi.org/10.1088/1757-899X/1056/1/012004

Bledzki, A. K., Jaszkiewicz, A., & Scherzer, D. (2009). Mechanical properties of PLA composites with man-made cellulose and abaca fibres. *Composites Part A: Applied Science and Manufacturing*, *40*(4), 404–412. https://doi.org/10.1016/j.compositesa.2009.01.002

Boubakri, A., Guermazi, N., Elleuch, K., & Ayedi, H. F. (2010). Study of UV-aging of thermoplastic polyurethane material. *Materials Science and Engineering: A*, *527*(7–8), 1649–1654. https://doi.org/10.1016/j.msea.2010.01.014

Bouzouita, A., Notta-Cuvier, D., Raquez, J.-M., Lauro, F., & Dubois, P. (2017). *Poly(Lactic Acid)-Based Materials for Automotive Applications* (pp. 177–219). Springer. https://doi.org/10.1007/12_2017_10

Brigham, C. (2018). Biopolymers. In *Green Chemistry* (pp. 753–770). Elsevier. https://doi.org/10.1016/B978-0-12-809270-5.00027-3

Cailloux, J., Santana, O. O., Franco-Urquiza, E., Bou, J. J., Carrasco, F., & Maspoch, M. L. (2014). Sheets of branched poly(lactic acid) obtained by one-step reactive extrusion – calendering process: Physical aging and fracture behavior. *Journal of Materials Science*, *49*(11), 4093–4107. https://doi.org/10.1007/s10853-014-8101-y

Celestine, A.-D. N., Agrawal, V., & Runnels, B. (2020). Experimental and numerical investigation into mechanical degradation of polymers. *Composites Part B: Engineering*, *201*, 108369. https://doi.org/10.1016/j.compositesb.2020.108369

Chen, F., Clough, A., Reinhard, B. M., Grinstaff, M. W., Jiang, N., Koga, T., & Tsui, O. K. C. (2013). Glass transition temperature of polymer – nanoparticle composites: Effect of polymer – particle interfacial energy. *Macromolecules*, *46*(11), 4663–4669. https://doi.org/10.1021/ma4000368

Chen, Y., Pan, M., Li, Y., Xu, J.-Z., Zhong, G.-J., Ji, X., Yan, Z., & Li, Z.-M. (2018). Core-shell nanoparticles toughened polylactide with excellent transparency and stiffness-toughness balance. *Composites Science and Technology*, *164*, 168–177. https://doi.org/10.1016/j.compscitech.2018.04.037

Chieng, B., Ibrahim, N., Yunus, W., & Hussein, M. (2013). Poly(lactic acid)/poly(ethylene glycol) polymer nanocomposites: Effects of graphene nanoplatelets. *Polymers*, *6*(1), 93–104. https://doi.org/10.3390/polym6010093

Csizmadia, R., Faludi, G., Renner, K., Móczó, J., & Pukánszky, B. (2013). PLA/wood biocomposites: Improving composite strength by chemical treatment of the fibers. *Composites Part A: Applied Science and Manufacturing*, *53*, 46–53. https://doi.org/10.1016/j.compositesa.2013.06.003

DeJong, L. E. S., DeBerardino, L. T. M., Brooks, D. E., & Judson, K. (2004). In vivo comparison of a metal versus a biodegradable suture anchor. *Arthroscopy: The Journal of Arthroscopic & Related Surgery*, *20*(5), 511–516. https://doi.org/10.1016/j.arthro.2004.03.008

Domenek, S., & Ducruet, V. (2016). Characteristics and applications of PLA. In *Biodegradable and Biobased Polymers for Environmental and Biomedical Applications* (pp. 171–224). Wiley. https://doi.org/10.1002/9781119117360.ch6

Ebrahimi, F., & Ramezani Dana, H. (2022). Poly lactic acid (PLA) polymers: From properties to biomedical applications. *International Journal of Polymeric Materials and Polymeric Biomaterials*, *71*(15), 1117–1130. https://doi.org/10.1080/00914037.2021.1944140

European Bioplastics. (2019). *Bioplastic Market Data Report*. https://docs.european-bioplastics.org/publications/market_data/Report_Bioplastics_Market_Data_2019.pdf

Fahim, I. S., Chbib, H., & Mahmoud, H. M. (2019). The synthesis, production & economic feasibility of manufacturing PLA from agricultural waste. *Sustainable Chemistry and Pharmacy*, *12*, 100142. https://doi.org/10.1016/j.scp.2019.100142

Farah, S., Anderson, D. G., & Langer, R. (2016). Physical and mechanical properties of PLA, and their functions in widespread applications – a comprehensive review. *Advanced Drug Delivery Reviews*, *107*, 367–392. https://doi.org/10.1016/j.addr.2016.06.012

Feng, J., Sun, Y., Song, P., Lei, W., Wu, Q., Liu, L., Yu, Y., & Wang, H. (2017). Fire-resistant, strong, and green polymer nanocomposites based on poly(lactic acid) and core – shell nanofibrous flame retardants. *ACS Sustainable Chemistry & Engineering*, *5*(9), 7894–7904. https://doi.org/10.1021/acssuschemeng.7b01430

Fredi, G., Rigotti, D., Bikiaris, D. N., & Dorigato, A. (2021). Tuning thermo-mechanical properties of poly(lactic acid) films through blending with bioderived poly(alkylene furanoate)s with different alkyl chain length for sustainable packaging. *Polymer*, *218*, 123527. https://doi.org/10.1016/j.polymer.2021.123527

Gazzotti, S., Adolfsson, K. H., Hakkarainen, M., Farina, H., Silvani, A., & Ortenzi, M. A. (2022). DOX mediated synthesis of PLA-co-PS graft copolymers with matrix-driven self-assembly tabiin PLA-based blends. *European Polymer Journal*, *170*, 111157. https://doi.org/10.1016/j.eurpolymj.2022.111157

Georgievna, P. T. (2021). *PLM-Systems Application in Aerospace Industry*. AIP Publishing. https://doi.org/10.1063/5.0035761

Gholampour, A., & Ozbakkaloglu, T. (2020). A review of natural fiber composites: Properties, modification and processing techniques, characterization, applications. *Journal of Materials Science*, *55*(3), 829–892. https://doi.org/10.1007/s10853-019-03990-y

Giv, A. N., Fu, Q., Yan, L., & Kasal, B. (2021). Durability and life prediction of fiber-reinforced polymer composites. In *Composite Materials* (pp. 179–203). Elsevier. https://doi.org/10.1016/B978-0-12-820512-9.00023-X

Guo, X., Yao, Y., Zhao, H., Chi, C., Zeng, F., Qian, F., Liu, Z., Huo, L., & Lv, Y. (2021). Environmental impacts of functional fillers in polylactide (PLA)-based bottles using life cycle assessment methodology. *Science of the Total Environment*, *788*, 147852. https://doi.org/10.1016/j.scitotenv.2021.147852

Hadasha, W., & Bezuidenhout, D. (2017). *Poly(lactic acid) as Biomaterial for Cardiovascular Devices and Tissue Engineering Applications* (pp. 51–77). https://doi.org/10.1007/12_2017_27

Hamad, K., Kaseem, M., Yang, H. W., Deri, F., & Ko, Y. G. (2015). Properties and medical applications of polylactic acid: A review. *Express Polymer Letters*, *9*(5), 435–455. https://doi.org/10.3144/expresspolymlett.2015.42

Harris, A. M., & Lee, E. C. (2010). Heat and humidity performance of injection molded PLA for durable applications. *Journal of Applied Polymer Science*, *115*(3), 1380–1389. https://doi.org/10.1002/app.30815

Ho, M., Wang, H., Lee, J.-H., Ho, C., Lau, K., Leng, J., & Hui, D. (2012). Critical factors on manufacturing processes of natural fibre composites. *Composites Part B: Engineering*, *43*(8), 3549–3562. https://doi.org/10.1016/j.compositesb.2011.10.001

Ilyas, R. A., Sapuan, S. M., Harussani, M. M., Hakimi, M. Y. A. Y., Haziq, M. Z. M., Atikah, M. S. N., Asyraf, M. R. M., Ishak, M. R., Razman, M. R., Nurazzi, N. M., & Norrahim, M. N. F. (2021). Polylactic acid (PLA) biocomposite: Processing, additive manufacturing and advanced applications. *Polymers*, *13*(8), 1326.

Ilyas, R., Zuhri, M., Aisyah, H., Asyraf, M. R. M., Hassan, S., Zainudin, E., Sapuan, S., Sharma, S., Bangar, S., Jumaidin, R., Nawab, Y., Faudzi, A., Abral, H., Asrofi, M., Syafri, E., & Sari, N. (2022). Natural fiber-reinforced polylactic acid, polylactic acid blends and their composites for advanced applications. *Polymers*, *14*(1), 202. https://doi.org/10.3390/polym14010202

Jeevitha, D., & Amarnath, K. (2013). Chitosan/PLA nanoparticles as a novel carrier for the delivery of anthraquinone: Synthesis, characterization and in vitro cytotoxicity evaluation. *Colloids and Surfaces B: Biointerfaces*, *101*, 126–134. https://doi.org/10.1016/j.colsurfb.2012.06.019

Jiang, N., Yu, T., Li, Y., Pirzada, T. J., & Marrow, T. J. (2019). Hygrothermal aging and structural damage of a jute/poly (lactic acid) (PLA) composite observed by X-ray tomography. *Composites Science and Technology*, *173*, 15–23. https://doi.org/10.1016/j.compscitech.2019.01.018

Joseph Arockiam, A., Subramanian, K., Padmanabhan, R. G., Selvaraj, R., Bagal, D. K., & Rajesh, S. (2022). A review on PLA with different fillers used as a filament in 3D printing. *Materials Today: Proceedings*, *50*, 2057–2064. https://doi.org/10.1016/j.matpr.2021.09.413

Kabirian, F., Ditkowski, B., Zamanian, A., Heying, R., & Mozafari, M. (2018). An innovative approach towards 3D-printed scaffolds for the next generation of tissue-engineered vascular grafts. *Materials Today: Proceedings*, *5*(7), 15586–15594. https://doi.org/10.1016/j.matpr.2018.04.167

Kaiser, M. R., Anuar, H. B., Samat, N. B., & Abdul Razak, S. B. (2013). Effect of processing routes on the mechanical, thermal and morphological properties of pla-based hybrid biocomposite. *Iranian Polymer Journal (English Edition)*, *22*(2), 123–131. https://doi.org/10.1007/s13726-012-0112-4

Kamarudin, S. H., Mohd Basri, M. S., Rayung, M., Abu, F., Ahmad, S., Norizan, M. N., Osman, S., Sarifuddin, N., Desa, M. S. Z. M., Abdullah, U. H., Mohamed Amin Tawakkal, I. S., & Abdullah, L. C. (2022). A review on natural fiber reinforced polymer composites (NFRPC) for sustainable industrial applications. *Polymers*, *14*(17), 3698. https://doi.org/10.3390/polym14173698

Kaseem, M., Ur Rehman, Z., Hossain, S., Singh, A. K., & Dikici, B. (2021). A review on synthesis, properties, and applications of polylactic acid/silica composites. *Polymers*, *13*(18), 3036. https://doi.org/10.3390/polym13183036

Khalid, M. Y., Al Rashid, A., Arif, Z. U., Ahmed, W., Arshad, H., & Zaidi, A. A. (2021). Natural fiber reinforced composites: Sustainable materials for emerging applications. *Results in Engineering*, *11*, 100263. https://doi.org/10.1016/j.rineng.2021.100263

Khan, T., Hameed Sultan, M. T. B., & Ariffin, A. H. (2018). The challenges of natural fiber in manufacturing, material selection, and technology application: A review. *Journal of Reinforced Plastics and Composites*, *37*(11), 770–779. https://doi.org/10.1177/0731684418756762

Kiely, E., Zwane, R., Fox, R., Reilly, A. M., & Guerin, S. (2021). Density functional theory predictions of the mechanical properties of crystalline materials. *CrystEngComm*, *23*(34), 5697–5710. https://doi.org/10.1039/D1CE00453K

Lv, S., Gu, J., Tan, H., & Zhang, Y. (2018). Enhanced durability of sustainable poly (lactic acid)-based composites with renewable starch and wood flour. *Journal of Cleaner Production*, *203*, 328–339. https://doi.org/10.1016/j.jclepro.2018.08.266

Lv, S., Zhao, X., Shi, L., Zhang, G., Wang, S., Kang, W., & Zhuang, X. (2018). Preparation and properties of sc-PLA/PMMA transparent nanofiber air filter. *Polymers*, *10*(9), 996. https://doi.org/10.3390/polym10090996

Mattana, G., Briand, D., Marette, A., Quintero, A. V., & De Rooij, N. F. (2015). Polylactic acid as a biodegradable material for all-solution-processed organic electronic devices. *Organic Electronics*, *17*, 77–86.

Mia, R., Islam, M. A., Ahmed, B., & Mojumdar, J. I. A. (2017). Woolenization of jute fibre. *European Scientific Journal*, *13*(30), 314. https://doi.org/10.19044/esj.2017.v13n30p314

Mohammed, A., & Abdullah, A. (2018). *Scanning Electron Microscopy*.

Monnier, X., Saiter, A., & Dargent, E. (2017). Physical aging in PLA through standard DSC and fast scanning calorimetry investigations. *Thermochimica Acta*, *648*, 13–22. https://doi.org/10.1016/j.tca.2016.12.006

Moudood, A., Rahman, A., Khanlou, H. M., Hall, W., Öchsner, A., & Francucci, G. (2019). Environmental effects on the durability and the mechanical performance of flax fiber/bio-epoxy composites. *Composites Part B: Engineering*, *171*, 284–293. https://doi.org/10.1016/j.compositesb.2019.05.032

Muller, J., González-Martínez, C., & Chiralt, A. (2017). Combination of poly(lactic) acid and starch for biodegradable food packaging. *Materials*, *10*(8), 952. https://doi.org/10.3390/ma10080952

Murariu, M., & Dubois, P. (2016). PLA composites: From production to properties. *Advanced Drug Delivery Reviews*, *107*, 17–46. https://doi.org/10.1016/j.addr.2016.04.003

Mustafa, Z., Razali, A. S., Rajendran, T. K., Fadzullah, S. H. S. M., Malingam, S. D., Abd Rahim, T., & Ratanawilai, T. (2022). Moisture absorption and thermal properties of unidirectional pineapple leaf fibre/polylactic acid composites under hygrothermal ageing conditions. *Journal of Advanced Research in Fluid Mechanics and Thermal Sciences*, *96*(2), 65–73. https://doi.org/10.37934/arfmts.96.2.6573

Notta-Cuvier, D., Odent, J., Delille, R., Murariu, M., Lauro, F., Raquez, J. M., Bennani, B., & Dubois, P. (2014). Tailoring polylactide (PLA) properties for automotive applications: Effect of addition of designed additives on main mechanical properties. *Polymer Testing*, *36*, 1–9. https://doi.org/10.1016/j.polymertesting.2014.03.007

Ogunsona, E. O., Misra, M., & Mohanty, A. K. (2017). Accelerated hydrothermal aging of biocarbon reinforced nylon biocomposites. *Polymer Degradation and Stability*, *139*, 76–88. https://doi.org/10.1016/j.polymdegradstab.2017.03.013

Oksiuta, Z., Jalbrzykowski, M., Mystkowska, J., Romanczuk, E., & Osiecki, T. (2020). Mechanical and thermal properties of polylactide (PLA) composites modified with Mg, Fe, and polyethylene (PE) additives. *Polymers*, *12*(12), 2939. https://doi.org/10.3390/polym12122939

Parvatareddy, H., Wang, J. Z., Dillard, D. A., Ward, T. C., & Rogalski, M. E. (1995). Environmental aging of high-performance polymeric composites: Effects on durability. *Composites Science and Technology*, *53*(4), 399–409. https://doi.org/10.1016/0266-3538(95)00029-1

Parveez, B., Kittur, M. I., Badruddin, I. A., Kamangar, S., Hussien, M., & Umarfarooq, M. A. (2022). Scientific advancements in composite materials for aircraft applications: A review. *Polymers*, *14*(22), 5007. https://doi.org/10.3390/polym14225007

Pina, S., Ribeiro, V. P., Marques, C. F., Maia, F. R., Silva, T. H., Reis, R. L., & Oliveira, J. M. (2019). Scaffolding strategies for tissue engineering and regenerative medicine applications. *Materials*, *12*(11), 1824. https://doi.org/10.3390/ma12111824

Puthumana, M., Santhana Gopala Krishnan, P., & Nayak, S. K. (2020). Chemical modifications of PLA through copolymerization. *International Journal of Polymer Analysis and Characterization*, *25*(8), 634–648. https://doi.org/10.1080/1023666X.2020.1830650

Qian, S., Sheng, K., Yu, K., Xu, L., & Fontanillo Lopez, C. A. (2018). Improved properties of PLA biocomposites toughened with bamboo cellulose nanowhiskers through silane modification. *Journal of Materials Science*, *53*(15), 10920–10932. https://doi.org/10.1007/s10853-018-2377-2

Raquez, J.-M., Habibi, Y., Murariu, M., & Dubois, P. (2013). Polylactide (PLA)-based nanocomposites. *Progress in Polymer Science*, *38*(10–11), 1504–1542. https://doi.org/10.1016/j.progpolymsci.2013.05.014

Rezvani Ghomi, E. R., Khosravi, F., Saedi Ardahaei, A. S., Dai, Y., Neisiany, R. E., Foroughi, F., Wu, M., Das, O., & Ramakrishna, S. (2021). The life cycle assessment for polylactic acid (PLA) to make it a low-carbon material. *Polymers*, *13*(11), 1854. https://doi.org/10.3390/polym13111854

Rostami, A., Vahdati, M., Alimoradi, Y., Karimi, M., & Nazockdast, H. (2018). Rheology provides insight into flow induced nano-structural breakdown and its recovery effect on crystallization of single and hybrid carbon nanofiller filled poly(lactic acid). *Polymer*, *134*, 143–154. https://doi.org/10.1016/j.polymer.2017.11.062

Ruz-Cruz, M. A., Herrera-Franco, P. J., Flores-Johnson, E. A., Moreno-Chulim, M. V., Galera-Manzano, L. M., & Valadez-González, A. (2022). Thermal and mechanical properties of PLA-based multiscale cellulosic biocomposites. *Journal of Materials Research and Technology*, *18*, 485–495. https://doi.org/10.1016/j.jmrt.2022.02.072

Sanivada, U. K., Mármol, G., Brito, F. P., & Fangueiro, R. (2020). PLA composites reinforced with flax and jute fibers – a review of recent trends, processing parameters and mechanical properties. *Polymers*, *12*(10), 2373. https://doi.org/10.3390/polym12102373

Sathish, S., Karthi, N., Prabhu, L., Gokulkumar, S., Balaji, D., Vigneshkumar, N., Ajeem Farhan, T. S., AkilKumar, A., & Dinesh, V. P. (2021). A review of natural fiber composites: Extraction methods, chemical treatments and applications. *Materials Today: Proceedings*, *45*, 8017–8023. https://doi.org/10.1016/j.matpr.2020.12.1105

Schick, C. (2009). Differential scanning calorimetry (DSC) of semicrystalline polymers. *Analytical and Bioanalytical Chemistry*, *395*(6), 1589–1611. https://doi.org/10.1007/s00216-009-3169-y

Siakeng, R., Jawaid, M., Ariffin, H., Sapuan, S. M., Asim, M., & Saba, N. (2019). Natural fiber reinforced polylactic acid composites: A review. *Polymer Composites*, *40*(2), 446–463. https://doi.org/10.1002/pc.24747

Simmons, H., Tiwary, P., Colwell, J. E., & Kontopoulou, M. (2019). Improvements in the crystallinity and mechanical properties of PLA by nucleation and annealing. *Polymer Degradation and Stability*, *166*, 248–257. https://doi.org/10.1016/j.polymdegradstab.2019.06.001

Siracusa, V., Rocculi, P., Romani, S., & Rosa, M. D. (2008). Biodegradable polymers for food packaging: A review. *Trends in Food Science & Technology*, *19*(12), 634–643. https://doi.org/10.1016/j.tifs.2008.07.003

Siriprom, W., Sangwaranatee, N., Herman, Chantarasunthon, K., Teanchai, K., & Chamchoi, N. (2018). Characterization and analyzation of the poly (L-lactic acid) (PLA) films. *Materials Today: Proceedings*, *5*(7), 14803–14806. https://doi.org/10.1016/j.matpr.2018.04.009

Sood, M., & Dwivedi, G. (2018). Effect of fiber treatment on flexural properties of natural fiber reinforced composites: A review. *Egyptian Journal of Petroleum*, *27*(4), 775–783. https://doi.org/10.1016/j.ejpe.2017.11.005

Sudamrao Getme, A., & Patel, B. (2020). A review: Bio-fiber's as reinforcement in composites of polylactic acid (PLA). *Materials Today: Proceedings*, *26*, 2116–2122. https://doi.org/10.1016/j.matpr.2020.02.457

Sujaritjun, W., Uawongsuwan, P., Pivsa-Art, W., & Hamada, H. (2013). Mechanical property of surface modified natural fiber reinforced PLA biocomposites. *Energy Procedia*, *34*, 664–672. https://doi.org/10.1016/j.egypro.2013.06.798

Sun, Z., Zhang, L., Liang, D., Xiao, W., & Lin, J. (2017). Mechanical and thermal properties of PLA biocomposites reinforced by coir fibers. *International Journal of Polymer Science*, *2017*(1), 2178329.

Tabi, T., Sajo, I. E., Szabo, F., Luyt, A. S., & Kovacs, J. G. (2010). Crystalline structure of annealed polylactic acid and its relation to processing. *Express Polymer Letters*, *4*(10), 659–668. https://doi.org/10.3144/expresspolymlett.2010.80

Toong, D. W. Y., Toh, H. W., Ng, J. C. K., Wong, P. E. H., Leo, H. L., Venkatraman, S., Tan, L. P., Ang, H. Y., & Huang, Y. (2020). Bioresorbable polymeric scaffold in cardiovascular applications. *International Journal of Molecular Sciences*, *21*(10), 3444. https://doi.org/10.3390/ijms21103444

Tripathi, N., Misra, M., & Mohanty, A. K. (2021). Durable polylactic acid (PLA)-based sustainable engineered blends and biocomposites: Recent developments, challenges, and opportunities. *ACS Engineering Au*, *1*(1), 7–38. https://doi.org/10.1021/acsengineeringau.1c00011

Tümer, E. H., & Erbil, H. Y. (2021). Extrusion-based 3D printing applications of PLA composites: A review. *Coatings*, *11*(4), 390. https://doi.org/10.3390/coatings11040390

Tyler, B., Gullotti, D., Mangraviti, A., Utsuki, T., & Brem, H. (2016). Polylactic acid (PLA) controlled delivery carriers for biomedical applications. *Advanced Drug Delivery Reviews*, *107*, 163–175. https://doi.org/10.1016/j.addr.2016.06.018

Varsavas, S. D., & Kaynak, C. (2018). Weathering degradation performance of PLA and its glass fiber reinforced composite. *Materials Today Communications*, *15*, 344–353. https://doi.org/10.1016/j.mtcomm.2017.11.008

Vasile, C., Tudorachi, N., Zaharescu, T., Darie-Nita, R. N., & Cheaburu-Yilmaz, C. N. (2020). Study on thermal behavior of some biocompatible and biodegradable materials based on plasticized PLA, chitosan, and rosemary ethanolic extract. *International Journal of Polymer Science*, *2020*, 1–18. https://doi.org/10.1155/2020/4269792

Wang, H., Chen, X., Ding, Y., Huang, D., Ma, Y., Pan, L., Zhang, K., & Wang, H. (2021). Combining novel polyether-based ionomers and polyethylene glycol as effective toughening agents for polylactide. *Polymer*, *229*, 123964. https://doi.org/10.1016/j.polymer.2021.123964

Xu, C., Sun, C., Wan, H., Tan, H., Zhao, J., & Zhang, Y. (2022). Microstructure and physical properties of poly(lactic acid)/polycaprolactone/rice straw lightweight bio-composite foams for wall insulation. *Construction and Building Materials*, *354*, 129216. https://doi.org/10.1016/j.conbuildmat.2022.129216

Xu, H., Xie, L., Jiang, X., Li, X.-J., Li, Y., Zhang, Z.-J., Zhong, G.-J., & Li, Z.-M. (2014). Toward stronger transcrystalline layers in poly(l-lactic acid)/natural fiber biocomposites with the aid of an accelerator of chain mobility. *The Journal of Physical Chemistry B*, *118*(3), 812–823. https://doi.org/10.1021/jp409021q

Yu, Y., Wang, H., Lu, F., Tian, G., & Lin, J. (2014). Bamboo fibers for composite applications: A mechanical and morphological investigation. *Journal of Materials Science*, *49*(6), 2559–2566. https://doi.org/10.1007/s10853-013-7951-z

Zakaria, Z., Islam, M. S., Hassan, A., Mohamad Haafiz, M. K., Arjmandi, R., Inuwa, I. M., & Hasan, M. (2013). Mechanical properties and morphological characterization of PLA/chitosan/epoxidized natural rubber composites. *Advances in Materials Science and Engineering*, *2013*(1), 629092.

Zare, Y., & Rhee, K. Y. (2019). Following the morphological and thermal properties of PLA/PEO blends containing carbon nanotubes (CNTs) during hydrolytic degradation. *Composites Part B: Engineering*, *175*, 107132. https://doi.org/10.1016/j.compositesb.2019.107132

9 Thermal Properties of Natural Fiber–PLA Composites

9.1 INTRODUCTION

Day by day, the usage of fuel-based polymer in everyday life is increasing and is taking a toll. The excess usage of polymer has caused serious issues to the environment such as air, land and water pollution. Millions of polymers are being manufactured every day and are disposed of to the environment. The usual kinds of polymers that are being used in the industries and daily life are polypropylene, polyethylene, polystyrene, polyvinyl chloride, and polyurethane (Zaaba et al., 2021; Geyer, 2020). These polymers are called plastic or nonbiodegradable polymers due to their property which takes a long time to degrade (Saliu et al., 2018). Since these polymers triggered environmental issues such as environmental contamination and greenhouse gas emissions, research is ongoing on alternatives to replace the polymers. In this context, biopolymers are effective alternatives (Geyer, 2020).

When it comes to biopolymers, PLA has led the way because of its exceptional qualities, renewability, biodegradability, and good attainability (McKeown & Jones, 2020). PLA can be processed using conventional plastic manufacturing techniques and has an excellent processing ability. It is also light, biodegradable, biocompatible, and has a reasonably high tensile strength. PLA has drawn interest for uses such as for material packaging, fiber manufacture, and to produce composite. However, PLA has poor toughness and impact strength and needs to be modified to overcome these shortcomings by making it into a composite material by mixing it with natural fibers (Mishra et al., 2021).

Presently, using natural fibers is the most practical approach to produce environmentally friendly composites, such as PLA composites. Animal fibers, cellulose fibers (plant-based fibers), and mineral fiber groups are the three main categories into which natural fibers can be classified (Thapliyal et al., 2023). Researchers are investigating exploiting plant fibers, such as pineapple leaf and bamboo fiber. These fibers are biodegradable and safe for the environment, enriching the ecosystem. Due to their flexibility, ecofriendly nature, inexpensive price, ability to renew, local availability, natural fibers are effective alternative materials in the composite sector (Park et al., 2019; Siakeng et al., 2019).

In the context of sustainable materials, adding natural fibers, specifically for reinforcement in composite materials, is regarded as an ideal approach for various applications. There are a lot of benefits in using natural fibers, including the ability to replace many synthetic fibers (Mahir et al., 2019). Green composite possesses

DOI: 10.1201/9781032678894-9

different benefits such as not being heavy, biodegradable, inexpensive, and showcasing good mechanical properties (Przybytek et al., 2018). Despite having a lot of benefits, there are also some setbacks such as poor water barrier properties, low toughness, and low glass transition temperature. PLA is mixed with a variety of natural fibers to improve the thermal, physical, mechanical, and degradability properties. Furthermore, adding natural fibers lowers the price of PLA products and aids in the production of competitively priced goods that are utilized across many industries. There are certain benefits of using biocomposites such as recyclability, renewability, low density, and low cost. Moreover, they are flexible and offer sufficient thermal and acoustic insulation. Due to these benefits, they are widely used in many different sectors such as automobiles and aviation (Rajeshkumar et al., 2021).

The use of plant natural fibers is also increasing. D-anhydroglucose (C6H11O5) repeating units make up the linear 1,4-β-glucan polymer known as cellulose, which enhances mechanical properties (Graninger, 2023). Hemicellulose controls the fibers' biodegradation, thermal breakdown, and moisture absorption. Plants use the phenolic chemical lignin as a structural support component. The high carbon concentration and low hydrogen level explain it. Its chemical structure is still unknown, despite the fact that most of the functional groups that comprise lignin molecules have been recognized. By varying the fiber length, Bisaria et al. investigated mechanical qualities and discovered that composites with a 15 mm length had the best flexural and tensile properties. Natural fibers are also used to reduce tool wear and tear during machining.

9.2 THERMAL BEHAVIOR OF POLYLACTIC ACID

9.2.1 PLA as a Thermoplastic Polymer

Lactic acid (LA) is the basic constitutional unit of PLA, an aliphatic polyester. PLA ($CH(CH_3)COO)n$). PLA is an excellent processing polymer that is recyclable, biodegradable, renewable, and reusable (Siakeng et al., 2019). PLA is biodegradable, sourced from nature, bioabsorbable, highly productive, easily accessible, and cheap. For the production of biodegradable bioplastic, PLA is a substitute material obtained using petrochemical sources (Ramezani Dana & Ebrahimi, 2023). Nevertheless, PLA has a number of clear disadvantages, including low thermal resistance, which severely restricts its widespread use in numerous industries; therefore, improving PLA's heat resistance has become important. Due to its limited applicability, PLA often blended with other thermoplastic polymers to increase the thermal properties of it. Isothermal heat treatment of glass fiber-reinforced PLA composites greatly increases their thermal stability, allowing them to be used in electronics and many other industries (Wang et al., 2019).

9.2.2 Thermal Transitions in PLA

PLA undergoes a number of variations in temperature when being processed and used. Heat resistance, which is often defined by the glass transition temperature (Tg), is to retain its exceptional mechanical and physical properties at the maximum service temperature. Tg is the temperature at which polymer chain segments change from being immobile to mobile. Material that is glassy and has limited stiffness, strength, hardness, and impact resistance below Tg becomes stretchy and highly deformable over Tg with a low modulus and strength (Zhao et al., 2022). For PLA, Tg is about 60 °C; in its flexible state, the material undergoes noticeable deformation from minor forces or even from its own weight. Some common applications for printed components, like motor flanges, have issues when the PLA is too soft (Suder et al., 2021). At their melting point, which for PLA is between 150 and 160 °C, thermoplastic materials turn liquid. The fact that thermoplastics don't significantly degrade when the heating and cooling process is done repeatedly is one of their main functional qualities. Thermoplastic materials such as PLA do not burn; instead, they liquefy, making injection molding and recycling of the material possible (Sreekumar et al., 2021).

PLA has a temperature range of 320–420 °C for thermal degradation, which can occur during thermal processing. The majority of polymers degrade thermally via a conventional radical chain mechanism, through initiation, propagation, and termination reactions (Faravelli et al., 2001). Random fracture and intramolecular ester exchange are the mechanisms by which PLA thermally cracks, releasing propylene glycol esters, cyclic oligomers, and linear oligomers. Propylene glycerides may be recovered at 350–400 °C, which is the ideal recovery temperature, thanks to thermal cracking (Li et al., 2023). Using a cutting-edge technique called fused filament fabrication (FFF), items are produced by layer-by-layer connecting polymer materials from three-dimensional computer models. FFF is distinguished from traditional manufacturing procedures by its layer-by-layer approach, which enables the fabrication of intricate and integrated parts at a fair cost and time (Gao et al., 2021).

The primary reason PLA performs poorly in terms of toughness and heat resistance is its low crystallization capacity. Therefore, improving PLA's crystallization is the main goal of PLA enhancement activities (Gao et al., 2021; Wang et al., 2019). Unfavorable thermal resistance and poor mechanical qualities with anisotropy up to 50% are typically seen in FFF-printed objects made of PLA because of PLA's inherently slow crystallization kinetics; the fast cooling that occurs during the FFF process causes the majority of PLA parts to be low crystalline (Gao et al., 2021).

9.2.3 Influence of Molecular Structure

L-lactic and D-lactic are the two optical isomers of the lactic acid monomer because the lactic acid molecule is chiral due to its asymmetric carbon atom. Figure 9.1 shows two carbon chains of the optical isomers of lactic acid. Consequently, PLA is classified as a copolymer of D-and L-blocks P(DL)LA, left-turning poly-l-lactide (PLLA), and right-turning poly-d-lactide (PDLA) polymer topologies. They differ in their chain arrangements, and their crystallinity is directly impacted by these variations. The optical purity of PLLA is also correlated with the polymer's ability to crystallize. Semi-crystalline PLA contains more than 93% L-lactic acid and between 50% and 93% L-lactic acid makes up PLA, which is completely amorphous (Zhao et al., 2022).

FIGURE 9.1 Carbon chain of two optical isomers of lactic acid monomer.

Source: Ojo and de Smidt (2023).

Solid-state polymerization (SSP) can increase a low-molecular-weight PLA by up to 67% when it is heated to 120 °C for 32 h in a nitrogen atmosphere. However, because degradation (such the formation of lactide and/or vinyl endings) might occur, it is crucial to thoroughly examine the SSP's circumstances. Thermal treatments provide advantages. Specifically, they don't call for the use of solvents or additives, nor do they involve complicated synthesis procedures. This makes them a potentially economical and non-polluting way to enhance mechanically recycled PLA's properties (Beltrán et al., 2020).

Due to various synthesizing methods, PLA has a wide range of molecular weights; PLA with low molecular weight exhibits high crystallinity and faster crystallization. The diffusion rate of the chains into lattice is affected by the poor mobility characteristics of PLA. This irregularity is caused by the formation of intramolecular or intermolecular hydrogen bonds between carbonyl groups on the chain (Zhao et al., 2022).

9.3 COMPOSITES FORMATION AND COMPOSITION

9.3.1 Composites Materials Overview

Natural fiber-reinforced PLA composite consists of the PLA, natural fibers, and fillers. PLA is produced from renewable sources such as starch and can be produced by different kinds of polymerization processes: polycondensation of lactic acid, ring-opening polymerization of lactides, and the direct methods azeotropic dehydration and enzymatic polymerization (Cristea et al., 2020).

Natural fibers are better than synthetic fibers in some ways, including flexibility, environmental friendliness, affordability, renewability, and local availability. These attributes make them viable substitute materials for the composite industry. They are classified into three groups: animal fibers, cellulose (plant) fibers, and mineral fibers. Among these natural fibers, plant fibers are the widely used in the fabrication green composite. Plant fibers contain cellulose, hemicellulose, and lignin. Cellulose is a linear 1,4-β-glucan polymer and is the strongest and stiffest component of the fibers. Figure 9.2 shows the structure of a part of β-glucan. Hemicellulose plays a role in thermal degradation and to absorb moisture. Meanwhile, lignin serves as a support structure for the plants (Rajeshkumar et al., 2021).

9.3.2 Fillers and Reinforcement

Fillers are widely used in synthesizing green composites, organic and inorganic fillers (Ramesh et al., 2022). Fillers can be expensive, however. To enhance the

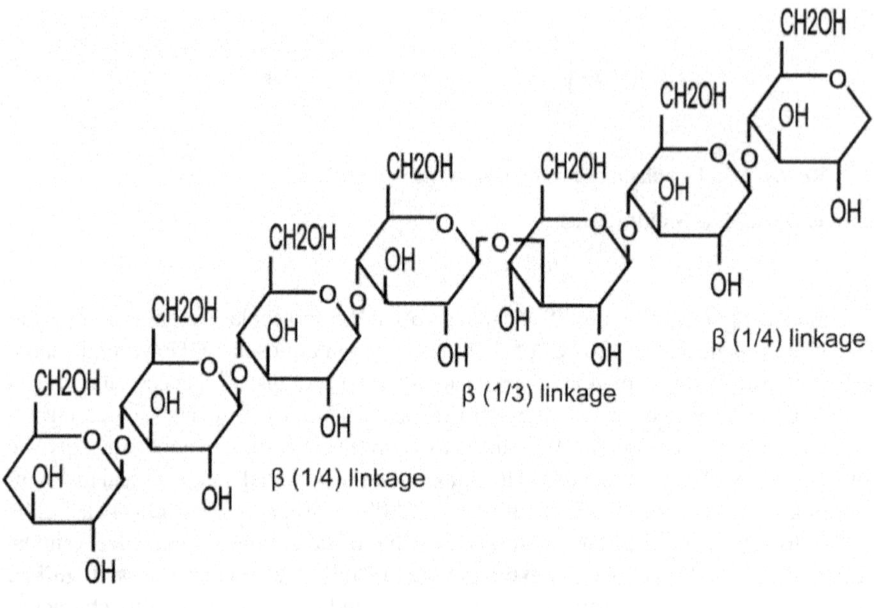

FIGURE 9.2 General structure of β-glucan.

Source: Hadiuzzaman et al. (2022).

effectiveness of the fillers and to optimize the cost, affordable fillers that are suitable for the composite should be used. The most frequently used fillers are talc and mica potassium titanate. These fillers are classified as inorganic and are commonly used due to their nucleating and reinforcing function (Gregorova et al., 2012). Organic fillers that are used in fiber-reinforced composites include core–shell rubber particles, rice husk powder, coconut shell particles, peanut shell powder, egg-shell powder, wood sawdust, and nanoclay.

Fillers play an important role in producing good-quality green composites. The addition of fillers produces a good surface finish preventing the coarse structure from forming and increasing the mechanical properties. Polymer composites with inorganic fillers have improved mechanical properties, excellent interfacial features, tribological behavior, and thermal stability. Factors such as size and shape of the fillers determine how these fillers affect the composites' qualities (Kundie et al., 2018; Ramesh et al., 2022). Fillers in polymer composites also enhance the processability.

Fiber–matrix contact is further enhanced by the filler, which transfers stress between the fiber and matrix, improving the specific surface area and stiffness of the composite. Filler concentration, filler particle size and shape, filler/polymer adhesion, filler dispersion in the polymer matrix, and the applied composite processing are some of the variables that affect how mineral fillers affect polymer composites (Gregorova et al., 2012). The effectiveness of the composite depends on a proper filler distribution in polymer matrices; excess filler can result in agglomeration of the fillers, thus reducing the bonding strength (Ramesh et al., 2022).

9.3.3 Composites Fabrication Methods

The preparation of PLA-based composites involves different methods based on the type of composite to be made and the type of reinforcement (Mishra et al., 2021). In addition to composite and reinforcement type, factors such as type and characteristics of matrix and nature of dispersed phases also need to be considered in choosing the right fabrication method (Murariu & Dubois, 2016). Many parameters need to be controlled during the fabrication. For example, long components can be produced via pultrusion, whereas medium-sized components can be produced by injection molding and hot pressing (Nasiri & Khosravani, 2020). Furthermore, choosing the right method is determined by factors like cost, ultimate product size, number of components needed, etc. There are several ways to produce PLA-based green composites (Murariu & Dubois, 2016), some of which we cover in depth here.

9.3.3.1 Injection molding

A common procedure is injection molding, which involves forcing a certain amount of liquid resin loaded with fibers into mold cavities. Pellets of PLA and shredded fibers are used to create PLA composites based on cellulose fiber. Initially, a hopper is filled with PLA pellets and chopped cellulose fibers that are then put into a barrel which is in the shape of cylindrical with heaters on top (Whyman, 2018). The heaters are maintained at a temperature that is at least as high as PLA's melting point. To create a homogenous combination, the barrel also has a revolving screw inside of it that mixes the chopped cellulose fibers and liquid PLA evenly. A little quantity of heat is dissipated by the friction. This heat is also utilized to melt the polymer pellets into a fluid form that is then forced into the mold cavity by the sprue nozzle, producing a portion of the necessary dimensions (Whyman, 2018; Saravana & Kandaswamy, 2019).

9.3.3.2 Extrusion Injection Molding

The process of extrusion involves softening and combining thermoplastic, which is often in the form of beads or pellets, with fiber that is moved by one or more rotating screws. The mixture is then squeezed and driven out of the chamber at a constant pace through a die. Excessive melt temperatures, air entrapment, and fiber breaking can all be caused by high screw speeds. On the other hand, low speeds result in inadequate wetting and poor mixing of the fibers. This technique can be applied alone or to create an IM precursor. It has been demonstrated that twin-screw systems outperform single-screw extruders in terms of mechanical performance and fiber dispersion (Pickering et al., 2016). The insufficient interfacial adhesion between the cellulose fibers and the PLA matrix is a disadvantage of PLA composites. Therefore, this two-step procedure of extrusion followed by injection molding increases the adhesion between the matrix and fiber (Rajeshkumar et al., 2021)

9.3.3.3 Extrusion Compression Molding

This procedure is just like extrusion injection molding. PLA pellets and cellulose fibers are first fed into a single- or twin-screw extruder, where they are transformed into PLA/fiber melt and moved to a mold to be heated to the appropriate temperature beforehand. After that, the mold is heatedly pushed within a compression molding machine. A study team used this method to create composites made of banana fibers and PLA. The mold was hot-pressed at 170 °C and 40 bars of pressure. When the temperature hit 50 °C, the final composite was taken out of the mold. Polyester sheets were utilized to make it simple to remove the composites (Rajeshkumar et al., 2021).

9.3.3.4 Film Stacking

Film stacking is a hot-pressing process where the fiber mats and films of the thermoplastic matrix are piled together (Valente et al., 2023). After that, pressure and heat are applied to solidify it. Although this procedure is clean, extra attention must be taken during fabrication to fully impregnate the fibers. Film stacking has only been applied in a small number of experiments to create PLA/cellulose fiber composites, but it was used in a significant study to create biodegradable composites made of PLA and sisal fibers. Initially, a compression molding machine was used to turn the PLA pellets into a film. After that, the fiber mats were divided into the proper parts based on the measurements of the mold; the fiber mats were always positioned in between two polymer sheets thanks to careful stacking. Teflon sheets were positioned at the top and bottom of the mold to stop the polymer films from adhering to the metallic mold. Teflon has a higher melting point than PLA. For eight minutes, the complete assembly was subjected to a hot press at 180 °C and 4 MPa of pressure. The assembly was then cooled under pressure after the pressure was increased to 6 MPa for a further 2 minutes. The composite was extracted from the mold after reaching 80 °C (Rajeshkumar et al., 2021; Valente et al., 2023).

9.4 THERMAL PROPERTIES OF PLA COMPOSITES

9.4.1 Melting Temperature

Melting temperature (Tm) plays a huge role in the performance of polymers and natural reinforced polymer composites such as PLA composites in different applications. This melting behavior of PLA polymer varies with its state (solid, amorphous, crystalline, molten). PLA exists as L-lactic acid and D-lactic acid (Siakeng et al., 2019) Since there are two enantiomers in PLA, the melting point depends on the copolymer ratio of lactic acid and lactide. The copolymer ratio influences the molecular weight and crystallinity (Murariu & Dubois, 2016). Tm for PLA increases if the molecular weight is higher: 178 °C when the copolymer ratio is 100/0 (L/D, L)-PLA; 164 °C for 95/5 (L/D,L)-PLA; 150 °C for 90/10 (L/D,L)-PLA; 140 °C for 85/15 (L/D,L)-PLA; and 125 °C for 80/20 (L/D,L)-PLA (Ilyas et al., 2022).

Tm for PLA reinforced with coir or pineapple leaf fiber and fabricated by melt blending is 290.07 °C. Starch-reinforced PLA composite made by extrusion recorded 165.32 °C. Mold-blended clay/RCF reinforced PLA has a Tm of 176.30 °C. Solution-casted graphene oxide/CNT-reinforced PLA melted at 154.00 °C. Lastly, hydroaxyapatite PLA composite made by air jet spinning has a melting temperature of 153.60 °C (Ilyas et al., 2022). Figure 9.3 shows the system.

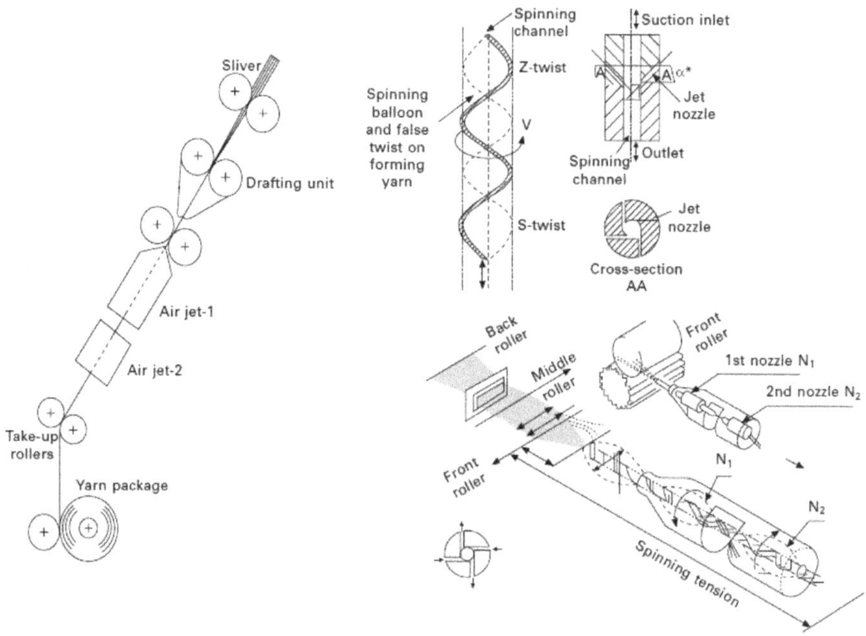

FIGURE 9.3 Air jet spinning system.

Source: Lawrence (2010).

9.4.2 Glass Transition Temperature

The glass transition temperature (Tg) of a polymer is the temperature at which the polymer changes from glassy to rubbery. Tg for PLA ranges from 55 to 60 °C (Zhao et al., 2022). Above Tg, PLA is rubbery; below Tg, it turns into a glass that can also creep until it cools to its transition temperature of around 45 °C, at which point it behaves like a brittle polymer. Factors such as molecular weight, molecular adhesion, and addition of fillers and natural fibers can influence the glass transition temperature (Panin et al., 2020). Different copolymer ratios of PLA cause variation in Tg. At 100/0(L/D,L)-PLA, Tg = 63 °C; at 95/5(L/D,L)-PLA, 59 °C; and at 90/10, 85/15, and 80/20, Tg = 56 °C each. Different natural fibers and different fabrication methods also affect Tg. PLA reinforced with PBSA/starch and fabricated by extrusion had a Tg of 54.01 °C, and PLA reinforced with chitosan/basalt and fabricated by reactive blending and injection molding had a Tg 63.32 °C (Ilyas et al., 2022).

9.4.3 Thermal Conductivity

PLA polymer reinforced with bamboo fiber was fabricated experimentally to study the thermal conductivity using molding. Evaluating the thermal conductivity involved both theoretical and experimental analysis, and a comparison was made between the thermal conductivity of wood and that of BFGC. With a 60% by weight bamboo fiber content, the PLA composite's thermal conductivity was determined using a rapid thermal conductivity meter (Showa Denko Co., Shotherm QTM-DII) that runs on the hot wire technique. The thermal conductivity of the PLA composite recorded for density of 0 is 0.20 W/(m · K). The thermal conductivity of the bamboo fiber reinforced PLA composite increases as the density increases (Takagi et al., 2007).

9.4.4 Thermal Expansion

The study of a sample's property change in response to an applied temperature change is known as thermal analysis. It was common practice to use thermal analysis techniques to describe the thermal behavior of polymeric nanoparticles (Leszczynska & Pielichowski, 2008). Dimensional stability is crucial for composites used in many different applications. Poor dimensional stability causes warping and other shape changes during the function. The coefficient of thermal expansion (CTE), a measure of a material's thermal stability, can be found using TMA testing, for instance of PLA composites reinforced with chicken feather fibers in varying quantities. CTE was higher for the CFF/PLA composites than for the pure PLA, possibly because of the low thermal stability of CFF or composite flaws like matrix and reinforcement fracture and interface debonding; both of these significantly affect the elasto-plastic behavior of the composites. Among the CFF/PLA composites with various CFF materials, the 5 wt.% CFF/PLA composite has the best thermal stability, which may be related to a good distribution of CFF in the matrix (Ilyas et al., 2022).

9.4.5 Heat Deflection Temperature

The temperature at which a polymeric material experiences deformation under a given load is known as the heat deflection temperature, also known as the heat distortion temperature (HDT). A plastic material's HDT is a crucial piece of data for product design. It appears for the material's maximum stability limit when under load and heat effect, with little to no physical deformation (mostly deflection) (Wong, 2003). This property, which is closely related to polymer crystallinity, is essential for the design and manufacture of thermoplastic components. A highly crystalline polymer's HDT is often higher than that of its amorphous equivalent (Aliotta et al., 2022).

A study was conducted to examine the impact of heat treatment and N-(4-carboxyphenyl) maleimide-alt-styrene (PCS) on the HDT of PLA. The HDT of PLA/PCS was approximately 10 °C higher than that of PLA when PCS and PLA were melt-blended in a certain ratio; that is, the HDTs were 62.8 °C and 53.3 °C, respectively. Following an 80 °C heat treatment, PLA/PCS-80's HDT was approximately 15 °C higher than PLA-80's, respectively, 118.7 °C and 103.2 °C. HDT was 65 °C higher for PLA/PCS-80 than for untreated pure PLA. PCS had a considerable impact on PLA's heat resistance when it was melt-blended into the material that increased with heat treatment (Jiang et al., 2021). Additionally, the HDT of clean PLA was 50.4 °C, whereas it was 100 °C for the composite because of the rise in crystallinity and the addition of reinforcement with a high aspect ratio (Roh & Lee, 2019).

9.5 MATERIAL TESTING AND CHARACTERIZATION

9.5.1 Differential Scanning Calorimetry

An analytical method that works well for describing a polymer's physical characteristics is differential scanning calorimetry, or DSC. Glass transition characterization, changes in heat capacity, and other effects indicating a latent heat can all be achieved by DSC. It also allows the determination of melting, crystallization, and mesomorphic transition temperatures, as well as the accompanying enthalpy and entropy changes (Schick, 2009). It is a method of thermal analysis that examines the relationship between temperature and a material's heat capacity. DSC is employed in research on biochemical transitions, also known as single-molecule changes in a molecule's conformation (Nurazzi et al., 2022). Researchers produced thermograms of cellulose-reinforced PLA composite films that show that adding cellulose considerably improved PLA's heat stability (Nurazzi et al., 2022). Table 9.1 summarizes and illustrates the DSC analysis based on the acquired values.

9.5.2 Thermogravimetric Analysis

Thermogravimetric analysis (TGA) is a widely used method to assess deterioration. It detects changes in mass using nonisothermal heating of the sample over a temperature range at a specific rate or isothermal heating of the sample at a specific temperature for a predetermined amount of time (Moseson et al., 2020). When calculating the mass loss rate, derivative thermogravimetry (DTG) is frequently used in conjunction with TGA. Thermogravimetry (TG)–DTG curves offer information in two forms: quantitative data, such as percentages of loss of mass and characteristic

TABLE 9.1
Thermal Characteristics and Crystallinity of PLA and Cellulose-Reinforced PLA Composite

	First Heating Scan			Second Heating Scan			
Samples	T_g (°C)	T_c (J/g)	T_m (°C)	T_g (°C)	T_c (J/g)	T_m (°C)	**Crystallinity (%)**
PLA	60.2	-	150	62.2	-	150.3	7.8
PLA/cellulose (2 wt.%)	59.8	115.0	149.9	62.2	125.0	150.6	13.0
PLA/cellulose (6 wt.%)	59.6	112.9	149.6	62.2	116.4	149.4	11.4
PLA/cellulose (10 wt.%)	57.8	111.0	143.6	62.2	117.2	149.4	10.5

Source: Nurazzi et al. (2022).

temperatures at crucial spots, and qualitative data, such as the identification of compounds (Gerassimidou et al., 2020).

For instance, Bhiogade et al. (2020), for microcrystalline cellulose (MCC), PLA, and PLA-based composites, produced a TGA curve that showed only one stage of degradation. Under 200 to 380 °C, all samples displayed an initial loss in weight. PLA began to decompose at 323 °C, but adding PEG sped decomposition to 233 °C (10 wt.%) and 240 °C (15 wt.%). PLA-PEG 10 wt% and PLA-PEG 15 wt% with 1 wt%, 3 wt%, and 5 wt% microcrystalline cellulose added showed breakdown at 267 °C, 262 °C, 243 °C, 240 °C, 237 °C, and 235 °C, respectively. These findings indicate that PLA is more thermally stable than the PLA-based composites. The thermal deterioration of PLA- and MCC-based composites, as determined by the TGA and DTG curves, is depicted in Figure 9.4.

9.5.3 Dynamic Mechanical Analysis

Dynamic mechanical analysis (DMA) determines viscoelastic properties, which correspond with temperature; the important properties in DMA analysis are storage modulus, loss modulus, and damping factor. Storage modulus (E') shows the stiffness of the composite. E' depends on the size of the fiber and type of matrix material used.

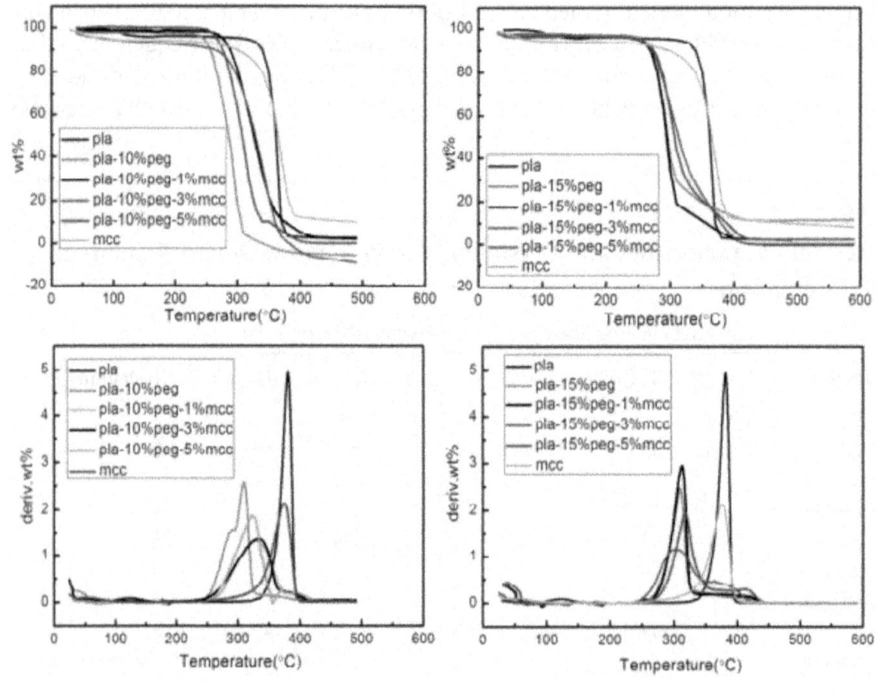

FIGURE 9.4 The TGA and DTG curves for composites based on PLA and MCC.

Source: Bhiogade et al. (2020).

TABLE 9.2
Dynamic Mechanical Properties of Different PLA Composites

Natural Fiber	Fabrication Method	Fiber Addition (Wt.%)	Best Combination (Wt.%)	Max. Storage Modulus (MPa)	Max. Loss Modulus (MPa)	Tg (for loss, °C)	Max. tan δ	Tg (for tan δ, °C)
Banana	Injection molding	10–30	20	3565	601	63.5	0.853	68
Hardwood	Film stacking	30–50	40	7900	585	61	0.17	64
Softwood	Film stacking	30–50	40	7850	595	62	0.14	67
Flax	Compression molding	30	-	4019	-	-	0.3	73
Jute	Extrusion injection	30	-	6200	1050	66	0.81	74

Source: Rajeshkumar et al. (2021).

Loss modulus (E") indicates the capacity to release energy that was stored under cyclic loading (Rajeshkumar et al., 2021). One of the dynamic analyzers that can be used for PLA composite is the Seiko instrument DMA 6100 (Gupta & Singh, 2018). DMA produces a complex modulus that is predicated on how the material responds to sinusoidal forcing and expressed as

$$(E) = E' + iE''$$

$$E' = E \cos \delta$$

$$E'' = E \sin \delta$$

where δ is the phase lag between stress and strain, the ratio of loss to storage modulus, damping factor, or tan. Factors such as viscoelasticity, phase, and grain boundaries influence δ (Rajeshkumar et al., 2021). Table 9.2 shows the properties of different composites.

9.6 FACTORS INFLUENCING THERMAL PROPERTIES

9.6.1 Type and Content of Fillers

Fillers influence the thermal properties of PLA-based composites, three types of fillers in particular: metals, ceramics, and carbon materials; metals and carbon materials are the most common fillers because they enhance heat transfer and heat dissipation but only for cases that do not require electrical insulation such as an exchanger. As contrast, carbon materials fillers are usually used only in areas that

TABLE 9.3
Fillers Used in PLA-Based Composites

Filler	Polymer	λ (W/mK)	Testing Method
30 wt% TA-GNPs	PLA	0.77	Hot disk
38 wt% Al_2O_3 þ 2 wt% AlN	PLA	0.72	Laser flash

Source: Guo et al. (2020).

require electrical insulation such as printed circuit boards. When creating thermally conductive polymer composites, the effectiveness of the thermally conductive fillers is crucial (Guo et al., 2020).

The filler properties such as thermal conductivity of natural fiber-reinforced PLA-based composites can be enhanced by the transportation of the phonon. The vibrations that affect its thermal conductivity are referred to as phonon transport. Thermoplastic materials have poor heat conductivity due to their low atomic density, structural inhomogeneities, and covalent bonding patterns, which naturally impede phonon transit. Although usage of fillers can increase the thermal conductivity, the properties will start to deteriorate if the filler has exceeded the threshold. Researchers studied a composite reinforced with aluminum nitride and found that when the fillers exceeded the threshold of 20 wt.%, the thermal properties of the composite started to deteriorate. As the filler content increased, the orientation that was present in the lower filler content aluminum nitride disappears when the particles agglomerated (Wolfsgruber et al., 2023). Guo et al. (2020) studied the effects of different types of fillers to PLA based composites and the corresponding effect on the thermal properties (Table 9.3).

9.6.2 Processing Conditions

Many manufacturing parameters have an impact on the thermal properties of PLA-based composites. The type of reinforcing material used and the intended end product determine the significant processing parameters for natural fiber PLA composites. The processing conditions differ with each natural fiber PLA composite as there are a variety of steps corresponding to different chemical, thermal and mechanical properties of the PLA and the natural fiber used to manufacture the composite.

Melt processing is a conventional polymer processing method that involves injection, extrusion, or blow molding. Both thermal and hydrolytic processes contribute to the degradation of commercial grade PLA in its molten form, and drying conditions along with duration have an impact on the material's rheological characteristics (Speranza et al., 2014). To prevent PLA from degrading thermally and guarantee that the reinforcing components are distributed uniformly, the production temperatures are adjusted constantly in melt processing. To illustrate this, prior to final production, scientists report compounding natural fibers and polymer pellets via extrusion. Nevertheless, it requires further procedures such as compounding by extrusion, pelletizing, and then drying the pellets at a high temperature in an oven to eliminate the moisture. High temperatures and shearing forces applied to the polymer pellets and natural fibers during compounding may cause the polymer and natural fibers to degrade later in the processing. Furthermore, the extra methods add to the process's time and energy requirements, raising the cost of processing (Komal et al., 2021).

The cooling rate of the natural fiber PLA composite during processing also influences the thermal properties. Cooling rate should be within the limit and requires constant monitoring so that it will not create defects in the composites. Crystallinity also decreases as cooling rate increases. Běhálek et al. (2020) compared PLA composites including cellulose fibers and crushed rice husks with pure PLA and found that the decrease was greater for pure PLA. After increasing the cooling rate 35 °C from 5 °C, crystallinity decreased to 89% for pure PLA and to 80% for composites using rice husk. At the same cooling rates, the PLA composites based on cellulose fiber only caused a 20% drop in crystallinity. The authors demonstrated that natural fiber PLA composites have better thermal resistance than pure PLA.

9.6.3 Thermal Stability

The thermal properties of PLA based composite are significantly influenced by thermal stability. According to Czerwinski (2020), thermal stability is a material attribute that characterizes changes during extended exposure to high temperatures. For example, PLA-cellulose nanomaterial produced did not lose weight below 260 °C, allowing for its potential use in 3D printing applications. Because of their strong connection, which protects the polymer from heat, the chemical modification of cellulose nanoparticles can also increase the thermal stability of the resulting composite materials. In addition, chemically changed materials showed a less noticeable decrease in thermal stability, which was explained by strong adhesion and contact. However, because these treatments incorporate functional groups into the material, extracting cellulose nanoparticles can change the thermal characteristics of the final composite material (Mokhena et al., 2018).

9.7 THERMAL PERFORMANCE IN APPLICATIONS

9.7.1 Automotive Industry

PLA is a very popular polymer that has good mechanical, physical, and thermal properties. Reinforcing PLA polymer with natural fibers enhances the properties to produce high-quality composites that can be used in automotive industries. To improve the composites' heat resistance and impact strength, Jae-won et al. used PLA with jute fiber. The composite jute/PLA treated with alkali showed good bonding between the fiber and matrix. Additionally, posttreatment annealing performed after the composites were manufactured revealed that the heat resistance of the treated green composites had increased noticeably. One application of PLA composites is in dashboard trims and door panels, both parts that are exposed to different temperatures such as high temperatures on sunny days and requires high thermal

TABLE 9.4
PLA Applications in the Automotive Industry

Sector	Application	Explanation
Automotive	Carpets Floor mats Trims Seat Fabric	PLA are used for interiors of the car by Ford Motor Company

Source: Rajeshkumar et al. (2021).

stability materials (Bouzouita et al., 2018). PLA composites are also used in engine covers and car spoilers as well (Ilyas et al., 2022). Table 9.4 below shows the automotive applications of PLA.

9.7.2 Packaging

PLA composites are also used widely in food packaging, although PLA's qualities have to be improved to increase its range of uses in packaging. Specifically, poly(hydroxybutyrate) (PHB) is highly recommended due to PHB's strong crystallinity and comparable melting temperature. Blends of PLA and PHB have proven to be very promising alternatives to the petroleum-based polymers now utilized in food packaging (Ilyas et al., 2022).

Aliphatic polyester PLA is extensively utilized in many different applications, including food packaging and paper coatings. One important aspect of packaging systems to consider is the barrier performance that prevents gases, water vapor, and aromatic organic compounds from penetrating. To prevent food from lipid oxidation, food packaging products typically need a strong oxygen barrier. One important factor improving barrier characteristics is the diffusion paths of the impermeable fillers in the nanocomposites form. To enhance the gas barrier qualities, the nanofillers must be distributed uniformly and aggregate less (Kim et al., 2020). Increasing the dispersion of graphene oxide carbon nanotubes in PLA increased the nanocomposite's crystallinity (Kim et al., 2020). Table 9.5 shows some PLA packaging applications.

9.7.3 Aviation

PLA composites also serve valuable purposes in aviation industries. Due to increasing concern of sustainability, aerospace and aviation industries incorporate PLA composites into their application where it replaces the conventional components and meets the sustainable demands. For example, the use of 3D printing technology in the aviation and space sectors has helped to lower fuel and material waste by replacing traditional structures with lighter, more flexible, and better geometric structures. Extrusion-based 3D printing in the aviation sector uses PLA composites based on cellulose. For instance, carbon fibers were fed into a PLA matrix and absorbed

TABLE 9.5
PLA Applications in the Packaging Industry

Sector	Application	Explanation
Packaging	Food packaging films	Food packaging films are suitable due to the twist retention characteristics
	Food containers	For example, Walmart uses food containers made of PLA.
	Food-contact articles	Cups, straws, etc.
	Bottle labels	S & B Foods uses PLA in its bottle labels.
	Grocery bags	Grocery bags use 45% PLA.
	Tea bags Coffee Pouches	PLA has high infusion capacity, making it suitable for tea bags and coffee pouches.

Source: Kim et al. (2020), Rajeshkumar et al. (2021).

during 3D printing, and the researchers monitored how the applied parameters affected the pressure and temperature during the process (Tümer & Erbil, 2021).

9.7.4 Biomedical

Natural fiber composites made of PLA have drawn interest for uses in the biomedical field because PLA has a long history of safety in human health; it is ecofriendly, biocompatible, and thermoplastic processable. The soluble oligomers with minimal immunogenicity are hydrolyzed and metabolized by cells in the bodies of humans and animals to degrade PLA. Ankle, knee, and hand interference screws, ligament attachment tacks and pins, spinal cages, soft-tissue implants and tissue engineering scaffolds are some of the use of biomedical PLA (Ramezani Dana & Ebrahimi, 2023).

PLA natural fiber composites are also utilized in drug delivery systems; it is possible to improve the mechanical qualities and regulate the rates of medication release by adding natural fibers. Investigators used the syringe test, a rheological test, at two distinct shear rates and temperatures, to evaluate the system's capacity for storage and the viability of local administration. Furthermore, research on the release of drugs from PLA nanoparticles, gelatin hydrogel networks, and the associated bionanocomposites have demonstrated how the gelatin network affects the kinetics of release. Similarly, the gelatin matrix was used to give the drug delivery system the mechanical characteristics it needed to first store the therapeutic agent locally and then easily distribute it using a syringe to minimize any negative effects on healthy cells (Moya-Lopez et al., 2022). Table 9.6 shows the biomedical applications of PLA.

9.7.5 Consumer Goods

Natural fiber composites made of PLA have grown in popularity for many uses, including consumer goods. In the fashion industry, textiles for apparel and accessories can be made with PLA natural fiber composites. PLA composites have drawn

TABLE 9.6
Biomedical Applications of PLA

Sector	Application	Explanation
Biomedical	Stent	-
	Drug delivery	PLA is used to deliver vaccines, proteins, and anesthetics.
	Screws and fixation pins	Employed in the foot, ankle, knee, wrist, and hand.
	Injectable microspheres	PLA microspheres are utilized in face reconstruction surgery as temporary fillers.
	Dermal fillers	Enhance the appearance of our faces by encouraging the body to produce more collagen.
	Bone scaffolds	As implants, osteogenic steam cells are positioned on PLA scaffolds to aid in the process of bone formation.
	3-D porous scaffolds	Used to manage disorders pertaining to the heart, joints, and nervous system.
	Spinal cages	Spinal infusion surgery has used PLA inter-body cages.

Source: Parbin et al. (2019).

interest in the textile sector because of their potential to replace conventional synthetic materials and being environmentally friendly. To illustrate this, Loh et al. (2021) tested their mechanical properties, various combinations of extrusion-printed polymer–textile composites were created utilizing PLA (printing material) and nylon and polyester as textile substrates. Their testing helps designers and researchers by offering valuable insights that will encourage more studies to be made to increase the quality of the adoption in the industry.

9.8 THERMAL MANAGEMENT AND HEAT DISSIPATION

9.8.1 Heat Transfer Mechanism

In PLA composites reinforced with natural fibers, heat is transferred via a number of methods, including conduction, convection, and radiation. The type of natural fibers used, how they are arranged within the composite, and the material's overall composition all affect the particular heat transmission properties. It is vital that the natural fiber and PLA matrix conduct heat effectively. Heat conduction is improved by a strong interfacial bond while a weak bond might result in thermal barriers. The level of compatibility of the fiber with the polymer and the surface treatment of the fiber can have an impact on the bond between the natural fiber and PLA matrix.

Heat transfer is also influenced by the thermal diffusivity of the natural fiber–PLA composite. Thermal diffusivity is also defined as the product of thermal conductivity and heat capacity. It demonstrates the rate at which temperature variations spread through the material itself. Neto et al. (2021) cited that for specimens containing 30 wt.% jute and 10 wt.% banana fibers, the hybrid composites' conductivities of thermal dropped by 44.35% and 34.98%, respectively. If the jute/banana hybrid composite increased, so did its thermal diffusivity and specific heat capacity. At a higher temperature, the combination of banana (10 wt.%) and jute (30 wt%) showed the ideal thermal stability.

9.8.2 Thermal Interface Materials

Thermal interface materials are essential for PLA green composites when efficient heat dissipation is required. Polymers are frequently used as thermal interface material matrices due to their exceptional mechanical qualities, handleability, and distinct flexibility. However, due to their lower thermal conductivity (~0.2 W·m−1·K−1), most polymer thermoelectric materials now in use are unable to meet the demands of thermal management for contemporary electronics with noticeably higher power consumption. To further enhance overall heat transfer characteristics, researchers added highly thermally conductive metals, carbon nanotubes, and low-dimensional inorganic materials as fillers to polymer matrices (Xing et al., 2022).

9.8.3 Cooling Solutions

Cooling solutions are crucial in the fabrication of natural fiber–PLA composites to streamline the production process and ensure the quality of finished products. There are a few cooling methods applied in manufacturing composites corresponding to the material and matrix of the composite. Air and water cooling are some of the notable methods used to cool the composite panel after fabrication that involves heating or hot pressing. Water cooling is where cold water at a specific temperature flows over the mold of the composite or the composite panel is immersed to begin the solidification process. Solidification quality depends on the cooling temperature and material ability to form instant bonds with the PLA.

9.9 CHALLENGES AND FUTURE DIRECTIONS

Given that PLA composite reinforced with natural fibers have fewer environmental consequences and mechanical strength that is equivalent to synthetic fibers, it is obvious that their usage in engineering applications is justifiable. However, employing PLA composites reinforced with natural fibers presents challenges and hurdles. PLA has not yet succeeded in the market due to its inadequate qualities and expensive production costs (Jem & Tan, 2020). Although research is being conducted, there are challenges in the homogenization of the fiber attributes, degree of polymerization, crystallization, and so on, and the price of PLA polymers is estimated to increase at a very fast rate (Tao et al., 2017).

A major technological obstacle to the widespread use of biopolymers is the problem of preserving their biodegradability while achieving mechanical and physical qualities similar to those of standard synthetic polymers. PLA's intrinsic stiffness, poor impact strength, and brittleness have been important bottlenecks for PLA-based biocomposites, impeding their widespread commercial applications. The strength of brittle PLA polymers is increased in many ways including amalgamation with hard polymers and rubber toughening. However, these processes significantly reduce the strength and modulus of the toughened PLA. Other basic weaknesses in PLA that impede its broader industrial application include its low heat resistance and restricted gas barrier property (Khalid et al., 2021).

It is anticipated that natural fiber composites based on PLA will eventually exhibit qualities comparable with those of synthetic composites. The future promises PLA composites made via inexpensive, widely accepted production techniques. The cost will most likely go down in light of consumer demands and the ongoing development of less expensive green polymers ((Pellis et al., 2021). Researchers continue to explore PLA-based composites with different natural fibers in different ratios and forms; for instance, nanocellulose in fiber-reinforced biocomposites improves their properties (Singh et al., 2023).

9.10 CONCLUSIONS

The usage of fuel-based polymers in everyday life is causing serious environmental issues. Research is being done on alternatives to replace these polymers, such as PLA. PLA has led the way due to its exceptional mechanical and physical qualities, renewability, biodegradability, and good availability. However, PLA has poor toughness and impact strength, so it needs to be modified by mixing it with reinforcements such as natural fibers. Natural fiber reinforcements in green PLA composites are lightweight, biodegradable, inexpensive, and showcasing good mechanical properties. Despite the many advantages, however, there are also hindrances such as poor water barrier properties, low toughness, low glass transition temperature, and hydrophilicity.

REFERENCES

Aliotta, L., Sciara, L. M., Cinelli, P., Canesi, I., & Lazzeri, A. (2022). Improvement of the PLA crystallinity and heat distortion temperature optimizing the content of nucleating agents and the injection molding cycle time. *Polymers*, *14*(5), 977. https://doi.org/10.3390/polym14050977

Běhálek, L., Borůvka, M., Brdlík, P., Habr, J., Lenfeld, P., Kroisová, D., Veselka, F., & Novák, J. (2020). Thermal properties and non-isothermal crystallization kinetics of biocomposites based on poly(lactic acid), rice husks and cellulose fibres. *Journal of Thermal Analysis and Calorimetry*, *142*(2), 629–649. https://doi.org/10.1007/s10973-020-09894-3

Beltrán, F. R., Climent-Pascual, E., de la Orden, M. U., & Martínez Urreaga, J. (2020). Effect of solid-state polymerization on the structure and properties of mechanically recycled poly(lactic acid). *Polymer Degradation and Stability*, *171*, 109045. https://doi.org/10.1016/j.polymdegradstab.2019.109045

Bhiogade, A., Kannan, M., & Devanathan, S. (2020). Degradation kinetics study of poly lactic acid (PLA) based biodegradable green composites. *Materials Today: Proceedings*, *24*, 806–814. https://doi.org/10.1016/j.matpr.2020.04.389

Bouzouita, A., Notta-Cuvier, D., Raquez, J. M., Lauro, F., & Dubois, P. (2018). Poly(lactic acid)-based materials for automotive applications. In *Advances in Polymer Science* (Vol. 282, pp. 177–219). Springer. https://doi.org/10.1007/12_2017_10

Cristea, M., Ionita, D., & Iftime, M. M. (2020). Dynamic mechanical analysis investigations of PLA-based renewable materials: How are they useful? *Materials*, *13*(22), 1–24. MDPI AG. https://doi.org/10.3390/ma13225302

Czerwinski, F. (2020). Thermal stability of aluminum alloys. *Materials*, *13*(15), 3441. https://doi.org/10.3390/ma13153441

Faravelli, T., Pinciroli, M., Pisano, F., Bozzano, G., Dente, M., & Ranzi, E. (2001). Thermal degradation of polystyrene. *Journal of Analytical and Applied Pyrolysis*, *60*(1), 103–121. https://doi.org/10.1016/S0165-2370(00)00159-5

Gao, X., Qi, S., Yang, B., Su, Y., Li, J., & Wang, D. (2021). Synergistic effect of plasticizer and nucleating agent on crystallization behavior of polylactide during fused filament fabrication. *Polymer*, *215*, 123426. https://doi.org/10.1016/j.polymer.2021.123426

Gerassimidou, S., Velis, C. A., Williams, P. T., & Komilis, D. (2020). Characterisation and composition identification of waste-derived fuels obtained from municipal solid waste using thermogravimetry: A review. *Waste Management & Research: The Journal for a Sustainable Circular Economy*, *38*(9), 942–965. https://doi.org/10.1177/0734242X20941085

Geyer, R. (2020). Production, use, and fate of synthetic polymers. In *Plastic Waste and Recycling* (pp. 13–32). Elsevier. https://doi.org/10.1016/b978-0-12-817880-5.00002-5

Graninger, G. (2023). *Cellulose as a Functional Additive to Develop Composite Materials* (Doctoral dissertation, Queen's University Belfast).

Gregorova, A., MacHovsky, M., & Wimmer, R. (2012). Viscoelastic properties of mineral-filled poly(lactic acid) composites. *International Journal of Polymer Science*, *2012*. https://doi.org/10.1155/2012/252981

Guo, Y., Ruan, K., Shi, X., Yang, X., & Gu, J. (2020). Factors affecting thermal conductivities of the polymers and polymer composites: A review. *Composites Science and Technology*, *193*, 108134.

Gupta, M. K., & Singh, R. (2018). Flexural and dynamic mechanical analysis (DMA) of polylactic acid (PLA) coated sisal fibre reinforced polyester composite. *Materials Today: Proceedings*, *5*. www.sciencedirect.comwww.materialstoday.com/proceedings2214-7853

Hadiuzzaman, M., Moniruzzaman, M., Shahjahan, M., Bai, S. C., Min, T., & Hossain, Z. (2022). β-Glucan: Mode of action and its uses in fish immunomodulation. *Frontiers in Marine Science*, *9*, 905986.

Ilyas, R. A., Zuhri, M. Y. M., Aisyah, H. A., Asyraf, M. R. M., Hassan, S. A., Zainudin, E. S., Sapuan, S. M., Sharma, S., Bangar, S. P., Jumaidin, R., Nawab, Y., Faudzi, A. A. M., Abral, H., Asrofi, M., Syafri, E., & Sari, N. H. (2022). Natural fiber-reinforced polylactic acid, polylactic acid blends and their composites for advanced applications. *Polymers*, *14*(1), 202. MDPI. https://doi.org/10.3390/polym14010202

Jem, K. J., & Tan, B. (2020). The development and challenges of poly (lactic acid) and poly (glycolic acid). *Advanced Industrial and Engineering Polymer Research*, *3*(2), 60–70.

Jiang, S., Liu, Y., Yan, W., He, M., Qin, S., & Yu, J. (2021). Enhanced thermostability of polylactic acid by N-(4-carboxyphenyl) maleimide – alt – styrene. *Polymers and Polymer Composites*, *29*(7), 953–960. https://doi.org/10.1177/0967391120951664

Khalid, M. Y., Al Rashid, A., Arif, Z. U., Ahmed, W., Arshad, H., & Zaidi, A. A. (2021). Natural fiber reinforced composites: Sustainable materials for emerging applications. *Results in Engineering*, *11*. Elsevier B.V. https://doi.org/10.1016/j.rineng.2021.100263

Kim, Y., Kim, J. S., Lee, S. Y., Mahajan, R. L., & Kim, Y. T. (2020). Exploration of hybrid nanocarbon composite with polylactic acid for packaging applications. *International Journal of Biological Macromolecules*, *144*, 135–142. https://doi.org/10.1016/j.ijbiomac.2019.11.239

Komal, U. K., Lila, M. K., & Singh, I. (2021). Processing of PLA/pineapple fiber based next generation composites. *Materials and Manufacturing Processes*, *36*(14), 1677–1692. https://doi.org/10.1080/10426914.2021.1942904

Kundie, F., Azhari, C. H., Muchtar, A., & Ahmad, Z. A. (2018). Effects of filler size on the mechanical properties of polymer-filled dental composites: A review of recent developments. *Journal of Physical Science*, *29*(1), 141–165. https://doi.org/10.21315/jps2018.29.1.10

Lawrence, C. A. (2010). Overview of developments in yarn spinning technology. In*: Advances in Yarn Spinning Technology* (pp. 3–41). Woodhead Publishing.

Leszczynska, A., & Pielichowski, K. (2008). Application of thermal analysis methods for characterization of polymer/montmorillonite nanocomposites. *Journal of Thermal Analysis and Calorimetry*, *93*(3), 677–687. https://doi.org/10.1007/s10973-008-9128-6

Li, X., Lin, Y., Liu, M., Meng, L., & Li, C. (2023). A review of research and application of polylactic acid composites. *Journal of Applied Polymer Science*, *140*(7). https://doi.org/10.1002/app.53477

Loh, G. H., Sotayo, A., & Pei, E. (2021). Development and testing of material extrusion additive manufactured polymer – textile composites. *Fashion and Textiles*, *8*(1), 2. https://doi.org/10.1186/s40691-020-00232-7

Mahir, F. I., Keya, K. N., Sarker, B., Nahiun, K. M., & Khan, R. A. (2019). A brief review on natural fiber used as a replacement of synthetic fiber in polymer composites. *Materials Engineering Research*, *1*(2), 88–99. https://doi.org/10.25082/MER.2019.02.007

McKeown, P., & Jones, M. D. (2020). The chemical recycling of PLA: A review. *Sustainable Chemistry*, *1*(1), 1–22. https://doi.org/10.3390/suschem1010001

Mishra, S. K., Dahiya, S., Gangil, B., Ranakoti, L., & Agrawal, N. (2021). Mechanical properties of fibre/filler based poly(lactic acid) (PLA) composites: A brief review. *Acta Innovations*, *41*, 5–18. https://doi.org/10.32933/ActaInnovations.41.1

Mokhena, T., Sefadi, J., Sadiku, E., John, M., Mochane, M., & Mtibe, A. (2018). Thermoplastic Processing of PLA/cellulose nanomaterials composites. *Polymers*, *10*(12), 1363. https://doi.org/10.3390/polym10121363

Moseson, D. E., Jordan, M. A., Shah, D. D., Corum, I. D., Alvarenga Jr., B. R., & Taylor, L. S. (2020). Application and limitations of thermogravimetric analysis to delineate the hot melt extrusion chemical stability processing window. *International Journal of Pharmaceutics*, *590*, 119916. https://doi.org/10.1016/j.ijpharm.2020.119916

Moya-Lopez, C., Juan, A., Donizeti, M., Valcarcel, J., Vazquez, J. A., Solano, E., Chapron, D., Bourson, P., Bravo, I., Alonso-Moreno, C., Clemente-Casares, P., Gracia-Fernández, C., Longo, A., Salloum-Abou-Jaoude, G., Ocaña, A., Piñeiro, M. M., Hermida-Merino, C., & Hermida-Merino, D. (2022). Multifunctional PLA/gelatin bionanocomposites for tailored drug delivery systems. *Pharmaceutics*, *14*(6), 1138. https://doi.org/10.3390/pharmaceutics14061138

Murariu, M., & Dubois, P. (2016). PLA composites: From production to properties. *Advanced Drug Delivery Reviews*, *107*, 17–46. Elsevier B.V. https://doi.org/10.1016/j.addr.2016.04.003

Nasiri, S., & Khosravani, M. R. (2020). Progress and challenges in fabrication of wearable sensors for health monitoring. *Sensors and Actuators, A: Physical*, *312*. Elsevier B.V. https://doi.org/10.1016/j.sna.2020.112105

Neto, J. S. S., de Queiroz, H. F. M., Aguiar, R. A. A., & Banea, M. D. (2021). A review on the thermal characterisation of natural and hybrid fiber composites. *Polymers*, *13*(24), 4425. https://doi.org/10.3390/polym13244425

Nurazzi, N. M., Abdullah, N., Norrrahim, M. N. F., Kamarudin, S. H., Ahmad, S., Shazleen, S. S., Rayung, M., Asyraf, M. R. M., Ilyas, R. A., & Kuzmin, M. (2022). Thermogravimetric analysis (TGA) and differential scanning calorimetry (DSC) of PLA/cellulose composites. In *Polylactic Acid-Based Nanocellulose and Cellulose Composites* (pp. 145–164). CRC Press. https://doi.org/10.1201/9781003160458-7

Ojo, A. O., & de Smidt, O. (2023). Lactic acid: A comprehensive review of production to purification. *Processes*, *11*(3), 688.

Panin, S. V., Buslovich, D. G., Kornienko, L. A., Aleksenko, V. O., Dontsov, Y. V., Ovechkin, B. B., & Shil'ko, S. V. (2020). Comparative analysis of tribological and mechanical properties of extrudable polymer–polymer UHMWPE composites fabricated by 3D printing and hot-pressing methods. *Journal of Friction and Wear*, *41*, 228–235.

Park, J. Y., Park, C. W., Han, S. Y., Kwon, G. J., Kim, N. H., & Lee, S. H. (2019). Effects of pH on nanofibrillation of TEMPO-oxidized paper mulberry bast fibers. *Polymers*, *11*(3), 414.

Pellis, A., Malinconico, M., Guarneri, A., & Gardossi, L. (2021). Renewable polymers and plastics: Performance beyond the green. *New Biotechnology*, *60*, 146–158. Elsevier B.V. https://doi.org/10.1016/j.nbt.2020.10.003

Pickering, K. L., Efendy, M. G. A., & Le, T. M. (2016). A review of recent developments in natural fibre composites and their mechanical performance. *Composites Part A: Applied Science and Manufacturing*, *83*, 98–112. Elsevier Ltd. https://doi.org/10.1016/j.compositesa.2015.08.038

Przybytek, A., Sienkiewicz, M., Kucińska-Lipka, J., & Janik, H. (2018). Preparation and characterization of biodegradable and compostable PLA/TPS/ESO compositions. *Industrial Crops and Products*, *122*, 375–383. https://doi.org/10.1016/j.indcrop.2018.06.016

Rajeshkumar, G., Arvindh Seshadri, S., Devnani, G. L., Sanjay, M. R., Siengchin, S., Prakash Maran, J., Al-Dhabi, N. A., Karuppiah, P., Mariadhas, V. A., Sivarajasekar, N., & Ronaldo Anuf, A. (2021). Environment friendly, renewable and sustainable poly lactic acid (PLA) based natural fiber reinforced composites – a comprehensive review. *Journal of Cleaner Production*, *310*. Elsevier Ltd. https://doi.org/10.1016/j.jclepro.2021.127483

Ramesh, M., Rajeshkumar, L. N., Srinivasan, N., Kumar, D. V., & Balaji, D. (2022). Influence of filler material on properties of fiber-reinforced polymer composites: A review. *E-Polymers*, *22*(1), 898–916. Walter de Gruyter GmbH. https://doi.org/10.1515/epoly-2022-0080

Ramezani Dana, H., & Ebrahimi, F. (2023). Synthesis, properties, and applications of polylactic acid-based polymers. *Polymer Engineering & Science*, *63*(1), 22–43. https://doi.org/10.1002/pen.26193

Roh, J. U., & Lee, W. I. (2019). Manufacture of continuous glass fiber/polylactic acid (PLA) composite and its properties. *EPI International Journal of Engineering*, *2*(1), 1–4. https://doi.org/10.25042/epi-ije.022019.01

Saliu, F., Montano, S., Garavaglia, M. G., Lasagni, M., Seveso, D., & Galli, P. (2018). Microplastic and charred microplastic in the Faafu Atoll, Maldives. *Marine Pollution Bulletin*, *136*, 464–471.

Saravana, S., & Kandaswamy, R. (2019). Investigation on the mechanical and thermal properties of PLA/calcium silicate biocomposites for injection molding applications. *Silicon*, *11*(2), 1143–1150. https://doi.org/10.1007/s12633-018-9926-9

Schick, C. (2009). Differential scanning calorimetry (DSC) of semicrystalline polymers. *Analytical and Bioanalytical Chemistry*, *395*(6), 1589–1611. https://doi.org/10.1007/s00216-009-3169-y

Siakeng, R., Jawaid, M., Ariffin, H., Sapuan, S. M., Asim, M., & Saba, N. (2019). Natural fiber reinforced polylactic acid composites: A review. *Polymer Composites*, *40*(2), 446–463. John Wiley and Sons Inc. https://doi.org/10.1002/pc.24747

Singh, A., Trivedi, A. K., Gupta, M. K., & Singh, H. (2023). Static, dynamic mechanical, and thermal analysis of coir/epoxy composites: Effect of hollow glass microspheres reinforcement. *Polymer Composites*, *44*(12), 8529–8540. https://doi.org/10.1002/pc.27717

Speranza, V., De Meo, A., & Pantani, R. (2014). Thermal and hydrolytic degradation kinetics of PLA in the molten state. *Polymer Degradation and Stability*, *100*, 37–41. https://doi.org/10.1016/j.polymdegradstab.2013.12.031

Sreekumar, K., Bindhu, B., & Veluraja, K. (2021). Perspectives of polylactic acid from structure to applications. *Polymers from Renewable Resources*, *12*(1–2), 60–74. https://doi.org/10.1177/20412479211008773

Suder, J., Bobovsky, Z., Mlotek, J., Vocetka, M., Zeman, Z., & Safar, M. (2021). Experimental analysis of temperature resistance of 3d printed PLA components. *MM Science Journal*, *2021*(1), 4322–4327. https://doi.org/10.17973/MMSJ.2021_03_2021004

Takagi, H., Kako, S., Kusano, K., & Ousaka, A. (2007). Thermal conductivity of PLA-bamboo fiber composites. *Advanced Composite Materials: The Official Journal of the Japan Society of Composite Materials*, *16*(4), 377–384. https://doi.org/10.1163/156855107782325186

Tao, Y., Wang, H., Li, Z., Li, P., & Shi, S. Q. (2017). Development and application of wood flour-filled polylactic acid composite filament for 3d printing. *Materials*, *10*(4). https://doi.org/10.3390/ma10040339

Thapliyal, D., Verma, S., Sen, P., Kumar, R., Thakur, A., Tiwari, A. K., Singh, D., Verros, G. D., & Arya, R. K. (2023). Natural fibers composites: Origin, importance, consumption pattern, and challenges. *Journal of Composites Science*, *7*(12), 506.

Tümer, E. H., & Erbil, H. Y. (2021). Extrusion-based 3D printing applications of PLA composites: A review. *Coatings*, *11*(4), 390.

Valente, M., Rossitti, I., & Sambucci, M. (2023). Different production processes for thermoplastic composite materials: Sustainability versus mechanical properties and processes parameter. *Polymers*, *15*(1), 242.

Wang, G., Zhang, D., Li, B., Wan, G., Zhao, G., & Zhang, A. (2019). Strong and thermal-resistance glass fiber-reinforced polylactic acid (PLA) composites enabled by heat treatment. *International Journal of Biological Macromolecules*, *129*, 448–459. https://doi.org/10.1016/j.ijbiomac.2019.02.020

Whyman, S. M. (2018). *Design and Development of a Small-Scale Pellet Extrusion System for 3D Printing Biopolymer Materials and Composites*. Massey University.

Wolfsgruber, N., Tanda, A., Archodoulaki, V. M., & Burgstaller, C. (2023). Influence of filler type and content on thermal conductivity and mechanical properties of thermoplastic compounds. *Polymer Engineering & Science*, *63*(4), 1094–1105.

Wong, A. C.-Y. (2003). Heat deflection characteristics of polypropylene and polypropylene/polyethylene binary systems. *Composites Part B: Engineering*, *34*(2), 199–208. https://doi.org/10.1016/S1359-8368(02)00080-X

Xing, W., Xu, Y., Song, C., & Deng, T. (2022). Recent advances in thermal interface materials for thermal management of high-power electronics. *Nanomaterials*, *12*(19), 3365. https://doi.org/10.3390/nano12193365

Zaaba, N. F., Jaafar, M., & Ismail, H. (2021). Tensile and morphological properties of nanocrystalline cellulose and nanofibrillated cellulose reinforced PLA bionanocomposites: A review. *Polymer Engineering and Science*, *61*(1), 22–38. John Wiley and Sons Inc. https://doi.org/10.1002/pen.25560

Zhao, X., Liu, J., Li, J., Liang, X., Zhou, W., & Peng, S. (2022). Strategies and techniques for improving heat resistance and mechanical performances of poly(lactic acid) (PLA) biodegradable materials. *International Journal of Biological Macromolecules*, *218*, 115–134. https://doi.org/10.1016/j.ijbiomac.2022.07.091

10 Life Cycle Assessment of Natural Fiber-Reinforced PLA Composites

10.1 INTRODUCTION

The growing awareness of environmental impacts associated with product life cycles has emphasized the need for sustainable production and disposal methods. Traditional materials such as metals and woods have been increasingly substituted with plastics due to their versatility and cost-effectiveness (Reddy et al., 2013). However, the environmental ramifications of such materials, particularly plastics, are profound and long-lasting, affecting ecosystems and human health globally. Life cycle assessment (LCA) is a critical methodology that evaluates the environmental impacts of a product from its inception (raw material extraction) to its disposal (end of life) (Klöpffer, 2012). This cradle-to-grave approach is essential as it allows manufacturers and consumers to understand the complete environmental burden associated with their products, guiding more informed decisions toward sustainability (Valero & Valero, 2013).

The significance of considering environmental impacts in the product life cycle is multifold. First, LCA identifies high-impact areas in the production process, enabling targeted improvements like energy efficiency or waste reduction. For instance, replacing petroleum-based plastics with bioplastics such as PLA can significantly reduce greenhouse gas emissions, as these are derived from renewable resources (Rezvani Ghomi et al., 2021). Moreover, these alternatives often offer end-of-life options that are less damaging to the environment, such as composting or recycling, reducing landfill use and resource depletion. Furthermore, LCA facilitates comparison between products, helping stakeholders choose materials with lower environmental impacts (Klöpffer & Grahl, 2014). This comparative analysis is not just limited to production but extends through the use phase and end-of-life management, thus promoting a circular economy.

In conclusion, the integration of environmental considerations throughout the product life cycle is not only beneficial for reducing ecological footprints but also essential for regulatory compliance and fostering consumer trust. Businesses adopting life cycle thinking will not only mitigate environmental risks but also enhance their market competitiveness in an increasingly ecoconscious global marketplace.

10.2 LCA OF PLA

The environmental impact assessment of PLA, a biodegradable polymer made from renewable resources such as corn and sugarcane, shows a nuanced picture

DOI: 10.1201/9781032678894-10

of its sustainability implications (Khan et al., 2023). Integrating findings from the reviewed studies provides a detailed overview of PLA's lifecycle from its production to its disposal, highlighting both the advantages and challenges in its environmental performance.

10.2.1 Raw Material Collection and Processing

The initial phase of PLA production involves harvesting bio-based resources like corn or sugarcane, which are then converted into lactic acid through fermentation (Castro-Aguirre et al., 2016). This step is noted for its efficiency and the ability to produce lactic acid with high optical purity, essential for the polymerization that follows. The production phase is notably energy intensive, with major energy inputs derived predominantly from natural gas and electricity, constituting a significant portion of the overall energy use (Guo & Crittenden, 2011).

10.2.2 Production and Applications

The conversion of lactic acid into PLA through polymerization dictates the polymer's characteristics such as crystallinity and molecular weight, which in turn influence its physical and mechanical properties (Marques et al., 2010). PLA is adaptable to various uses, including medical devices, packaging materials, and agricultural films, employing methods like injection molding and extrusion (Sin & Tueen, 2019). While PLA is versatile, in higher proportions, it shows brittleness under weight (Murariu & Dubois, 2016).

10.2.3 Comparative Environmental Impact

Research generally show that PLA has lower nonrenewable energy usage (NREU) and global warming potential (GWP) compared with traditional petrochemical plastics when comparisons are made on a weight basis (Ramesh & Vinodh, 2020). However, when additional material is necessary for certain applications, such as packaging, the environmental benefits may be lessened. Some reports even indicate higher impacts in categories like acidification potential and eutrophication potential, particularly when taking into account indirect emissions from land use changes (Menglei et al., 2022).

10.2.4 Disposal Options

PLA disposal options include recycling, composting, and landfilling (Rezvani Ghomi et al., 2021). Each method has its own set of challenges and advantages. Recycling processes, both mechanical and chemical, are currently constrained by infrastructure limitations and often result in recycled PLA of lower quality (Maga et al., 2019). Composting and landfilling do not substantially alleviate its environmental impact due to PLA's slow degradation rate, which contributes little to methane and carbon dioxide emissions.

In sum, the life cycle of PLA offers a promising yet complex alternative to conventional plastics. Its advantages in terms of lower NREU and GWP are noteworthy, but these benefits can be diminished by greater impacts in other environmental categories and practical challenges in waste management. The studies underscore the need for a multifaceted approach to assessing the sustainability of biopolymers like PLA. Future enhancements in production processes, recycling technologies, and waste management systems could improve PLA's environmental standing, making it a more viable sustainable material in the future (Figure 10.1).

10.2.5 Environmental Impacts of PLA Production

Morão and De Bie (2019) provide a comprehensive analysis of the environmental footprint of PLA production at the Total Corbion PLA site in Rayong, Thailand,

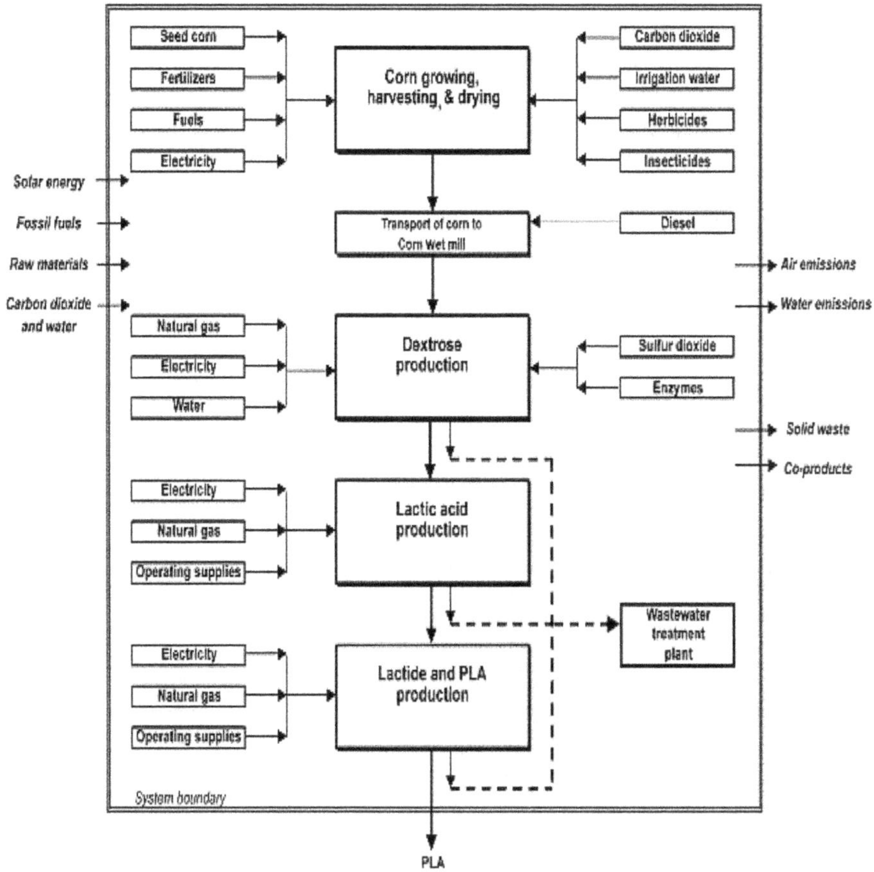

FIGURE 10.1 Simplified flow diagram and boundary of a PLA production system.

Source: Vink et al. (2003).

through an LCA following ISO 14040/44 standards. The authors focused on cradle-to-gate impacts, from the cultivation of sugarcane to the production of PLA. Key stages of PLA production include sugarcane cultivation, its processing into sugar, the fermentation of sugar to produce lactic acid (by Corbion), and the conversion of lactic acid to lactide and then to PLA. The environmental impacts of these processes were meticulously assessed, considering factors like GWP, water usage, eutrophication, acidification, and land use.

The study authors found that the primary environmental impacts of PLA production were associated with the agricultural production of sugarcane and the PLA manufacturing, which both entail intensive use of water, chemicals, and energy. However, the sustainability risk assessment showed low risks of significant land transformation and water-related impacts in the sourcing areas. They also highlight the substantial potential for reducing the environmental impacts of PLA. Recommendations for improvement include enhancing sugarcane yield through better agricultural practices, optimizing fertilizer applications to reduce emissions (NOx and SOx), improving energy efficiency at the sugar mill, and increasing the use of renewable energy in the production processes. Such improvements could significantly reduce the CO_2 footprint of PLA production from 501 kg CO_2 eq/ton to potentially as low as −909 kg CO_2 eq/ton. Overall, this LCA study not only details the environmental impacts associated with each step of PLA production but also emphasizes the potential for significant reductions in these impacts through targeted improvements in production practices and technologies.

10.2.6 Greenhouse Gas Emission of PLA

Benavides et al. (2020) provide an in-depth analysis of the greenhouse gas (GHG) emissions associated with the lifecycle of PLA and compare it with bio-polyethylene

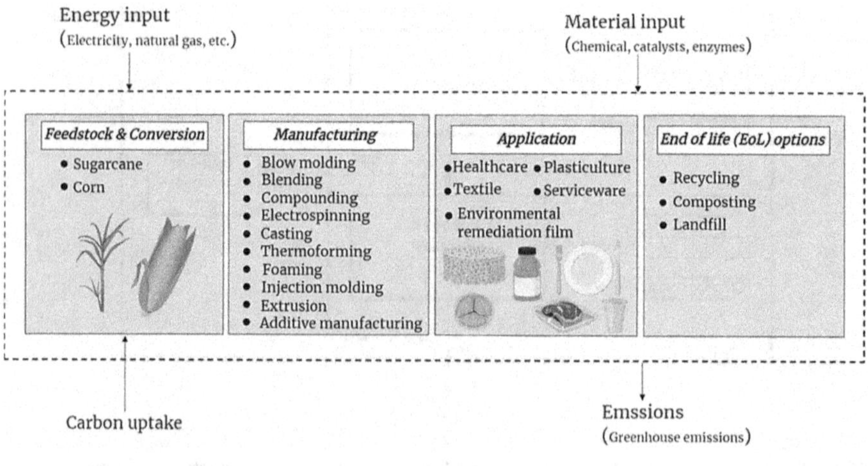

FIGURE 10.2 Life cycle of PLA illustrating potential emissions at various stages.

Source: With permission from Rezvani Ghomi et al. (2021).

(bio-PE) and conventional fossil-based high-density polyethylene (HDPE) and low-density polyethylene (LDPE). Employing an LCA, the authors assessed environmental effects from feedstock production to end-of-life (EOL) scenarios of these plastics. Initially, bio-PE and PLA exhibited lower GHG emissions than fossil-based plastics, with PLA showing emissions of 1.7 kg CO_2e per kg if biodegradation is not considered. However, the biodegradability of PLA significantly impacts its GHG profile; when EOL scenarios like landfilling and composting are considered, PLA's GHG emissions increase significantly – ranging from an increase of 16% to 163% – due to the release of carbon dioxide and methane during the biodegradation process. In landfill conditions, the biodegradation of PLA can lead to considerable GHG emissions, particularly methane, depending on landfill management practices such as the efficiency of methane collection systems, which can vary widely. Composting, while resulting in lower methane emissions, still contributes to CO_2 emissions.

Benavides et al. (2020) also highlight the influence of the regional electricity mix on the GHG emissions of PLA production; a renewable-based electricity mix can significantly reduce these emissions. Additionally, the choice of plastic manufacturing technique, such as less energy-intensive methods like extrusion, impacts the overall GHG emissions, with potential reductions in GHG outputs. While PLA offers a renewable alternative to fossil-based plastics, its end-of-life biodegradability can lead to significant GHG emissions, especially under anaerobic conditions such as those found in landfills. These emissions can offset the initial advantages of lower GHG emissions during the production phase. Effective management of EOL scenarios, improved landfill gas collection systems, and the use of renewable energy sources in production processes are critical factors that can enhance the environmental profile of PLA and other bioplastics.

10.3 CASE STUDIES OF LCA OF PLA BIOCOMPOSITES

10.3.1 Coffee Jar Lid from PLA–Banana Fiber Biocomposites

Rodríguez et al. (2020) presented an environmental LCA of coffee jar lids made from HDPE, PLA, and banana fiber composites, evaluating and comparing the environmental performance of these materials. The LCA followed ISO standards and considered all life cycle phases from goal and scope to interpretation. The functional unit for the study was a cover for a glass jar that would preserve coffee freshness during one year of storage. The system boundaries encompassed banana agriculture, fiber production, lid production, lid use, and disposal stages. Sensitivity scenarios were considered to assess the influence of factors such as the source of banana fiber feedstock and pretreatment conditions.

Data for the foreground system were collected from local agricultural associations, producers, companies, laboratory analysis, and bibliographic resources; the background system data were obtained from the Ecoinvent database. Rodríguez et al. (2020) conducted the uncertainty analysis using the Pedigree matrix to estimate uncertainties in input and output data and performed the LCA using ReCiPe midpoint 2016 methodology. The results indicate that lids made from a blend of 40% banana fiber, HDPE, and PLA performed better in most impact categories than lids made

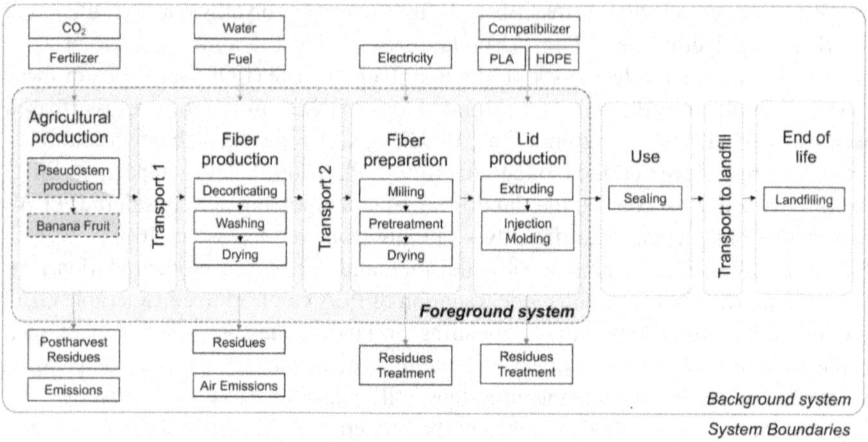

FIGURE 10.3 The system boundaries involve in the LCA for producing disposable lids.

Source: Rodríguez et al. (2020).

from 100% PLA or HDPE. However, lids containing a high content of PLA required optimized production systems to achieve comparable environmental performance to HDPE lids. Rodriguez et al. concluded that banana fiber is an environmentally sound biomaterial for making biocomposite jar lids as a substitute for PLA. Recommendations for future work include comparing the environmental performance of other formulations of biobased composite materials for various food packaging items.

10.3.2 PLA–Chicken Feather Biocomposites

Molins et al. (2018) assessed the environmental impacts of new biocomposite materials made from PLA and chicken feathers (CFs). The aim is to compare these impacts with those of virgin PLA. Two stabilization methods for CFs, autoclave and surfactant, were analyzed to prioritize the most environmentally friendly option. The authors defined a functional unit as a composite plate of specific dimensions. The materials and methods section outlines the goal, functional unit, and scope of the study, including the CFs stabilization process, crushing, and composite plate manufacturing. Data sources for the study were carefully selected, and life cycle inventories (LCI) were conducted for different proportions of CFs in the composite plates.

Two stabilization processes for CFs were compared: surfactant treatment and autoclave treatment. The surfactant process involved washing CFs with a cationic benzalkonium surfactant solution, while the autoclave process used steam. Energy consumption and wastewater analysis were conducted for both processes. The crushing reduced CFs to a maximum size of 1 mm, followed by drying to eliminate remaining humidity. The manufacturing involved compounding CFs with the PLA matrix and hot-pressing to form composite plates. Energy consumption data for the manufacturing process were collected. Allocation rules were established to exclude environmental impacts attributed to chicken meat production from the analysis of CFs. LCA was conducted focusing on selected impact categories such as global warming potential, human toxicity, and aquatic ecotoxicity.

Molins et al. (2018) concluded that CFs, when properly stabilized, can be an alternative raw material for producing biocomposites with lower environmental impact than that of pure PLA. The autoclave stabilization was found to be more environmentally friendly due to lower resource consumption and wastewater loads. Therefore, it is recommended for industrial production of CFs/PLA biocomposites. Table 10.1 presents the characterization data for 1 kg of stabilized chicken feathers processed with autoclave and surfactant, utilizing the modified CML-IA EU25 method and excluding infrastructure processes and long-term emissions.

10.3.3 PLA-Based Nanocomposite Active Packaging

Lorite et al. (2017) developed PLA-based active packaging for ready-to-eat fresh-cut fruits, aiming to improve protection and extend shelf life. Nanoclays and surfactants were added to the PLA formulation to enhance its properties, and they evaluated the packaging using fresh-cut melons compared with pristine PLA and conventional PET packaging. Physicochemical properties such as weight loss, visual appearance, pH, color, and firmness were assessed, along with microbial profiles

TABLE 10.1
Experimental Characterization of Chicken Feather–PLA Biocomposites

"Impact Category	Unit	Autoclave Stabilized CFs	Surfactant Stabilized CFs
Abiotic depletion	kg Sb eq	3.57E–08	9.65E–07
Abiotic depletion (fossil fuels)	MJ	0.60E+01	1.03E+02
Global warming (GWP100a)	kg CO_2 eq	4.10E+01	0.59E+01
Ozone layer depletion (ODP)	kg CFC-11 eq	6.41E–08	8.06E–07
Human toxicity	kg 1,4-DB eq	3.84E–02	5.72E–01
Fresh water aquatic ecotox	kg 1,4-DB eq	4.94E–03	1.07E–01
Marine aquatic ecotoxicity	kg 1,4-DB eq	2.74E+02	3.42E+03
Terrestrial ecotoxicity	kg 1,4-DB eq	2.09E–04	2.59E–03
Photochemical oxidation	kg C_2H_4 eq	9.70E–05	1.88E–03
Acidification	kg SO_2 eq	2.42E–03	3.53E–02
Eutrophication	kg PO_4^{-3} eq	7.58E–04	4.59E–02
Agricultural land occupation	m^2 year	5.26E–03	6.21E–02
Urban land occupation	m^2 year	1.47E–03	1.75E–02
Natural land transformation	m^2 year	9.29E–06	1.10E–04"

Source: Molins et al. (2018)

via microbiological assays. The study authors also included a LCA to compared the environmental impact of PLA-based packaging with that of PET. Materials included PLA pellets, nanoclays, surfactants, and PET packaging. The PLA-based packaging was produced using a twin-screw extruder containing 3.5% OnCap™ Bio in its formulation. Fresh-cut melons were prepared and stored in climatic chambers under specified conditions for 2, 5, and 7 days.

Physicochemical analyses included weight loss, visual appearance, pH, color, and firmness, while microbiological assays covered various microorganisms. Statistical analysis was performed using ANOVA, and LCA was conducted from cradle to grave using SimaPro 8™ software. Results showed that PLA-based packaging with nanoclays and surfactants improved performance, resembling that of PET packaging, and exhibited the best microbial growth prevention. LCA results indicated that while PLA-based packaging with additives required more energy for production, it was competitive with PET in terms of environmental impact and more environmentally friendly with lower health impacts. Figure 10.4 displays the impact of energy, transport, waste, and materials on the environmental performance of innovative packaging using nanoclay and PET during their life cycles. Both were modeled assuming a 15-day shelf life and current waste management practices (Lorite et al., 2017).

Overall, PLA-based nanoclay composite packaging showed potential for fresh food packaging applications, potentially reducing food loss by extending shelf life. These findings suggest further development of PLA nanoclay composites for improved packaging solutions.

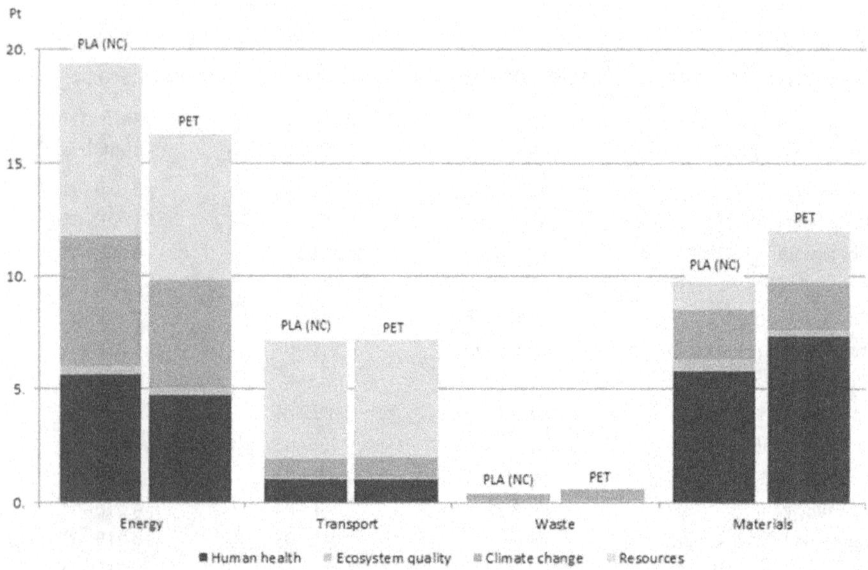

FIGURE 10.4 The different impacts of different variables involved in the life cycle of experimentally produced innovative food packaging using nanoclay and PET.

Source: Lorite et al. (2017).

10.3.4 Wood Fiber–PLA and PLA–Thermoplastic Starch Biocomposites

Mahalle et al. (2014) conducted a detailed LCA of biocomposite formulations, where wood fiber is extruded with either PLA or a combination of PLA and thermoplastic starch (TPS), comparing these with the conventional petroleum-based polypropylene (PP). The assessment strictly adheres to the International Organization for Standardization (ISO) guidelines, specifically ISO 14040 and ISO 14044. The environmental impacts evaluated in this study span several categories, including global warming potential, ozone layer depletion, acidification of land and water, eutrophication, smog formation, and various health impacts such as respiratory, carcinogenic, and noncarcinogenic effects.

The study revealed that PLA, despite its biobased origin, contributed significantly to fossil fuel consumption, acidification, and smog, largely due to the energy consumed and emissions released during its transportation from production sites, often located far from the point of use. In contrast, using locally sourced TPS in conjunction with PLA significantly mitigated these impacts, enhancing the environmental performance of the biocomposite formulations. This substitution not only decreases the dependency on fossil fuels but also lessens the overall environmental footprint of the manufacturing process. When compared directly with PP, the biocomposite formulations exhibited superior environmental performance in nearly all the categories assessed, with the notable exception of eutrophication, where PP still performed better. This superior performance was particularly prominent when the biocomposites were manufactured using hydroelectric power, suggesting that the choice of energy source plays a critical role in determining the environmental impact of biocomposites (Mahalle et al., 2014).

The study, which was based on a prototype produced at lab scale, indicated that scaling up the production of these biocomposites could lead to even lower per-unit environmental impacts due to efficiencies gained in larger-scale operations. Despite this promising outlook, Mahalle et al. (2014) caution that their findings (Figure 10.5)

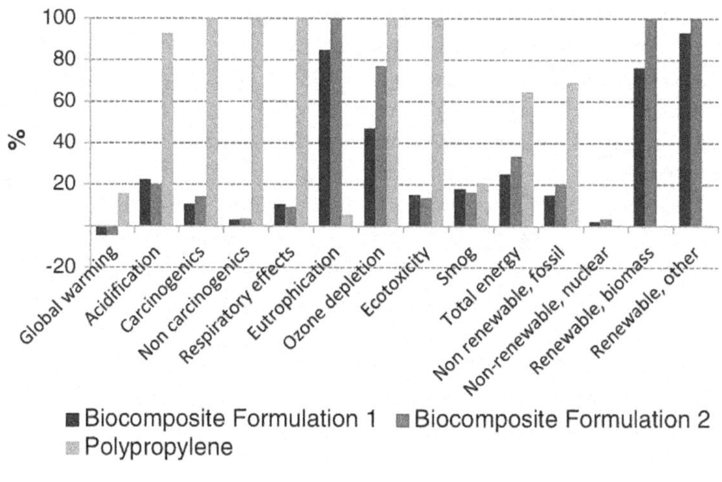

FIGURE 10.5 Comparison of wood fiber–PLA (formulation 1) and PLA–TPS (formulation 2) with polypropylene, in percentages.

Source: Mahalle et al. (2014).

were based on preliminary prototype analysis and determined that comprehensive LCAs including EOL scenarios are essential for fully ascertaining the long-term environmental benefits of these biocomposite materials.

In conclusion, while Mahalle et al. (2014) highlighted the potential for biocomposites to significantly reduce environmental impacts compared with traditional petroleum-based polymers, they underscore the need for further research into their full life cycle impacts. This includes assessing their performance during use and at end-of-life, to ensure that biocomposites can truly offer a more sustainable alternative in the materials sector. This thorough understanding is crucial for moving toward the commercialization and wider adoption of environmentally friendly biocomposite materials.

10.3.5 PLA-Based Food Packaging

Menglei et al. (2022) conducted an ISO 14040 LCA to analyze the energy consumption and environmental impacts of PLA plastic packaging, focusing specifically on PLA food box packaging with 1 ton as the functional unit. The LCA examines five lifecycle stages: raw material acquisition, transportation, production, usage, and disposal, identifying the production phase as the most environmentally impactful. The environmental impact is quantified in seven categories: GWP, abiotic depletion potential of fossil fuels (ADP), human toxicity potential (HTP), acidification potential, terrestrial ecotoxicity potential, eutrophication potential, and photochemical ozone formation potential. The LCA results indicate that GWP, ADP, and HTP are the top three environmental impact categories, contributing 32.63%, 24.83%, and 14.01%, respectively.

Corn, sugarcane, and sugar beet, with corn being the primary, were the raw materials used in PLA packaging (Menglei et al., 2022). The process involved extracting corn starch, fermenting it to glucose, processing glucose to lactic acid, and polymerizing the lactic acid to form PLA. This PLA was then molded into packaging via extrusion and blow molding. The energy consumption during the production of one ton of PLA from corn was reported at 7.35×10^3 MJ, with a production cost of 1,830 yuan. Menglei et al. (2022) suggested mitigating environmental impacts by reducing electricity consumption, increasing the use of renewable energy, and improving material conversion rates in the production process. Replacing conventional fertilizers with organic and water-soluble alternatives, using ecofriendly fuels, and establishing a robust recycling system are also recommended.

The final disposal phase involved landfill and incineration, where PLA products naturally degrade into non-toxic byproducts like water and carbon dioxide. However, the natural degradation efficiency in soil is relatively low, with only 1% of the material degrading over 100 years. Overall, Menglei et al. (2022) highlighted the need for improved production technologies and policies surrounding biodegradable PLA plastics in China. They emphasize the potential shift toward second-generation biomass materials to further reduce environmental impacts and energy consumption.

10.4 ENVIRONMENTAL BENEFITS OF PLA VERSUS FOSSIL-BASED PLASTICS

10.4.1 Reduced Environmental Footprint

Composites made from PLA, particularly those with organic fillers like dried distiller grains with solubles (DDGS), wood, and rice husks, generally exhibit much lower environmental impacts compared to traditional petroleum-based plastics and PLA variants with inorganic fillers like glass (Haylock & Rosentrater, 2018). This is evident in their reduced contributions to global warming, lower energy requirements, and fewer emissions in categories such as air acidification and eutrophication.

10.4.2 Biodegradability

PLA stands out due to its ability to biodegrade under conditions found in industrial composting facilities (Li et al., 2024). This property significantly helps mitigate the environmental issues associated with waste disposal and landfill usage.

10.4.3 Renewable Source Material

Derived from annually renewable sources like corn starch or sugarcane, PLA offers a sustainable alternative to petroleum-based plastics, which depend on finite resources that contribute to higher extraction and processing emissions (Taib et al., 2023).

10.5 DRAWBACKS OF PLA COMPARED WITH FOSSIL-BASED PLASTICS

10.5.1 Higher Production Cost and Complexity

The manufacture of PLA and related composites can be costlier and more complex than those of standard plastics (Sin & Tueen, 2019). For instance, natural fillers often need extra processing like moisture reduction, which increases both cost and energy use.

10.5.2 Performance Constraints in Specific Uses

Despite their overall effectiveness, PLA composites can have mechanical limitations such as lower tensile and flexural strengths compared with conventional plastics, limiting their application in demanding environments unless they are specially treated or mixed with other materials (Dhinesh et al., 2021).

10.6 CHALLENGES WITH END-OF-LIFE PROCESSING

Although biodegradable, PLA requires particular conditions for effective degradation, such as those provided by industrial composting facilities. Without access to such facilities, PLA may not break down more effectively than traditional plastics, presenting similar disposal challenges (McKeown & Jones, 2020).

10.6.1 Resource Competition

Using crops like corn and sugarcane for PLA production can lead to conflicts with food supply, especially in areas where these are primary food sources. This could potentially affect food prices and security if not managed responsibly.

10.6.2 Recycling PLA

PLA stands out not only for its origin from renewable resources but also for its commendable properties such as biodegradability and industrial compostability. The demand for PLA is ever increasing driven by a global shift toward sustainability and a reduction in fossil resource dependence (Taib et al., 2023). Therefore, the focus intensifies on its end-of-life scenarios and recycling strategies. The intricate journey of PLA recycling involves multiple pathways, each with its own set of challenges and opportunities. From mechanical to chemical, and solvent-based recycling, these technologies represent a crucial frontier in the quest to minimize the environmental footprint of plastics (Maga et al., 2019). The development and refinement of these technologies are still in progress, highlighting a dynamic field ripe for innovation and improvement.

Maga et al. (2019) conducted an attributional LCA of various recycling technologies for PLA waste in Germany, comparing mechanical, solvent-based, and chemical recycling methods with thermal treatment. Mechanical recycling processes PLA waste from both industrial production and consumer waste. Postindustrial waste is processed into granulates through sorting, milling, and extrusion. Postconsumer waste undergoes additional steps like near-infrared sorting and intensive purification before similar processing. Solvent-based recycling, tailored for contaminated post-consumer waste, uses solvents to separate PLA, requiring significant energy for solvent recovery. Chemical recycling converts waste back into monomers, integrating into commercial PLA production, and is still predominantly experimental.

All recycling methods demonstrated considerable environmental advantages over incineration, significantly reducing impacts on global warming, energy demand, and resource depletion (Figure 10.6). The reduced agricultural land occupation due to avoided biomass cultivation for virgin PLA was a notable benefit across recycling methods. However, challenges persist in scalability and direct comparisons due to variations in output quality and technology maturity (Maga et al., 2019).

Enhancements in recycling technologies could substantially improve the environmental performance of PLA products. Future research should focus on scaling chemical and solvent-based recycling and exploring systemic impacts of increased recycling through consequential LCA. Overall, more research should emphasize the potential of recycling to mitigate the environmental burdens associated with PLA production and disposal, highlighting the importance of policy and market developments to support these technologies (Maga et al., 2019).

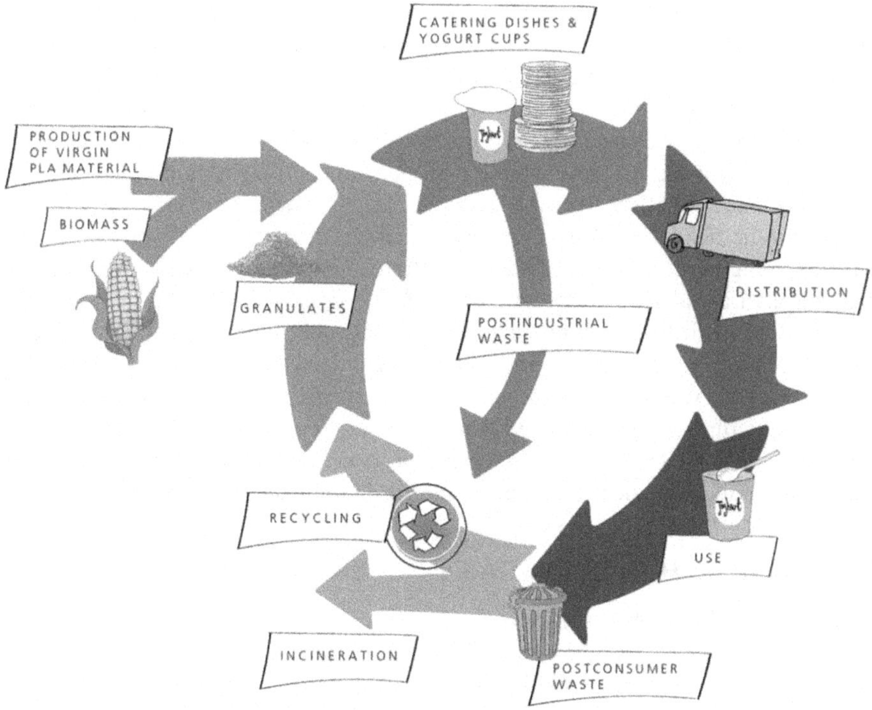

FIGURE 10.6 Recycling options for PLA.

Source: Reprinted with permission from Maga et al. (2019).

10.7 COMPARATIVE LCA OF ORGANIC AND INORGANIC FILLERS IN PLA COMPOSITES

Haylock and Rosentrater (2018) offer a comprehensive analysis comparing the use of organic versus inorganic fillers in PLA composites, with specific emphasis on their economic, environmental, and mechanical impacts. Economically, it was revealed that composites with glass fillers were the most expensive to produce, maintaining a higher life cycle cost, while those with organic fillers like DDGS and wood were the least costly, with costs reducing by \$1–\$2 per kg when recycling was implemented as the EOL treatment. This trend was consistent across all EOL options, with landfilling emerging as the most expensive choice.

From an environmental perspective, the research showed that while glass filler composites paired with recycling produced the lowest CO_2 emissions (CO_{2eq}), organic fillers such as DDGS, rice husks, and wood, particularly with a 1 kg part weight and recycling, resulted in the lowest emissions across multiple categories including global warming potential, energy intensity, and air acidification. Haylock and Rosentrater quantitatively demonstrated that organic fillers are more beneficial in terms of reducing environmental impact, with inorganic fillers like glass and talc proving to be more energy and emission-intensive, especially when smaller part weights (0.01 kg) and incineration are used.

Mechanical properties were also significantly influenced by the choice of filler. Glass, wood pulp, and flax increased the tensile strength of PLA composites, with tensile strength being higher for these materials compared with pure PLA. On the other hand, organic fillers like DDGS showed improvements in elongation at break, enhancing the flexibility of the composites. However, although 10% talc improved flexural strength, it also reduced the crystallinity, leading to a weaker overall composite structure (Haylock & Rosentrater, 2018).

Overall, organic fillers are preferred to inorganic ones for PLA composites due to their lower cost and reduced environmental impact. However, mechanical properties and commercial viability need to be considered as organic fillers have tradeoffs in terms of structural integrity and performance. Therefore, organic fillers are economically and environmentally preferable, but the selection of filler must be balanced with considerations of mechanical performance and market applicability.

10.8 RESEARCH GAPS IN LCA OF PLA BIOCOMPOSITES

There are several notable research gaps in the LCA of PLA biocomposites. First, EOL scenarios for PLA and its composites need comprehensive evaluation, particularly in terms of environmental impacts across different geographical regions and under varied environmental settings (Bher et al., 2022). While recycling, landfilling, and incineration are commonly studied, the impacts of composting, particularly for PLA composites with organic fillers, are less understood and require deeper exploration (Tolga et al., 2020). Second, climate change and other specific environmental impacts of PLA, especially during its conversion process from lactic acid, are areas where more detailed studies could provide insights into reducing emissions and optimizing energy use (X. Guo et al., 2021). Additionally, there's a lack of comparative LCA studies that include PLA, other bioplastics, and traditional petroleum-based plastics, which would help in clearly positioning PLA within a broader environmental context. Finally, the influence of novel additives and fillers on the environmental performance of PLA composites needs further investigation to validate their benefits and identify potential environmental trade-offs.

10.9 CONCLUSIONS AND FUTURE RESEARCH DIRECTIONS

Future research directions in LCA research of PLA should focus on enhancing the integration of PLA into circular economy models. This involves not only improving recycling technologies to accommodate PLA waste more effectively but also exploring systemic changes necessary for increasing recycling rates and minimizing environmental impacts. Research should also aim to expand LCA studies to include a broader range of bio-based and synthetic fillers and additives, thereby assessing their full lifecycle impacts in PLA composites. Developing more complex economic models and advanced statistical methods as part of techno-economic analyses could provide deeper insights into the economic viability and scalability of PLA applications.

Additionally, regional-specific LCAs could help tailor PLA usage to local environmental conditions and regulations, ensuring more globally sustainable applications of PLA. These efforts would collectively enhance the understanding and optimization of PLA's environmental profile, aiding in its broader adoption across various industries.

REFERENCES

Benavides, P. T., Lee, U., & Zarè-Mehrjerdi, O. (2020). Life cycle greenhouse gas emissions and energy use of polylactic acid, bio-derived polyethylene, and fossil-derived polyethylene. *Journal of Cleaner Production*, *277*, 124010. https://doi.org/10.1016/j.jclepro.2020.124010

Bher, A., Castro-Aguirre, E., & Auras, R. (2022). End-of-life scenarios for poly (lactic acid). *Poly (Lactic Acid) Synthesis, Structures, Properties, Processing, Applications, and End of Life*, 559–579.

Castro-Aguirre, E., Iniguez-Franco, F., Samsudin, H., Fang, X., & Auras, R. (2016). Poly (lactic acid) – mass production, processing, industrial applications, and end of life. *Advanced Drug Delivery Reviews*, *107*, 333–366.

Dhinesh, S. K., Arun, P. S., Senthil, K. K. L., & Megalingam, A. (2021). Study on flexural and tensile behavior of PLA, ABS and PLA-ABS materials. *Materials Today: Proceedings*, *45*, 1175–1180.

Guo, Q., & Crittenden, J. C. (2011). An energy analysis of polylactic acid (PLA) produced from corn grain and corn stover integrated system. *Proceedings of the 2011 IEEE International Symposium on Sustainable Systems and Technology*, 1–5.

Guo, X., Yao, Y., Zhao, H., Chi, C., Zeng, F., Qian, F., Liu, Z., Huo, L., & Lv, Y. (2021). Environmental impacts of functional fillers in polylactide (PLA)-based bottles using life cycle assessment methodology. *Science of the Total Environment*, *788*, 147852.

Haylock, R., & Rosentrater, K. A. (2018). Cradle-to-grave life cycle assessment and techno-economic analysis of polylactic acid composites with traditional and bio-based fillers. *Journal of Polymers and the Environment*, *26*, 1484–1503.

Khan, A., Sapuan, S. M., Yusuf, J., Siddiqui, V. U., Zainudin, E. S., Zuhri, M. Y. M., Baharuddin, B. T. H. T., Ansari, M. A., & Rahman, A. A. A. (2023). An examination of cutting-edge developments in bamboo-PLA composite research: A comprehensive review. *Renewable and Sustainable Energy Reviews*, *188*, 113832.

Klöpffer, W. (2012). The critical review of life cycle assessment studies according to ISO 14040 and 14044: Origin, purpose and practical performance. *The International Journal of Life Cycle Assessment*, *17*, 1087–1093.

Klöpffer, W., & Grahl, B. (2014). *Life cycle assessment (LCA): A guide to best practice*. John Wiley & Sons.

Li, Y., Wang, S., Qian, S., Liu, Z., Weng, Y., & Zhang, Y. (2024). Depolymerization and re/upcycling of biodegradable PLA plastics. *ACS Omega*, *9*(12), 13509–13521.

Lorite, G. S., Rocha, J. M., Miilumäki, N., Saavalainen, P., Selkälä, T., Morales-Cid, G., Gonçalves, M. P., Pongrácz, E., Rocha, C. M. R., & Toth, G. (2017). Evaluation of physicochemical/microbial properties and life cycle assessment (LCA) of PLA-based nanocomposite active packaging. *LWT*, *75*, 305–315. https://doi.org/10.1016/j.lwt.2016.09.004

Maga, D., Hiebel, M., & Thonemann, N. (2019). Life cycle assessment of recycling options for polylactic acid. *Resources, Conservation and Recycling*, *149*, 86–96.

Mahalle, L., Alemdar, A., Mihai, M., & Legros, N. (2014). A cradle-to-gate life cycle assessment of wood fibre-reinforced polylactic acid (PLA) and polylactic acid/thermoplastic starch (PLA/TPS) biocomposites. *The International Journal of Life Cycle Assessment*, *19*, 1305–1315.

Marques, D. A. S., Jarmelo, S., G. Baptista, C. M. S., & Gil, M. H. (2010). Poly (lactic acid) synthesis in solution polymerization. *Macromolecular Symposia*, *296*(1), 63–71.

McKeown, P., & Jones, M. D. (2020). The chemical recycling of PLA: A review. *Sustainable Chemistry*, *1*(1), 1–22.

Menglei, Z., Zeng, Y., Jingnan, Z., Yan, W., Xiaolei, M. A., & Jian, G. U. O. (2022). Life cycle assessment of biodegradable polylactic acid (PLA) plastic packaging products – taking Tianjin, China as a case study. *Journal of Resources and Ecology*, *13*(3), 428.

Molins, G., Álvarez, M. D., Garrido, N., Macanás, J., & Carrillo, F. (2018). Environmental impact assessment of polylactide (PLA)/chicken feathers biocomposite materials. *Journal of Polymers and the Environment*, *26*, 873–884.

Morão, A., & De Bie, F. (2019). Life cycle impact assessment of polylactic acid (PLA) produced from sugarcane in Thailand. *Journal of Polymers and the Environment*, *27*(11), 2523–2539.

Murariu, M., & Dubois, P. (2016). PLA composites: From production to properties. *Advanced Drug Delivery Reviews*, *107*, 17–46.

Ramesh, P., & Vinodh, S. (2020). State of art review on life cycle assessment of polymers. *International Journal of Sustainable Engineering*, *13*(6), 411–422.

Reddy, M. M., Vivekanandhan, S., Misra, M., Bhatia, S. K., & Mohanty, A. K. (2013). Biobased plastics and bionanocomposites: Current status and future opportunities. *Progress in Polymer Science*, *38*(10–11), 1653–1689.

Rezvani Ghomi, E., Khosravi, F., Saedi Ardahaei, A., Dai, Y., Neisiany, R. E., Foroughi, F., Wu, M., Das, O., & Ramakrishna, S. (2021). The life cycle assessment for polylactic acid (PLA) to make it a low-carbon material. *Polymers*, *13*(11), 1854.

Rodríguez, L. J., Fabbri, S., Orrego, C. E., & Owsianiak, M. (2020). Comparative life cycle assessment of coffee jar lids made from biocomposites containing poly (lactic acid) and banana fiber. *Journal of Environmental Management*, *266*, 110493.

Sin, L. T., & Tueen, B. S. (2019). *Polylactic acid: A practical guide for the processing, manufacturing, and applications of PLA*. William Andrew.

Taib, N.-A. A. B., Rahman, M. R., Huda, D., Kuok, K. K., Hamdan, S., Bakri, M. K. B., Julaihi, M. R. M. B., & Khan, A. (2023). A review on poly lactic acid (PLA) as a biodegradable polymer. *Polymer Bulletin*, *80*(2), 1179–1213.

Tolga, S., Kabasci, S., & Duhme, M. (2020). Progress of disintegration of polylactide (PLA)/poly (butylene succinate)(PBS) blends containing talc and chalk inorganic fillers under industrial composting conditions. *Polymers*, *13*(1), 10.

Valero, A., & Valero, A. (2013). From grave to cradle. *Journal of Industrial Ecology*, *17*(1), 43–52.

Vink, E. T. H., Rábago, K. R., Glassner, D. A., & Gruber, P. R. (2003). Applications of life cycle assessment to NatureWorksTM polylactide (PLA) production. *Polymer Degradation and Stability*, *80*(3), 403–419. https://doi.org/10.1016/S0141-3910(02)00372-5

11 Fabricating Natural Fiber–PLA Composites Components

11.1 INTRODUCTION

The growing environmental concerns and the need for sustainable materials have driven the development of natural fiber-reinforced polymer composites as alternatives to traditional synthetic fiber-based composites. Natural fibers are renewable and biodegradable and have high specific strength and stiffness, making them attractive reinforcements for polymer matrices. PLA, a biodegradable and biocompatible polymer derived from renewable resources, is a particularly suitable matrix material for these natural fiber composites. PLA-based natural fiber composites are entirely biobased and offer promising biodegradability and mechanical properties, positioning them as viable substitutes for petroleum-based products.

Natural fiber-reinforced PLA composites have found a wide range of applications in industries including automotive, construction, packaging, and consumer goods. In the automotive industry, they offer a lightweight and environmentally friendly alternative to traditional materials, and in construction, PLA-based natural fiber composites are used for structural and nonstructural applications such as wall panels, floors, and roofing materials. Furthermore, the biodegradable nature of these composites makes them suitable for packaging applications, including food containers, disposable tableware, and compost bags (Rajeshkumar et al. 2021).

In this chapter, we explore the processes involved in fabricating components from natural fiber-reinforced PLA composites. We investigate preprocessing natural fibers, different composite fabrication methods, and the impacts of processing parameters on the final properties of the components. Preprocessing techniques such as chemical, physical, and biological treatments are essential for modifying the inherent variability of natural fibers and enhancing their compatibility with the PLA matrix. Chemical treatments like alkali treatment, silane treatment, and acetylation improve fiber–matrix adhesion and enhance mechanical properties (Faiz Norrrahim et al. 2021; Syafri et al. 2022). Physical treatments, including mechanical fibrillation and heat treatment, modify the fibers' physical properties, improving bonding with the PLA matrix and enhancing dimensional stability. Biological treatments, using enzymes or microorganisms, can selectively remove lignin or hemicellulose, further improving fiber–matrix adhesion. These preprocessing techniques significantly influence the final properties of the composites, impacting aspects such as surface morphology, aspect ratio, hydrophilicity, and thermal stability, all of which play crucial roles in determining the overall performance of the natural fiber-reinforced PLA composites.

DOI: 10.1201/9781032678894-11

11.2 NATURAL FIBERS

Natural fibers can be obtained from plants, animals, or the environment. Figures 11.1 and 11.2 show basic information about the branches of biocomposites and natural fibers classification. The advantages of natural fibers over synthetic fibers are their recyclability and biodegradability (Ilyas et al. 2021; Sabaruddin et al. 2020; Ayu

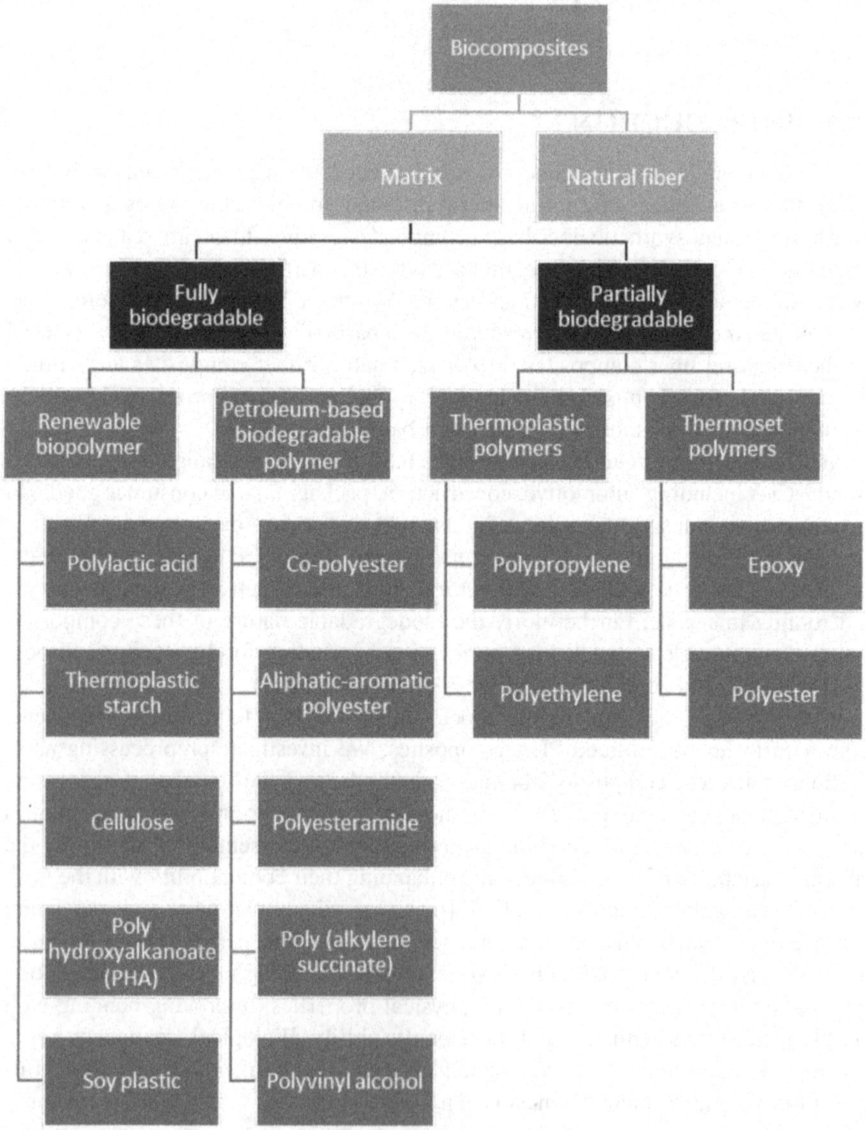

FIGURE 11.1 Classification of biocomposites.

Source: Adapted with copyright permission from Mohanty, Misra, and Dreal (2001).

et al. 2020). The natural fiber market and production are drastically progressing, shifting science and engineering focus to PLA composites. Nowadays, natural fibers are well known in reinforced polymeric materials for industrial developments, e.g. glass fiber as matrix material (Wambua, Ivens, and Verpoest 2003). The polymer is obtained from the fermentation of corn, potato, sugar, beet, and other agricultural sources. Despite their biodegradability properties, natural fibers exhibit several main drawbacks that hinder their developments, including differences in consistency, sensitive to moisture intake due to their hydrophilic nature, and low thermal stability (Oksman, Skrifvars, and Selin 2003; Jumaidin, Ilyas, et al. 2019).

Studies on the processing effects and natural fibers properties show improved mechanical properties parallel to National Policy on Industry 4.0 (Industry 4WRD). Natural fibers have sparked great interest among researchers and industry players for their applications in the military (Nurazzi et al. 2021), automotive (Nurazzi et al. 2020; Aisyah et al. 2019), industrial (Syafri et al. 2019; Abral, Atmajaya, et al. 2020; Asyraf et al. 2022), furniture (Mazani et al. 2019), civil (Asyraf et al. 2020, 2019), membrane (Norfarhana, Ilyas, and Ngadi 2022; Mubarak et al. 2024; Othman et al. 2024), packaging (Punia Bangar et al. 2023; Low et al. 2024; Rahmadiawan et al. 2024), fire retardant (Suriani et al. 2021), and biomedical (Alam, Maniruzzaman, and Morshed 2014) applications. Global production of natural fibers exceeded 105 million metric tons in 2018 (Chemiefaser 2018). Migneault et al. (2008) studied the applications of wood-plastic composites that are typically manufactured as hybrid products by combining wood flour with recycled plastic and observed enhanced mechanical properties with increasing fiber content. Nonwood fibers include plant straw, leaf, bast, fruit, seed, and grass; straw fibers that originate from resources such as maize, wheat, and rice hull are comparatively solid, rigid, low density, and sustainable. These natural fiber-reinforced composites are used as deck boards in housing construction materials (Mukherjee and Kao 2011).

Biopolymer PLA has long been a matter of considerable concern in multiple sectors because of the biocomposites' mechanical performance. Mukherjee and Kao (2011) showed that cellulose content and microfibril angle govern the mechanical properties of natural fiber. The better the mechanical properties of fiber, the better the mechanical properties of the biocomposites. The mechanical properties of natural fibres depend on the cellulose content in them. Table 11.1 shows the chemical compositions of different natural fibers, reflecting that their compositions and cell structures are complex and different. Depending on the cellulose crystallinity, the physical, chemical, and mechanical behaviors of the lignocellulosic fibers vary from one another (Cosgrove 2005). Type of fiber is an important parameter in determining their chemical composition. The primary natural fibers components comprise cellulose, hemicellulose, and lignin (Martins, Kiyohara, and Joekes 2004; Ilyas et al. 2018; Ferreira et al. 2018).

Each fiber exists as a composite by nature, where the rigid cellulose microfibrils are protected in the amorphous matrix containing hemicellulose and lignin (Abral, Ariksa, et al. 2020; Ahmad Ilyas et al. 2019). Thus, natural fibers can also be identified as cellulosic or lignocellulosic fibers. Generally, the main natural fibers' constituent is cellulose with 30%–80% followed by hemicellulose of 7%–40%, and 3%–33% lignin as shown in Table 11.1. Besides that, the surface structure of the

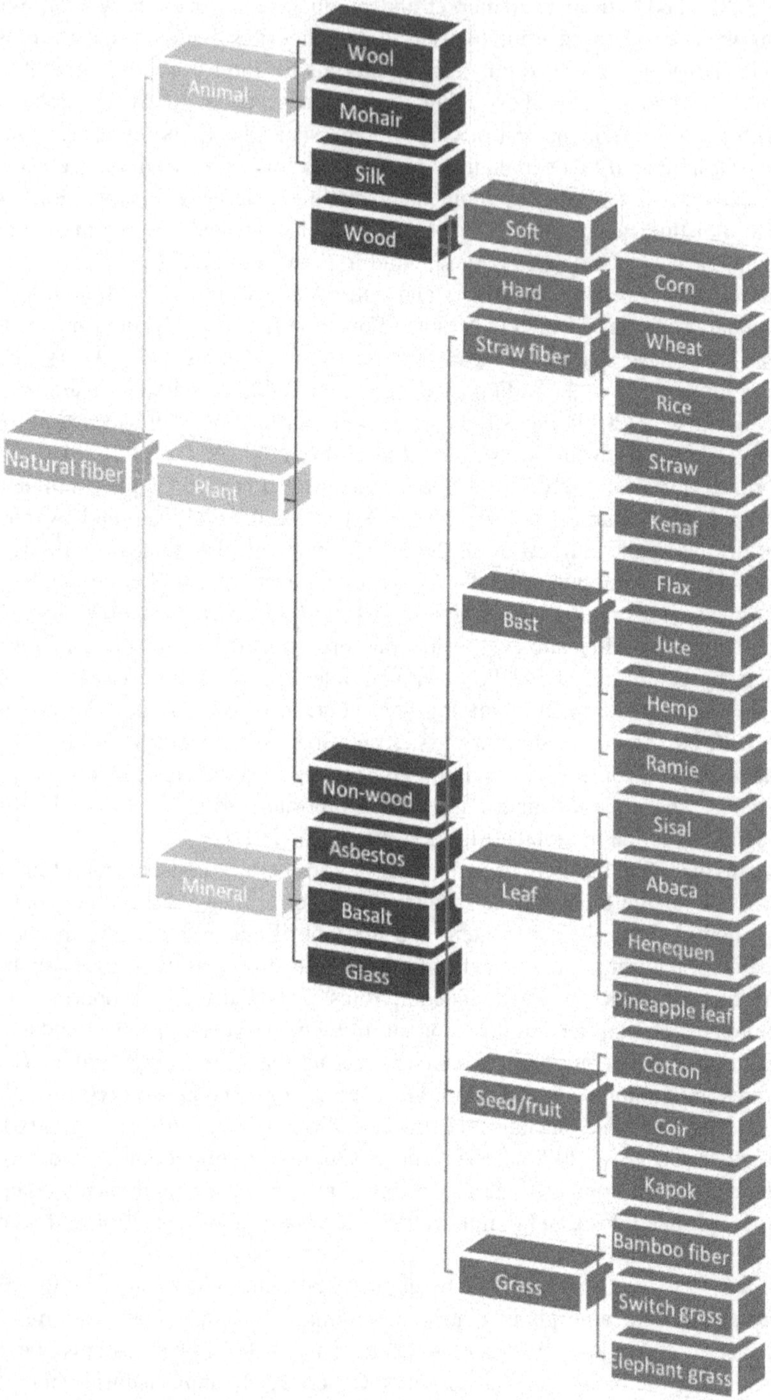

FIGURE 11.2 Classification of natural fibers.

TABLE 11.1
Chemical Compositions of Common Natural Fibers

Fibres	Holocellulose (wt%) Cellulose (wt%)	Hemicellulose (wt%)	Lignin (wt%)	Ash (wt%)	Extractives (wt%)	Crystallinity (%)	Ref.
Sugar palm fiber	43.88	7.24	33.24	1.01	2.73	55.8	(R. A. Ilyas, Sapuan, and Ishak 2018)
Wheat straw fiber	43.2 ± 0.15	34.1 ± 1.2	22.0 ± 3.1	-	-	57.5	(Alemdar and Sain 2008b)
Soy hull fiber	56.4 ± 0.92	12.5 ± 0.72	18.0 ± 2.5	-	-	59.8	(Alemdar and Sain 2008b)
Areca nut husk fiber	34.18	20.83	31.60	2.34	-	37	(Julie Chandra, George, and Narayanankutty 2016)
Helicteres isora	71 ± 2.6	3.1 ± 0.5	21 ± 0.9	-	-	38	(Chirayil et al. 2014)
Pineapple leaf fiber	81.27 ± 2.45	12.31 ± 1.35	3.46 ± 0.58	-	-	35.97	(Cherian et al. 2010)
Ramie fiber	69.83	9.63	3.98	-	-	55.48	(Syafri et al. 2018)
Oil palm mesocarp fiber	28.2 ± 0.8	32.7 ± 4.8	32.4 ± 4.0	-	6.5 ± 0.1	34.3	(Megashah et al. 2018)
Oil palm empty fruit bunch	37.1 ± 4.4	39.9 ± 0.75	18.6 ± 1.3	-	3.1 ± 3.4	45.0	(Megashah et al. 2018)
Oil palm frond	45.0 ± 0.6	32.0 ± 1.4	16.9 ± 0.4	-	2.3 ± 1.0	54.5	(Megashah et al. 2018)
Oil palm empty fruit bunch fiber	40 ± 2	23 ± 2	21 ± 1	-	2.0 ± 0.2	40	(Jonoobi et al. 2011)
Rubber wood	45 ± 3	20 ± 2	29 ± 2	-	2.5 ± 0.5	46	(Jonoobi et al. 2011)
Curaua fiber	70.2 ± 0.7	18.3 ± 0.8	9.3 ± 0.9	-	-	64	(Corrêa et al. 2010)
Banana fiber	7.5	74.9	7.9	0.01	9.6	15.0	(Tibolla, Pelissari, and Menegalli 2014)
Sugarcane bagasse	43.6	27.7	27.7	-	-	76	(Teixeira et al. 2011)
Kenaf bast	63.5 ± 0.5	17.6 ± 1.4	12.7 ± 1.5	2.2 ± 0.8	4.0 ±1.0	48.2	(Jonoobi et al. 2009)
***Phoenix dactylifera* palm leaflet**	33.5	26.0	27.0	6.5	-	50	(Bendahou et al. 2009)
***Phoenix dactylifera* rachis**	44.0	28.0	14.0	2.5	-	55	(Bendahou et al. 2009)
Kenaf core powder	80.26		23.58	-	-	48.1	(Chan et al. 2013)
Water hyacinth fiber	42.8	20.6	4.1	-	-	59.56	(Abral et al. 2018)
Wheat straw	43.2 ± 0.15	34.1 ± 1.2	22.0 ± 3.1	-	-	57.5	(Alemdar and Sain 2008a)
Sugar beet fiber	44.95 ± 0.09	25.40 ± 2.06	11.23 ± 1.66	17.67 ± 1.54	-	35.67	(M. Li et al. 2014)
Mengkuang leaves	37.3 ± 0.6	34.4 ± 0.2	24 ± 0.8		2.5 ± 0.02	55.1	(Sheltami et al. 2012)

natural fiber plays a vital role in the adhesion behavior of the matrix polymer and the fibers. The higher and stronger the fibers and matrix adhesion, the better the natural fibers' mechanical properties. The composite intensity depends on the various level of the fibers in use and approximately equal to its matrix binding mechanism.

11.3 PREPROCESSING OF NATURAL FIBERS

The use of natural fibers in polymer composites has gained significant attention due to their environmental benefits and desirable mechanical properties. However, natural fibers exhibit high moisture absorption, poor interfacial adhesion with hydrophobic polymer matrices, and variability in fiber quality. To address these issues, preprocessing techniques including chemical, physical, and biological treatments are applied to enhance the performance of natural fibers in composite materials.

11.3.1 Chemical Treatments

Various studies have assessed the efficacy of reinforced natural fiber polymer composite in a variety of fields. In recent years, a number of treatments were developed to enhance interfacial bonding and to improve mechanical properties and water-resistance characteristics of natural fibers (Bangar et al. 2022). The most critical downside of these natural fibers in polymer composites is their hydrophilicity, which creates weak interfacial bonding between fibers and matrix. Numerous physical impurities and hydroxyl groups are found on the surface of the fibers that affect their performance as reinforcing agents. Thus, several surface modifications and treatments have been explored in order to obtain the desired properties of polymer composites.

To identifying a knowledge gap in the field of study in natural fibers reinforced PLA composites, Sawpan, Pickering, and Fernyhough (2011) examined improving performance of industrial hemp fiber reinforced polylactide biocomposites was conducted. Interestingly, aligned PLA composites reinforced with long hemp fiber (0 wt.%–40 wt.%) were studied, and 30% alkali-treated fiber-reinforced PLA composite (PLA/ALK) had the highest mechanical strength with a tensile strength of more than 70 MPa, Young's modulus of more than 8 GPa module, and flexural toughness of 2.64 kJ/m^2.

In a previous literature review, Li, Tabil, and Panigrahi (2007) focused on the effect of chemical treatments toward natural fibers for the application of natural fiber-reinforced composites. The authors identified poor fiber–matrix compatibility and high moisture absorption as the key drawbacks of natural fibers in composites and for the study explored modifying the fiber surface properties using chemical treatments. Other authors found that chemical treatments with alkali, silane, acetylation, benzoylation, etc. improved fiber–matrix surface adhesion and reduce the water absorption properties of the fibers (Asyraf et al. 2021).

Reinforcing natural fiber with polymer composite was a focus of a study on natural fiber surface modifications and the performance of the subsequent biocomposites, Mohanty, Misra, and Dreal (2001) stated that adequate adhesion between the surface of hydrophilic natural fiber cellulose and the polymer matrix resin is needed to ensure the desirable properties of biocomposites. Useful methods of enhancing fiber–matrix adhesion in natural fiber composites are alkali, peroxide, coupling agents, isocyanate treatments, dewaxing, bleaching, vinyl grafting, and acetylation. Replacing the hydroxyl group with some chemical functionality can break the crystalline structure of cellulose. This process of decrystallization enhances the thermoplasticity of the cellulose because the role of the plasticizer is performed by the substituted groups. Essentially, Mohanty et al. established that surface modifications of hydrophilic natural fibers can create superior interfaces but that modifying the surfaces needs to be cheaper for biocomposites to replace glass fiber composites.

Chemical treatments play a crucial role in modifying the surface properties of natural fibers to improve their compatibility with the PLA matrix. One of the most common chemical treatments is alkali treatment, also known as mercerization. This process involves treating fibers with a sodium hydroxide (NaOH) solution that removed impurities such as lignin, pectin, and hemicellulose from the fiber surface; removing these noncellulosic components exposes the cellulose, leading to increased surface roughness and a higher aspect ratio (Ilyas et al. 2017). This, in turn, enhances the mechanical interlocking between the fiber and the polymer matrix, resulting in improved tensile strength and stiffness of the composite.

For instance, mercerization of jute fibers significantly improved fiber–matrix adhesion and tensile strength. Ilehuv et al. (2014) prepared natural fiber-reinforced biodegradable polymer composites using short jute fibers treated with varying concentrations of NaOH and H_2O_2 as reinforcement in a PLA matrix. They found that composites with fibers treated with 10% NaOH and H_2O_2 at 20% fiber loading exhibited a 7.5% increase in tensile strength and a significant enhancement in tensile modulus compared with neat PLA and untreated fiber composites. SEM analysis revealed better interfacial bonding in the treated fiber composites.

Other researchers investigated the effect of pretreatment on ramie fiber-reinforced PLA composites using alkali and silane treatments to improve fiber–matrix compatibility. They examined nine samples to study the influence of temperature, fiber volume ratio, molding pressure, and time on composite fabrication, and they identified the optimal processing parameters for the best tensile, bending, and impact strengths. The authors also explored the degradation process of these composites through underground burial experiments, emphasizing their ecofriendly nature (Debeli, Qin, and Guo 2018).

Silane treatment is another widely used chemical method where silane coupling agents are employed to improve the bond between the hydrophilic fiber surface and the hydrophobic polymer matrix. The silane molecules react with the hydroxyl groups on the fiber surface, forming a covalent bond that enhances interfacial adhesion. This treatment is particularly beneficial in applications requiring strong mechanical properties, such as automotive interior components. For example, in flax fiber-reinforced PLA composites, silane treatment increased the tensile strength and reduced water absorption, making the composite more durable and suitable for use in moisture-prone environments (Debeli, Qin, and Guo 2018).

Debeli, Qin, and Guo (2018) examined the effects of silane treatment on polylactic acid/short bamboo fiber (PLA/SBF) composites focusing on water absorption and mechanical properties. The silane treatment significantly reduced water absorption and enhanced mechanical strength, with a 72% increase in tensile strength and a

30% increase in Young's modulus compared with untreated composites. The treatment also improved interfacial adhesion, leading to better water-proofing and overall performance of the PLA/SBF composites. The authors also looked at improving the matrix/fiber interfacial bonding in unidirectional flax fiber–reinforced PLA composites, a key challenge for their use in outdoor applications. They treated flax fibers with varying amounts of amino–silane to optimize the composite's mechanical properties, particularly their flexural response and long-term performance. The findings revealed that while silane treatment improved the composites' mechanical performance, excessive silane loading led to deterioration of these properties (Debeli, Qin, and Guo 2018).

Acetylation is a chemical treatment that introduces acetyl groups into the fiber structure to replace some of the hydroxyl groups; this process reduces the hydrophilicity of the fibers. Acetylated fibers exhibit improved dimensional stability and resistance to fungal attack, making them ideal for applications in construction materials where durability is critical. In bamboo fiber-reinforced PLA composites, acetylation significantly enhanced the thermal stability and mechanical properties of the composite, making it a viable alternative to for many applications (Fang et al. 2022). The PLA-based composites demonstrated significant improvements in performance, with increases of 33.10%, 72.29%, 44.38%, and 40.80% in bending strength, tensile strength, impact strength, and elongation-at-break, respectively. Their thermal stability also improved, as the onset temperature (Tonset) rose from 290.6 °C to 331.9 °C, and water resistance improved with a reduction in water absorption from 12% to 8%. These results, achieved using a self-supplied biocoupling agent, highlight a promising pathway for more sustainable, stronger, and environmentally friendly biocomposites (Fang et al. 2022).

11.3.2 Physical Treatments

Physical treatments are applied to natural fibers to modify their structural properties, enhancing their performance in composites. In mechanical fibrillation, for instance, fiber bundles are separated into finer microfibrils, increasing the surface area available for bonding with the polymer matrix. This process has been used effectively on natural fibers, where the increased surface area after fibrillation led to better dispersion in the PLA matrix and improved tensile strength of the composite. Another physical treatment, heat treatment, involves subjecting fibers to controlled heating to remove moisture, volatile compounds, and internal stresses. This treatment not only enhances the dimensional stability of the fibers but also improves their thermal stability, making them suitable for high-temperature applications. For example, heat-treated sisal fibers used in PLA composites have shown increased resistance to thermal degradation. Heat treatment (HT) of sisal fibers (SFs) can significantly improve the mechanical properties of SF/PLA composites by reducing fiber hydrophilicity and enhancing interfacial adhesion. Optimal HT conditions, such as 220 °C for 10 minutes, result in a tensile strength increase of 21%, while a treatment at 150 °C for 15 minutes boosts flexural strength by 53% compared to untreated composites. However, improper HT can degrade fiber strength due to hemicellulose and lignin breakdown, potentially compromising composite performance (Huang et al. 2015).

11.3.3 Biological Treatments

Biological treatments offer ecofriendly alternatives to chemical treatments by using enzymes or microorganisms to modify the fiber surface. Enzymatic treatments, such as the use of cellulases, can selectively remove noncellulosic components like hemicellulose and lignin, resulting in cleaner fiber surfaces and better fiber-matrix adhesion. These treatments are particularly useful in applications where environmental sustainability is a priority. For instance, enzymatically treated flax fibers have been used in PLA composites for biodegradable packaging materials, where the treatment improved the mechanical properties and biodegradability of the composite without the need for harsh chemicals.

Huang et al. (2015) improved the interfacial adhesion between PLA and Alfa short fibers through various treatments, enhancing the mechanical properties of biocomposites. Enzymatic treatment effectively decomposed lignin, pectin, and hemicellulose, reducing fiber diameter and increasing water resistance, which led to a 27% increase in tensile strength compared with untreated samples. These treated biocomposites, demonstrating improved fiber–matrix adhesion, have potential applications in construction, automotive, and packaging industries. In another study, Coskun et al. (2021) investigated the impacts of various enzymatic treatments on the mechanical properties of coir fiber-reinforced PLA composites. Lipase and lactase enhanced tensile strength, while pectinase and cellulase had minimal effect; all treatments improved impact strength and elastic modulus below the glass transition temperature. Overall, enzymatic treatments enhanced interfacial adhesion by removing waxes and fatty acids or increasing surface roughness, though flexural properties remained unaffected.

Bendourou et al. (2021) used cellulose microfibers (CMFs) extracted from hemp fiber and pulp and paper solid waste as reinforcing agents in recycled polylactic acid (rPLA) biocomposites. Various physicochemical and enzymatic treatments, including laccase and cellulase, were applied to CMFs to enhance their interfacial adhesion with rPLA and improved the mechanical properties and biodegradability. Among the biocomposites tested, those incorporating hemp fiber achieved the highest Young's modulus (324.53 MPa), impact strength (27.61 kJ/m^2), and biodegradation rate (1.97%). The use of enzymes is emerging as a promising technology for the modification of the structure and surface of natural fibers, as it is environmentally friendly.

11.3.4 Impact on Fiber Properties

The impact on fiber properties due to These preprocessing techniques is significant and directly influences the performance of the resulting composites. Chemical treatments such as mercerization and silane treatment not only improve fiber–matrix adhesion but also enhance the mechanical properties of the composites, including tensile strength, stiffness, and impact resistance (Nurazzi et al. 2021). Physical treatments like mechanical fibrillation and heat treatment improve the structural integrity and thermal stability of the fibers, making the composites more robust and durable. Biological treatments, while less harsh, still provide substantial improvements in fiber compatibility and composite performance, particularly in applications where biodegradability is essential.

For example, in the automotive industry, natural fiber-reinforced PLA composites treated with a combination of chemical and physical treatments have been successfully used to produce lightweight, high-strength components like door panels and dashboard parts. These composites offer a sustainable alternative to traditional petroleum-based materials, reducing the overall carbon footprint of the vehicle. Similarly, in the construction industry, acetylated bamboo fiber-reinforced PLA composites have been used to create durable, moisture-resistant panels and insulation materials, offering both environmental and performance advantages over conventional materials.

In summary, the preprocessing of natural fibers through chemical, physical, and biological treatments is essential for enhancing their properties and making them suitable for high-performance composite applications. These treatments not only improve the mechanical and thermal properties of the fibers but also increase their compatibility with the PLA matrix, resulting in composites that are strong, durable, and environmentally friendly (Khalid et al. 2021; Jagadeesh et al. 2021). The continued development and optimization of these preprocessing techniques will be crucial for expanding the use of natural fiber-reinforced PLA composites in various industrial applications, contributing to a more sustainable future.

11.4 FABRICATION TECHNIQUES

11.4.1 Injection Molding

11.4.1.1 Process Overview

Injection molding is a widely used manufacturing process for producing parts by injecting molten materials into a mold. When it comes to natural fiber-reinforced PLA, this technique allows for efficient production of complex geometries while leveraging the mechanical properties of both the natural fibers and the biodegradable PLA matrix. The process begins with the blending of PLA pellets and natural fibers, which can include materials such as kenaf, hemp, jute, sugar palm, and date palm. This blend is then heated until it becomes a viscous fluid, after which it is injected under high pressure into a predesigned mold.

The cooling phase is crucial, as it solidifies the material into the desired shape. Once cooled, the mold is opened, and the finished part is ejected. This method is particularly effective for high-volume production runs, as it offers rapid cycle times and the ability to produce consistent and repeatable parts. The versatility of injection molding also allows for the incorporation of various additives to enhance the properties of the composite material.

11.4.1.2 Advantages and Limitations

One of the primary advantages of injection molding with natural fiber-reinforced PLA is its efficiency. The ability to produce many parts in a short time is particularly beneficial for manufacturers aiming to meet high demand. Additionally, the method allows for precise control over part dimensions, resulting in minimal post-processing requirements.

However, there are limitations to consider. The initial setup costs for molds can be substantial, making injection molding less viable for small production runs. Furthermore, the processing temperature and pressure must be carefully controlled to prevent thermal degradation of PLA, which can negatively affect the material properties. Lastly, the dispersion of natural fibers within the PLA matrix can be challenging, impacting the uniformity and mechanical performance of the final product.

11.4.2 Compression Molding

11.4.2.1 Process Overview

Compression molding is another method for fabricating natural fiber-reinforced PLA composites. This technique involves placing a premeasured amount of the composite material into an open mold, which is then closed under heat and pressure. The heat activates the PLA, causing it to flow and fill the mold cavity. After a designated time, the mold is cooled, and the solidified part is removed.

Compression molding is particularly advantageous for producing larger parts or components with thicker cross-sections. The process allows for a high fiber content, which can enhance the mechanical properties of the PLA composite. The simplicity of the equipment required for compression molding makes it accessible for small-scale manufacturers and laboratories.

11.4.2.2 Advantages and Limitations

One of the key advantages of compression molding is its ability to process materials with high fiber loadings, which can lead to improved mechanical performance. The method is also relatively cost-effective for low to medium production volumes as it does not require complex tooling. Additionally, the process allows for incorporating additives to enhance specific properties, such as flame resistance or UV stability. On the downside, compression molding cycle times are longer than those for injection molding, particularly for larger parts. The requirement for heating the mold and allowing adequate cooling can extend the overall production time. Moreover, the uniformity of fiber distribution can be affected, leading to variability in mechanical properties, and the process can require additional finishing steps to achieve a desired surface quality.

11.4.3 Extrusion

11.4.3.1 Process Overview

Extrusion is a continuous manufacturing process used to create long shapes and profiles from natural fiber-reinforced PLA. In this method, the composite materials are fed into an extruder, where they are heated and mixed thoroughly. The molten material is then forced through a die to form a continuous profile, which can be cut into specific lengths or cooled to form sheets. The flexibility of extrusion allows for the production of a variety of shapes, including films, rods, and sheets. Extrusion offers significant advantages in terms of scalability, as it can be easily adapted for large production runs. The process also allows for incorporating various additives and fillers to modify the properties of the composite material. The resulting products can be further processed through methods such as injection molding or 3D printing, making extrusion a versatile option in the manufacturing workflow.

11.4.3.2 Advantages and Limitations

The primary advantages of extrusion are its efficiency and versatility. It can produce consistent and high-quality materials at a rapid pace. The continuous nature of the process makes it particularly suitable for applications requiring large quantities of material. Additionally, extrusion can facilitate the blending of multiple components, which can enhance the overall performance of the final product.

However, there are limitations to the extrusion process. The need for high temperatures poses a challenge for the thermal stability of natural fibers, potentially leading to degradation. Furthermore, achieving uniform dispersion of fibers within the PLA matrix can be difficult, which affects the mechanical properties of the extruded material. Finally, the initial investment in extrusion equipment can be significant, making it less feasible for smaller operations or projects with limited budgets.

11.4.4 3D Printing

11.4.4.1 Process Overview

Additive manufacturing, commonly called 3D printing, has gained popularity for producing natural fiber-reinforced PLA components due to its flexibility and design freedom. In this process, a digital model is created using computer-aided design (CAD) software that is then sliced into thin layers. The printer extrudes the composite filament, composed of PLA and natural fibers, layer by layer to build the final object.

There are various 3D printing technologies available, including fused deposition modeling (FDM), selective laser sintering (SLS), and stereolithography (SLA). Each technology has its unique advantages and applications (Azlin et al. 2022; Ilyas et al. 2021). FDM is the most commonly used method for PLA-based composites due to its accessibility and cost-effectiveness. The ability to produce intricate designs and prototypes quickly is one of the primary draws of 3D printing.

11.4.5 Advantages and Limitations

One of the key advantages of 3D printing with natural fiber-reinforced PLA is the ability to create highly customized parts with complex geometries that may be challenging to achieve through traditional manufacturing methods. This process also allows for rapid prototyping, enabling designers to test and iterate on designs quickly. The use of natural fibers can enhance the sustainability of 3D printed products, aligning with the growing demand for ecofriendly materials.

However, there are limitations associated with 3D printing. The mechanical properties of printed parts vary depending on the printing parameters, such as temperature, speed, and layer adhesion. Additionally, layer-by-layer construction can result in anisotropic properties, where strength differs along different axes. Another challenge is ensuring consistent quality and performance of the filament, as variations in fiber content or moisture levels can impact the final product's integrity.

11.5 SOLVENT CASTING

11.5.1 Process Overview

Solvent casting involves dissolving PLA in a suitable solvent to create a homogeneous solution. Natural fibers are then added to this solution and mixed thoroughly. The mixture is poured into a mold and allowed to evaporate, forming a solid composite material as the solvent dissipates. This method is particularly useful for creating films, coatings, or flexible components with specific properties tailored to the intended application. Solvent casting allows for excellent control over the fiber distribution and alignment within the PLA matrix. This method is particularly advantageous for applications requiring high flexibility and surface smoothness. The simplicity of the process also makes it accessible for small-scale production and laboratory settings.

11.5.2 Advantages and Limitations

One of the primary advantages of solvent casting is its ability to produce high-quality films and coatings with uniform properties. The process allows for precise control over the thickness and composition of the resulting material, making it suitable for applications where specific mechanical or barrier properties are required. Additionally, the low equipment costs and straightforward setup make it an appealing option for researchers and small manufacturers.

However, solvent casting does come with limitations. The reliance on solvents can raise environmental and safety concerns, particularly if volatile organic compounds are involved. Furthermore, the evaporation phase can be time-consuming, leading to longer production times than with other methods. There are also challenges associated with scaling the process for larger production volumes, as maintaining consistent quality can become more complex with increased batch sizes. Finally, the compatibility of natural fibers with solvents must be considered, as some fibers may degrade or lose strength during the casting process.

11.6 PROCESSING METHOD DEVELOPMENTS OF NATURAL FIBER/PLA COMPOSITES

Natural fiber reinforcement with PLA composite has been the subject of a great deal of previous natural fiber studies. In one study, Khan et al. (2016) introduced an ecofriendly reinforced PLA biocomposite to examine its mechanical properties in woven jute fabrics as an alternative to the nonbiodegradable synthetic fiber composite. They used hot press molding to prepare the PLA fabric and found that the average tensile strength, tensile modulus, flexural strength, flexural modulus, and impact strength of raw warp-directed woven jute composite increased by 103%, 211%, 95.2%, 42.4%, and 85.9%, respectively, and the strain at maximum tensile stress increased by 11.7% after the reinforcement. The authors concluded that jute composites based on PLA could be suitable synthetic fiber composite alternatives even for high load-bearing applications.

The value of natural fiber-reinforced polymer composites has been illustrated by several studies. In an engaging study, Ogin et al. (2016) described the components, design, and generic degradation of composite materials; the fundamental components of composite materials; the generic defects arising from the manufacturing process; and the external loading of the material. Due to resin shrinkage, the authors reported observing common process-related defects include porosity, shrinkage cracking, and fiber matrix debonding. The curing kinetics – temperature cycles, gelation mechanisms, and resin shrinkage – resulted in internal curing and temperature stresses owing to the difference in thermal expansion coefficients between the various basic components, which could lead to material defects too. The study authors also emphasized the importance of optimizing the material's design by reducing weight and avoiding unnecessary strength. They focused on ensuring that the material was only strong in the areas and directions that bear stress.

Bergeret and Benezet (2011) investigated natural fiber-reinforced biofoams made of starches and PLA and prepared via melt extrusion to produce starch foams. The natural fiber starch foams were 33% less dense and showed a void content of 48% for cellulose to increase mechanical properties of fiber-reinforced PLA foams. Jauhari et al. (2015) studied natural fiber-reinforced composite laminates by assembling long and short bundles of natural fibers. In polymeric composite terminology, the authors created a flat sheet up to tens of layers of fibers thick. Different methods were proposed to classify the interlock of the fibers themselves with a binder in mechanical properties. The authors modeled a PLA drop-off laminate and analyzed it using ANSYS software. The results of this investigation showed that the layers were held together to keep these materials together.

Lim et al. (2008) studied PLA processing technology for mitigating waste management issues and reducing dependency on synthetic plastic for food packaging because of the biodegradability and ecofriendly properties of PLA-based composite. The authors discussed some of the many PLA processing approaches, injection molding, extrusion, and thermoforming fiber spinning. Overall, the existing data highlighted the importance of the developments of sustainable natural sources in order to address possible alternative raw materials in the food supply.

In one study, Nechwatal et al. (2003) characterized natural fiber properties for composites and found that reinforcement with natural fiber had increased the composites' tensile strength and Young's modulus. Moreover, they proposed a new system for processing long fiber-reinforced thermoplastic granules using regular plastic equipment. A significant alternative to the current processes, namely, pultrusion or extrusion is processes that involve long fiber granules. The new long fiber granules have a helical structure that can permit long fibers for composites.

Van De Velde and Kiekens (2001) investigated the thermoplastic pultrusion of natural fiber-reinforced composites. As a reinforcement material for composites, natural flax fiber offered good mechanical characteristics and environmental advantage. One method to satisfy continuous demand for composites is pultrusion, and the authors determined that thermoplastic pultrusion shows the capacity to become a major new force in the pultrusion industry and showed that thermoplastic pultruded composites reinforced with flax fiber might respond to the need for constantly reinforced profiles that are environmentally friendly.

11.6.1 Natural Fiber-Reinforced PLA

Most green composites are made using the same processes as conventional synthetic FRP matrix composites, divided into an open and closed mold. The fabrication methods and processing parameters for PLA-based green composites are summarized in Table 11.2. Open mold processes include hand layup, spray up, tape layup, filament winding, and the autoclave system. Closed mold processes include compression molding, injection molding, and transfer molding. Table 11.2 presents other methods.

Lee et al. (2009) used carding, treatment with a 3-glycidoxypropyl trimethoxy silane, and hot pressing to fabricate kenaf fiber-reinforced polylactide biocomposites. They used carding to create a uniform mixture of the two fibers followed by needle punching, prepressing, and eventually hot pressing to create the composite material.

TABLE 11.2
Fabrication Methods and Processing Parameters for PLA-Based Green Composites

Fiber	Process	Temperature	Pressure	Time of Heating	Reference
Kenaf	Wet impregnation	Room temperature	Process under vacuum	No heating, 24h drying	(Nishino et al. 2003)
Kenaf	Hot pressing	160 °C	10 MPa	10 min	(Ochi 2008)
Flax	Twin-screw extrusion + injection molding	250 °C	70 MPa Screw speed 250 rpm	-	(Oksman et al. 2010)
Chicken feather	Twin-screw extrusion + injection molding	180 °C	Screw speed 100 r/min	10 min	(Cheng et al. 2009)
Bamboo	Twin-screw extrusion + injection molding	180 °C	50–60 MPa	-	(Tokoro et al. 2008)
Bamboo	Compounding + injection molding	170 °C	Screw speed-150 r/min	-	(Gamon, Evon, and Rigal 2013)
Treated ramie	Two-roll plastics mill + hot pressing	140–170 °C	20 MPa	4 min (hot press)	(Yu et al. 2010)
Short ramie	Two-roll plastics mill + hot pressing	140–170 °C	5 MPa	4 min (hot press)	(Yu, Li, and Ren 2009)
Ramie and jute	Two-roll plastics mill + hot pressing	140–170 °C	20 MPa (of hot press)	4 min (hot press)	(Yu, Li, and Ren 2009)

The mat's thickness was reduced by pressing the PLA/kenaf nonwoven web formed after the carding process. They applied a silane coupling agent to the prepressed nonwoven web in quantities of 1, 3, and 5 parts per hundred (pph) of the prepressed composite material. For two hours, the silane was allowed to penetrate and respond with the pre-pressed mat. Finally, the silane-treated prepressed mat was hot-pressed at 200 °C for 5 minutes at 0.7 MPa pressure.

Using film stacking, Plackett et al. (2003) created a PLA/jute biodegradable composite containing around 40% jute fiber by weight. Using a single-screw extruder, PLA was first converted into a 0.2 mm thick film for this analysis. Sections of jute fiber mats were stacked with multiple PLA film layers on either side inside a metal frame to create layups. At the top and bottom of the frame, Teflon sheets were used. The layups went through a rapid press consolidation process that included precompression, touch heating under vacuum, rapid transfer to a press for consolidating and cooling, and removal of the finished portion from the press (Bajpai, Singh, and Madaan 2013).

Based on Table 11.3, another list of some PLA-based composites from the literature, PLA/NBR19 has the highest tensile strength at 51.57 MPa, and PLA/PP has the lowest tensile strength, 33.71 MPa. Results from dynamic mechanical analysis and rheological analysis of blends suggested better compatibility between PLA and NBR19 as reflected in the more homogenous and tougher blend, which was in good agreement with the interfacial tension data and the fine morphological structure. PLA/PP/Cloisite30B had the highest tensile modulus at 3.10 GPa, and PLA/PA had the lowest, 1.20 MPa, because when stress is imposed on a nanocomposite, incorporated nanoparticles can act as stress concentration points, leading to the progress of crazes at the interphase and happening of the failure.

PLA/ABS/SAN-GMA had flexural strength of 62.9 MPa, the highest, and PLA/ABS had the lowest with 45.6 MPa. PLA/ABS/SAN-GMA also had the highest flexural modulus at 2.30 GPa and PLA/ABS the lowest at 1.96 GPa. The composite with the highest impact was PLA/PA at 276 kJ/m^2, whereas the composite with the lowest impact was PLA/PP/Cloisite30B, 3.6 kJ/m^2. The impact is related to the more homogeneous microstructure and apparently a better interfacial adhesion since the PA dispersed phase was covered with the matrix.

PLA bioblends with predominantly biosourced PA10.10 in the composition range 10–50 wt.% were prepared by melt blending in order to overcome the brittleness of PLA. According to Cailloux et al. (2018), the viscosity ratio affects the morphology and mechanical properties of PLA/PA10.10 bioblends, resulting in a consistent two-phase structure regardless of the formulation. Sea-island morphology confirmed the immiscible nature of the PLA/PA10.10 bioblends, where the PLA/PA composite was prepared by melt blending and furthered by compression molding. PLA/PA engineered blends are promising technical progress that opens new perspectives for PLA-based products in technical application.

PLA/ABS/SAN-GMA replacement is beneficial in the cost aspect because the price of ABS is higher than that of PLA. However, the incompatibility between PLA and ABS as well as the brittleness of PLA make it difficult to achieve the desired material properties. Jo et al. (2012a) reported that using compatibilizers in ABS/PLA composites improved their mechanical properties. Adding the compatibilizers

TABLE 11.3
Mechanical Properties of PLA Blend Composites

Polymers	Fibers	Processing Technique	Property: Tensile Strength (MPa)	Tensile Modulus (GPa)	Flexural Strength (MPa)	Flexural Modulus (GPa)	Impact (kJ/m²)	Reference
PLA/ABS	-	Injection molding	37.3	-	45.6	1.96	-	(Jo et al. 2012b)
PLA/ABS/ SAN-GMA	-	Injection molding	50.9	-	62.9	2.30	-	(Jo et al. 2012a)
PLA/NBR19	-	Melt blending	49.63–51.57	2.65–3.15	-	-	-	(Maroufkhani et al. 2017)
PLA/NBR33	-	Melt blending	47.62–50.44	2.51–2.97	-	-	-	(Maroufkhani et al. 2017)
PLA/NBR51	-	Melt blending	44.74–49.92	2.71–3.23	-	-	-	(Maroufkhani et al. 2017)
PLA/PP	–	Melt blending	33.71–35.09	1.93–2.03	–	–	7.8–8.6	(Ebadi-Dehaghani et al. 2015)
PLA/PP/PTW	–	Melt blending	37.53–38.27	2.20–2.50	–	–	32.8–34.6	(Ebadi-Dehaghani et al. 2015)
PLA/PP	Cloisite 30B nanocomposites	Melt blending	36.94–40.66	2.85–3.10	–	–	3.4–3.6	(Ebadi-Dehaghani et al. 2015)
PLA/PP/PTW	Cloisite30B nanocomposites	Melt blending	39.15–39.45	2.50–2.61	-	-	4.3–4.9	(Ebadi-Dehaghani et al. 2015)
PLA/PA	-	Melt blending	47.0–49.0	1.20–1.40	-	-	166–276	(Cailloux et al. 2018)
PLA/TPS	Sugar palm crystalline nanocellulose	Melt blending	-	-	-	-	1.07–3.25	(Nazrin et al. 2021)

increased the polybutadiene content in ABS. Introducing the correct compatibilizer, along with heat and stabilizers, further enhanced the mechanical properties of ABS/PLA composites. ABS/PLA was prepared by extradition and injection molding. SAN-GMA exhibited tensile strength higher than 40 MPa, the target value for car console boxes. The impact strength also was increased due to the addition of SAN-GMA. Composites composed of ABS and PLA were prepared to develop ABS-based automobile console boxes with improved environmental friendliness. PLA/PP is one of the simple approaches to overcome the main problems of PLA, its low impact strength and elongation at break. Ebadi-Dehaghani et al. (2015) conducted experimental and theoretical analyses of mechanical properties of PP/PLA/clay nanocomposites and showed that PLA/PP requires compatibilizer to achieve better tensile properties.

11.6.2 PLA Hybrid Composites

PLA and its copolymers have been extensively used in the biomedical field, specifically in orthopedic applications. However, the main problem associated with PLA-based polymers is that they have low mechanical properties, making it difficult to achieve stiffness equivalent to metallic implants. As a result, problems such as increased implant size, longer degradation periods, and adverse effects from acidic species often emerge. Researchers have established that using biodegradable reinforcements in fabricating fully absorbable composite materials will solve these problems. Hybrid composites are used when investigators desire a combination of properties or when longitudinal as well as lateral mechanical performances are required (Alias et al. 2021). Table 11.4 shows the mechanical properties of various PLA hybrid composites as established from research.

Asaithambi, Ganesan, and Ananda Kumar (2014) reported that untreated hybrid fiber added to virgin PLA provided support by acting as a stress bearer for the PLA matrix. Treating the fiber with alkali improved the adhesive characteristics of fiber surface due to the elimination of natural and artificial impurities. The authors found that the flexural properties of PLA composites improved due to the mercerization process that transformed crude cellulose structure I into a refined cellulose structure II with reactive groups, resulting in short crystallites that formed intimate bonds with the PLA matrix. The alkali treatment increased the strength of chemical bonding between fiber and PLA matrix in the composites.

Meanwhile, Pappu, Pickering, and Thakur (2019) reported that fiber length/orientation, matrix properties (kinetics and crystallinity), and processing techniques all influence the properties of natural fiber-reinforced PLA composites. It is also worth noting that fiber surface roughness aids mechanical interlocking with matrices. In terms of flexural strength, PLA/banana/sisal fiber obtained the highest value at 125.00 MPa, and PLA/polycaprolactone/oil palm mesocarp achieved the lowest at 21.45 MPa. Surprisingly, as for flexural modulus, PLA/montmorillonite nanoclay/short kenaf achieved the highest value at 7.50 GPa while PLA/flax/jute achieved the lowest value at 2.25 GPa.

TABLE 11.4
Mechanical Properties of PLA Hybrid Composites

			Property					
Polymers	**Fibers**	**Processing Technique**	**Tensile Strength (MPa)**	**Tensile Modulus (GPa)**	**Flexural Strength (MPa)**	**Flexural Modulus (GPa)**	**Impact**	**Reference**
PLA	Banana/sisal	Injection molding	79.00	4.10	125.00	5.60	47.80 kJ/m^2	(Asaithambi, Ganesan, and Ananda Kumar 2014)
PLA	Flax/jute	Compression molding	49.35	2.80	80.50	2.25	61.46 J/m	(Manral, Ahmad, and Chaudhary 2019)
PLA	Polycaprolactone/oil palm mesocarp	Melt blending	33.48	0.88	21.45	2.43	95.44 J/m	(Eng et al. 2014)
PLA	Montmorillonite nanoclay/ short kenaf	Double extrusion	37.00	2.80	50.00	7.50	82.00 kJ/m^2	(Kaiser et al. 2013)
PLA	Corn stover/wheat straw/soy stalks	Extrusion + injection molding	58.00	5.55	80.00	6.90	23 J/m	(Sanchez-Garcia and Lagaron 2010; Nyambo, Mohanty, and Misra 2010)
PLA	Hemp/sisal	Injection Molding	46.25	6.10	94.83	6.04	10.29 kJ/m^2	(Pappu, Pickering, and Thakur 2019)
PLA	Coir/pineapple leaf	Melt mixing	18.00	5.00	33.00	5.00	4.3 kJ/m^2	(Siakeng et al. 2019)
PLA	Banana/kenaf	Molding	50.00	-	61.00	-	16.00 kJ/m^2	(Thiruchitrambalam et al. 2009)
PLA	Cotton gin waste/flax	Extrusion + Melt Blending	–	–	13.99	3.97	–	(Bajracharya, Bajwa, and Bajwa 2017)
PLA	Bamboo/microfibrillated cellulose	Milling	–	4.81	53.80	–	–	(Okubo, Fujii, and Thostenson 2009)
PLA	Hemp/yarn	Compression molding/prepreg	62.00	6.50	122.00	9.00	25.00 kJ/m^2	(Baghaei, Skrifvars, and Berglin 2013)
PLA	MMT clay/aloe vera	Extrusion	56.00	3.20	100.00	6.70	55.00 kJ/m^2	(Ramesh, Prasad, and Narayana 2020)
PLA	Softwood flour/ cellulose	Injection molding	70.00	56.00	-	-	-	(Bledzki, Franciszczak, and Meljon 2015)

11.7 CHALLENGES AND SOLUTIONS IN FABRICATION

Fabricating components from natural fiber-reinforced PLA composites presents several challenges that can impact the quality and performance of the final product. One of the most common issues is fiber breakage during processing, which can occur due to the mechanical stresses applied during mixing, extrusion, or molding. Fiber breakage reduces the reinforcing potential of the fibers, leading to composites with lower mechanical strength and stiffness.

Another significant challenge is poor adhesion between the hydrophilic natural fibers and the hydrophobic PLA matrix. This lack of adhesion can result in weak interfacial bonding, leading to composite materials with inferior mechanical properties and poor durability. Additionally, the variability in fiber quality, such as differences in fiber length, diameter, and surface characteristics, can lead to inconsistencies in composite performance, making it difficult to achieve uniformity in mass production.

To overcome these challenges, several strategies have been developed. Chemical treatments, such as silane coupling agents, are often used to improve the adhesion between the fibers and the PLA matrix by modifying the fiber surface to create a stronger bond. Physical treatments, including mechanical fibrillation and controlled heat treatments, can enhance fiber quality and reduce the likelihood of fiber breakage during processing. Optimizing processing parameters such as temperature, pressure, and mixing speed can also help minimize fiber damage and improve the overall quality of the composite. Moreover, the development of hybrid composites, where natural fibers are combined with synthetic fibers or additives, has shown promise in addressing the limitations of single-fiber composites, offering improved mechanical properties and processing stability.

Innovations in fabrication techniques continue to drive advancements in the production of natural fiber-reinforced PLA composites. Additive manufacturing, or 3D printing, has emerged as a cutting-edge technique for fabricating complex composite structures with precise control over fiber orientation and distribution. This technology allows for customizing composite properties, enabling the production of components with tailored mechanical and thermal characteristics. Additionally, the use of nano-reinforcements, such as nanocellulose or graphene, in combination with natural fibers, has shown potential in enhancing the strength and toughness of PLA composites without compromising their biodegradability. These innovations are paving the way for more efficient and sustainable fabrication processes, making natural fiber-reinforced PLA composites increasingly viable for industrial applications.

11.8 CONCLUSIONS

In summary, fabricating components from natural fiber-reinforced PLA composites involves overcoming challenges including fiber breakage, poor adhesion, and variability in fiber quality. Chemical and physical treatments can mitigate these challenges, as can optimizing the processing parameters, leading to composites with enhanced mechanical properties and durability. Innovations such as additive manufacturing and the use of nano-reinforcements further contribute to the advancement of this field, offering new possibilities for producing high-performance, sustainable materials.

The implications of these findings are significant for both industry and research. For the industry, the ability to fabricate strong, durable, and environmentally friendly composites opens new opportunities for applications in automotive, construction, and packaging sectors. For research, the ongoing exploration of new materials, treatments, and fabrication techniques will be essential in further improving the performance and sustainability of these composites. Future work should focus on the continued development of hybrid composites, the integration of advanced manufacturing techniques, and the exploration of new natural fibers and biobased polymers. By addressing these areas, the potential of natural fiber-reinforced PLA composites can be fully realized, contributing to a more sustainable and innovative material landscape.

REFERENCES

Abral, Hairul, Jeri Ariksa, Melbi Mahardika, Dian Handayani, Ibtisamatul Aminah, Neny Sandrawati, Angga Bahri Pratama, Nural Fajri, S.M. Sapuan, and R.A. Ilyas. 2020. "Transparent and Antimicrobial Cellulose Film from Ginger Nanofiber." *Food Hydrocolloids* 98 (January): 105266. https://doi.org/10.1016/j.foodhyd.2019.105266.

Abral, Hairul, Arief Atmajaya, Melbi Mahardika, Fadli Hafizulhaq, Kadriadi, Dian Handayani, S.M. Sapuan, and R.A. Ilyas. 2020. "Effect of Ultrasonication Duration of Polyvinyl Alcohol (PVA) Gel on Characterizations of PVA Film." *Journal of Materials Research and Technology* 9 (2): 2477–86. https://doi.org/10.1016/j.jmrt.2019.12.078.

Abral, Hairul, Maro Hagabean Dalimunthe, Joko Hartono, Rice Putra Efendi, Mochamad Asrofi, Eni Sugiarti, S.M. Sapuan, Ji Won Park, and Hyun Joong Kim. 2018. "Characterization of Tapioca Starch Biopolymer Composites Reinforced with Micro Scale Water Hyacinth Fibers." *Starch/Staerke* 70 (7–8): 1–8. https://doi.org/10.1002/star.201700287.

Ahmad Ilyas, Rushdan, Salit Mohd Sapuan, Rushdan Ibrahim, Hairul Abral, M.R. Ishak, E.S. Zainudin, Mochamad Asrofi, et al. 2019. "Sugar Palm (Arenga Pinnata (Wurmb.) Merr) Cellulosic Fibre Hierarchy: A Comprehensiveapproach from Macro to Nano Scale." *Journal of Materials Research and Technology* 8 (3): 2753–66. https://doi.org/10.1016/j.jmrt.2019.04.011.

Aisyah, H.A., M.T. Paridah, S.M. Sapuan, A. Khalina, O.B. Berkalp, S.H. Lee, C.H. Lee, et al. 2019. "Thermal Properties of Woven Kenaf/Carbon Fibre-Reinforced Epoxy Hybrid Composite Panels." *International Journal of Polymer Science* 2019 (December): 1–8. https://doi.org/10.1155/2019/5258621.

Alam, M.M., M. Maniruzzaman, and M.M. Morshed. 2014. "Application and Advances in Microprocessing of Natural Fiber (Jute) Based Composites." In *Comprehensive Materials Processing*, edited by Saleem Hashmi, Gilmar Ferreira Batalha, Chester J. Van Tyne, and Bekir Yilbas, 1st ed., 243–60. Elsevier. https://doi.org/10.1016/B978-0-08-096532-1.00714-7.

Alemdar, Ayse, and Mohini Sain. 2008a. "Biocomposites from Wheat Straw Nanofibers: Morphology, Thermal and Mechanical Properties." *Composites Science and Technology* 68 (2): 557–65. https://doi.org/10.1016/j.compscitech.2007.05.044.

Alemdar, Ayse, and Mohini Sain. 2008b. "Isolation and Characterization of Nanofibers from Agricultural Residues – Wheat Straw and Soy Hulls." *Bioresource Technology* 99 (6): 1664–71. https://doi.org/10.1016/j.biortech.2007.04.029.

Alias, Aisyah Humaira, Mohd Nurazzi Norizan, Fatimah Athiyah Sabaruddin, Muhammad Rizal Muhammad Asyraf, Mohd Nor Faiz Norrahim, Ahmad Rushdan Ilyas, Anton M. Kuzmin, et al. 2021. "Hybridization of MMT/Lignocellulosic Fiber Reinforced Polymer Nanocomposites for Structural Applications: A Review." *Coatings* 11 (11): 1355. https://doi.org/10.3390/coatings11111355.

Asaithambi, B., G. Ganesan, and S. Ananda Kumar. 2014. "Bio-Composites: Development and Mechanical Characterization of Banana/Sisal Fibre Reinforced Poly Lactic Acid (PLA) Hybrid Composites." *Fibers and Polymers* 15 (4): 847–54. https://doi.org/10.1007/s12221-014-0847-y.

Asyraf, Muhammad Rizal Muhammad, M.R. Ishak, S.M. Sapuan, and N. Yidris. 2019. "Conceptual Design of Creep Testing Rig for Full-Scale Cross Arm Using TRIZ-Morphological Chart-Analytic Network Process Technique." *Journal of Materials Research and Technology* 8 (6): 5647–58. https://doi.org/10.1016/j.jmrt.2019.09.033.

Asyraf, Muhammad Rizal Muhammad, M.R. Ishak, S.M. Sapuan, N. Yidris, R.A. Ilyas, M. Rafidah, and M.R. Razman. 2020. "Potential Application of Green Composites for Cross Arm Component in Transmission Tower: A Brief Review." Edited by Yulin Deng. *International Journal of Polymer Science* 2020 (December): 1–15. https://doi.org/10.1155/2020/8878300.

Asyraf, Muhammad Rizal Muhammad, M. Rafidah, A. Azrina, and M.R. Razman. 2021. "Dynamic Mechanical Behaviour of Kenaf Cellulosic Fibre Biocomposites: A Comprehensive Review on Chemical Treatments." *Cellulose* (February). https://doi.org/10.1007/s10570-021-03710-3.

Asyraf, Muhammad Rizal Muhammad, Agusril Syamsir, Nazirul Mubin Zahari, Abu Bakar Mohd Supian, Mohamad Ridzwan Ishak, Salit Mohd Sapuan, Shubham Sharma, et al. 2022. "Product Development of Natural Fibre-Composites for Various Applications: Design for Sustainability." *Polymers* 14 (5): 920. https://doi.org/10.3390/polym14050920.

Ayu, Rafiqah S., Abdan Khalina, Ahmad Saffian Harmaen, Khairul Zaman, Tawakkal Isma, Qiuyun Liu, R.A. Ilyas, and Ching Hao Lee. 2020. "Characterization Study of Empty Fruit Bunch (EFB) Fibers Reinforcement in Poly(Butylene) Succinate (PBS)/Starch/Glycerol Composite Sheet." *Polymers* 12 (7): 1571. https://doi.org/10.3390/polym12071571.

Azlin, M.N.M., R.A. Ilyas, M.Y.M. Zuhri, S.M. Sapuan, M.M. Harussani, Shubham Sharma, A.H. Nordin, N.M. Nurazzi, and A.N. Afiqah. 2022. "3D Printing and Shaping Polymers, Composites, and Nanocomposites: A Review." *Polymers* 14 (1): 180. https://doi.org/10.3390/polym14010180.

Baghaei, Behnaz, Mikael Skrifvars, and Lena Berglin. 2013. "Manufacture and Characterisation of Thermoplastic Composites Made from PLA/Hemp Co-Wrapped Hybrid Yarn Prepregs." *Composites Part A: Applied Science and Manufacturing* 50: 93–101. https://doi.org/10.1016/j.compositesa.2013.03.012.

Bajpai, Pramendra Kumar, Inderdeep Singh, and Jitendra Madaan. 2013. "Tribological Behavior of Natural Fiber Reinforced PLA Composites." *Wear* 297 (1–2): 829–40. https://doi.org/10.1016/j.wear.2012.10.019.

Bajracharya, Rohan M., Dilpreet S. Bajwa, and Sreekala G. Bajwa. 2017. "Mechanical Properties of Polylactic Acid Composites Reinforced with Cotton Gin Waste and Flax Fibers." *Procedia Engineering* 200: 370–76. https://doi.org/10.1016/j.proeng.2017.07.052.

Bangar, Sneh Punia, M.M. Harussani, R.A. Ilyas, Adeleke Omodunbi Ashogbon, Arashdeep Singh, Monica Trif, and Seid Mahdi Jafari. 2022. "Surface Modifications of Cellulose Nanocrystals: Processes, Properties, and Applications." *Food Hydrocolloids* 130 (September): 107689. https://doi.org/10.1016/j.foodhyd.2022.107689.

Bendahou, Abdelkader, Youssef Habibi, Hamid Kaddami, and Alain Dufresne. 2009. "Physico-Chemical Characterization of Palm from Phoenix Dactylifera-L, Preparation of Cellulose Whiskers and Natural Rubber-Based Nanocomposites." *Journal of Biobased Materials and Bioenergy* 3 (1): 81–90. https://doi.org/10.1166/jbmb.2009.1011.

Bendourou, Fatima Ezzahra, Gayatri Suresh, Mohamed Amine Laadila, Pratik Kumar, Tarek Rouissi, Gurpreet S. Dhillon, Khiari Zied, Satinder K. Brar, and Rosa Galvez. 2021. "Feasibility of the Use of Different Types of Enzymatically Treated Cellulosic Fibres for Polylactic Acid (PLA) Recycling." *Waste Management* 121 (February): 237–47. https://doi.org/10.1016/j.wasman.2020.11.058.

Bergeret, Anne, and Jean Charles Benezet. 2011. "Natural Fibre-Reinforced Biofoams." *International Journal of Polymer Science*. https://doi.org/10.1155/2011/569871.

Bledzki, A.K., P. Franciszczak, and A. Meljon. 2015. "High Performance Hybrid PP and PLA Biocomposites Reinforced with Short Man-Made Cellulose Fibres and Softwood Flour." *Composites Part A: Applied Science and Manufacturing* 74 (July): 132–39. https://doi.org/10.1016/j.compositesa.2015.03.029.

Cailloux, J., T. Abt, V. García-Masabet, O. Santana, M. Sánchez-Soto, F. Carrasco, and M. Ll Maspoch. 2018. "Effect of the Viscosity Ratio on the PLA/PA10.10 Bioblends Morphology and Mechanical Properties." *Express Polymer Letters* 12 (6): 569–82. https://doi.org/10.3144/expresspolymlett.2018.47.

Chan, Chi Hoong, Chin Hua Chia, Sarani Zakaria, Ishak Ahmad, and Alain Dufresne. 2013. "Production and Characterisation of Cellulose and Nano- Crystalline Cellulose from Kenaf Core Wood." *BioResources* 8 (1): 785–94. https://doi.org/10.15376/biores.8.1.785-794.

Chemiefaser, Industrievereinigung. 2018. *Worldwide Production Volume of Chemical and Textile Fibers from 1975 to 2018*. Statista Inc.

Cheng, Sha, Kin Tak Lau, Tao Liu, Yongqing Zhao, Pou Man Lam, and Yansheng Yin. 2009. "Mechanical and Thermal Properties of Chicken Feather Fiber/PLA Green Composites." *Composites Part B: Engineering* 40 (7): 650–54. https://doi.org/10.1016/j.compositesb.2009.04.011.

Cherian, Bibin Mathew, Alcides Lopes Leão, Sivoney Ferreira de Souza, Sabu Thomas, Laly A. Pothan, and M. Kottaisamy. 2010. "Isolation of Nanocellulose from Pineapple Leaf Fibres by Steam Explosion." *Carbohydrate Polymers* 81 (3): 720–25. https://doi.org/10.1016/j.carbpol.2010.03.046.

Chirayil, Cintil Jose, Jithin Joy, Lovely Mathew, Miran Mozetic, Joachim Koetz, and Sabu Thomas. 2014. "Isolation and Characterization of Cellulose Nanofibrils from Helicteres Isora Plant." *Industrial Crops and Products* 59: 27–34. https://doi.org/10.1016/j.indcrop.2014.04.020.

Corrêa, Ana Carolina, Eliangela de Morais Teixeira, Luiz Antonio Pessan, and Luiz Henrique Capparelli Mattoso. 2010. "Cellulose Nanofibers from Curaua Fibers." *Cellulose* 17 (6): 1183–92. https://doi.org/10.1007/s10570-010-9453-3.

Cosgrove, Daniel J. 2005. "Growth of the Plant Cell Wall." *Nature Reviews Molecular Cell Biology* 6 (11): 850–61. https://doi.org/10.1038/nrm1746.

Coskun, Kubra, Aysenur Mutlu, Mehmet Dogan, and Ebru Bozacı. 2021. "Effect of Various Enzymatic Treatments on the Mechanical Properties of Coir Fiber/Poly(Lactic Acid) Biocomposites." *Journal of Thermoplastic Composite Materials* 34 (8): 1066–79. https://doi.org/10.1177/0892705719864618.

Debeli, Dereje Kebebew, Zhao Qin, and Jiansheng Guo. 2018. "Study on the Pre-Treatment, Physical and Chemical Properties of Ramie Fibers Reinforced Poly (Lactic Acid) (PLA) Biocomposite." *Journal of Natural Fibers* 15 (4): 596–610. https://doi.org/10.1080/15440478.2017.1349711.

Ebadi-Dehaghani, Hassan, Hossein Ali Khonakdar, Mehdi Barikani, and Seyed Hassan Jafari. 2015. "Experimental and Theoretical Analyses of Mechanical Properties of PP/PLA/Clay Nanocomposites." *Composites Part B: Engineering* 69: 133–44. https://doi.org/10.1016/j.compositesb.2014.09.006.

Eng, Chern Chiet, Nor Azowa Ibrahim, Norhazlin Zainuddin, Hidayah Ariffin, Wan Md Zin Wan Yunus, and Yoon Yee Then. 2014. "Enhancement of Mechanical and Dynamic Mechanical Properties of Hydrophilic Nanoclay Reinforced Polylactic Acid/Polycaprolactone/Oil Palm Mesocarp Fiber Hybrid Composites." *International Journal of Polymer Science*. https://doi.org/10.1155/2014/715801.

Faiz Norrahim, Mohd Nor, Rushdan Ahmad Ilyas, Norizan Mohd Nurazzi, Mohd Saiful Asmal Rani, Mahamud Siti Nur Atikah, and Siti Shazra Shazleen. 2021. "Chemical Pretreatment of Lignocellulosic Biomass for the Production of Bioproducts: An Overview." *Applied Science and Engineering Progress* (July): 1–18. https://doi.org/10.14416/j.asep.2021.07.004.

Fang, Xiaoyang, Yanchen Li, Junqi Zhao, Jianuo Xu, Chuanchuan Li, Jiaxun Liu, Yi Liu, and Hongwu Guo. 2022. "Improved Interfacial Performance of Bamboo Fibers/Polylactic Acid Composites Enabled by a Self-Supplied Bio-Coupling Agent Strategy." *Journal of Cleaner Production* 380 (December): 134719. https://doi.org/10.1016/j.jclepro.2022.134719.

Ferreira, F.V., M. Mariano, S.C. Rabelo, R.F. Gouveia, and L.M.F. Lona. 2018. "Isolation and Surface Modification of Cellulose Nanocrystals from Sugarcane Bagasse Waste: From a Micro- to a Nano-Scale View." *Applied Surface Science* 436 (April): 1113–22. https://doi.org/10.1016/j.apsusc.2017.12.137.

Gamon, G., Ph Evon, and L. Rigal. 2013. "Twin-Screw Extrusion Impact on Natural Fibre Morphology and Material Properties in Poly(Lactic Acid) Based Biocomposites." *Industrial Crops and Products* 46: 173–85. https://doi.org/10.1016/j.indcrop.2013.01.026.

Huang, Can-Jun, Xiang-Li Li, Ye-Qing Zhang, Yan-Hong Feng, Jin-Ping Qu, He-Zhi He, and Han-Zhi Shen. 2015. "Properties of Heat-Treated Sisal Fiber/Polylactide Composites." *Journal of Thermoplastic Composite Materials* 28 (6): 777–90. https://doi.org/10.1177/0892705713495221.

Ilehuv, Wuhdwhg, D.W. Kljkhu, Ileuh Ordglqj, and Z.D.V. Dqg. 2014. "Tensile Properties of Successive Alkali Treated Short Jute Fiber Reinforced PLA Composites." *Procedia Materials Science* 5: 2188–96. https://doi.org/10.1016/j.mspro.2014.07.425.

Ilyas, R.A., S.M. Sapuan, M.S.N. Atikah, M.R.M. Asyraf, S. Ayu Rafiqah, H.A. Aisyah, N. Mohd Nurazzi, and M.N.F. Norrrahim. 2021. "Effect of Hydrolysis Time on the Morphological, Physical, Chemical, and Thermal Behavior of Sugar Palm Nanocrystalline Cellulose (Arenga Pinnata (Wurmb.) Merr)." *Textile Research Journal* 91 (1–2): 152–67. https://doi.org/10.1177/0040517520932393.

Ilyas, R.A., S.M. Sapuan, and M.R. Ishak. 2018. "Isolation and Characterization of Nanocrystalline Cellulose from Sugar Palm Fibres (Arenga Pinnata)." *Carbohydrate Polymers* 181 (February): 1038–51. https://doi.org/10.1016/j.carbpol.2017.11.045.

Ilyas, R.A., S.M. Sapuan, M.R. Ishak, and E.S. Zainudin. 2017. "Effect of Delignification on the Physical, Thermal, Chemical, and Structural Properties of Sugar Palm Fibre." *BioResources* 12 (4): 8734–54. https://doi.org/10.15376/biores.12.4.8734-8754.

Ilyas, R.A., S.M. Sapuan, M.L. Sanyang, M.R. Ishak, and E.S. Zainudin. 2018. "Nanocrystalline Cellulose as Reinforcement for Polymeric Matrix Nanocomposites and Its Potential Applications: A Review." *Current Analytical Chemistry* 14 (3): 203–25. https://doi.org/10.2174/1573411013666171003155624.

Jagadeesh, Praveenkumara, Madhu Puttegowda, Sanjay Mavinkere Rangappa, and Suchart Siengchin. 2021. "A Review on Extraction, Chemical Treatment, Characterization of Natural Fibers and Its Composites for Potential Applications." *Polymer Composites* 42 (12): 6239–64. https://doi.org/10.1002/pc.26312.

Jauhari, Nitin, Raghvendra Mishra, and Harischchandra Thakur. 2015. "Natural Fibre Reinforced Composite Laminates – A Review." *Materials Today: Proceedings*. https://doi.org/10.1016/j.matpr.2015.07.304.

Jo, Mi Yeong, Yeon Jong Ryu, Jun Hee Ko, and Jin San Yoon. 2012a. "Effects of Compatibilizers on the Mechanical Properties of ABS/PLA Composites." *Journal of Applied Polymer Science* 125 (Suppl. 2). https://doi.org/10.1002/app.36732.

Jo, Mi Yeong, Yeon Jong Ryu, Jun Hee Ko, and Jin San Yoon. 2012b. "Effects of Compatibilizers on the Mechanical Properties of ABS/PLA Composites." *Journal of Applied Polymer Science* 125 (Suppl. 2). https://doi.org/10.1002/app.36732.

Jonoobi, Mehdi, Jalaludin Harun, Alireza Shakeri, Manjusri Misra, and Kristiina Oksmand. 2009. "Chemical Composition, Crystallinity, and Thermal Degradation of Bleached and Unbleached Kenaf Bast (Hibiscus Cannabinus) Pulp and Nanofibers." *BioResources* 4 (2): 626–39. https://doi.org/10.15376/biores.4.2.626-639.

Jonoobi, Mehdi, Abolghasem Khazaeian, Paridah Md Tahir, Syeed Saiful Azry, and Kristiina Oksman. 2011. "Characteristics of Cellulose Nanofibers Isolated from Rubberwood and Empty Fruit Bunches of Oil Palm Using Chemo-Mechanical Process." *Cellulose* 18 (4): 1085–95. https://doi.org/10.1007/s10570-011-9546-7.

Julie Chandra, C.S., Neena George, and Sunil K. Narayanankutty. 2016. "Isolation and Characterization of Cellulose Nanofibrils from Arecanut Husk Fibre." *Carbohydrate Polymers* 142: 158–66. https://doi.org/10.1016/j.carbpol.2016.01.015.

Jumaidin, Ridhwan, RA. Ilyas, Mohamed Saiful, Firdaus Hussin, and M.T. Mastura. 2019. "Water Transport and Physical Properties of Sugarcane Bagasse Fibre Reinforced Thermoplastic Potato Starch Biocomposite." *Journal of Advanced Research in Fluid Mechanics and Thermal Sciences* 61 (2): 273–81.

Kaiser, Mohammad R., Hazleen B. Anuar, Noorasikin B. Samat, and Shamsul B. Abdul Razak. 2013. "Effect of Processing Routes on the Mechanical, Thermal and Morphological Properties of PLA-Based Hybrid Biocomposite." *Iranian Polymer Journal (English Edition)* 22 (2): 123–31. https://doi.org/10.1007/s13726-012-0112-4.

Khalid, Muhammad Yasir, Ramsha Imran, Zia Ullah Arif, Naveed Akram, Hassan Arshad, Ans Al Rashid, and Fausto Pedro García Márquez. 2021. "Developments in Chemical Treatments, Manufacturing Techniques and Potential Applications of Natural-Fibers-Based Biodegradable Composites." *Coatings* 11 (3): 293. https://doi.org/10.3390/coatings11030293.

Khan, G.M. Arifuzzaman, M. Terano, M.A. Gafur, and M. Shamsul Alam. 2016. "Studies on the Mechanical Properties of Woven Jute Fabric Reinforced Poly(L-Lactic Acid) Composites." *Journal of King Saud University – Engineering Sciences* 28 (1): 69–74. https://doi.org/10.1016/j.jksues.2013.12.002.

Lee, Byoung Ho, Hee Soo Kim, Sena Lee, Hyun Joong Kim, and John R. Dorgan. 2009. "Bio-Composites of Kenaf Fibers in Polylactide: Role of Improved Interfacial Adhesion in the Carding Process." *Composites Science and Technology* 69 (15–16): 2573–79. https://doi.org/10.1016/j.compscitech.2009.07.015.

Li, Meng, Li Jun Wang, Dong Li, Yan Ling Cheng, and Benu Adhikari. 2014. "Preparation and Characterization of Cellulose Nanofibers from De-Pectinated Sugar Beet Pulp." *Carbohydrate Polymers* 102 (1): 136–43. https://doi.org/10.1016/j.carbpol.2013.11.021.

Li, Xue, Lope G. Tabil, and Satyanarayan Panigrahi. 2007. "Chemical Treatments of Natural Fiber for Use in Natural Fiber-Reinforced Composites: A Review." *Journal of Polymers and the Environment* 15 (1): 25–33. https://doi.org/10.1007/s10924-006-0042-3.

Lim, L.T., R. Auras, and M. Rubino. 2008. "Processing Technologies for Poly(Lactic Acid)." *Progress in Polymer Science (Oxford)* 33 (8): 820–52. https://doi.org/10.1016/j.progpolymsci.2008.05.004.

Low, L.Q., R.A. Ilyas, R. Jalil, H.S.N. Hawanis, R. Ibrahim, H.A.A. Azriena, M.Y.M. Zuhri, et al. 2024. "Physical and Mechanical Properties Enhancement of Beaten Oil Palm Trunk Pulp and Paper by Optimizing Starch Addition: Towards Sustainable

Packaging Solutions." *Industrial Crops and Products* 221 (December): 119232. https://doi.org/10.1016/j.indcrop.2024.119232.

Manral, Ankit, Furkan Ahmad, and Vijay Chaudhary. 2019. "Static and Dynamic Mechanical Properties of PLA Bio-Composite with Hybrid Reinforcement of Flax and Jute." *Materials Today: Proceedings* 2–5. https://doi.org/10.1016/j.matpr.2019.07.240.

Maroufkhani, Mahshid, Ali Asghar Katbab, Wangcheng Liu, and Jinwen Zhang. 2017. "Polylactide (PLA) and Acrylonitrile Butadiene Rubber (NBR) Blends: The Effect of ACN Content on Morphology, Compatibility and Mechanical Properties." *Polymer* 115: 37–44. https://doi.org/10.1016/j.polymer.2017.03.025.

Martins, Maria Alice, Pedro Kunihiro Kiyohara, and Inés Joekes. 2004. "Scanning Electron Microscopy Study of Raw and Chemically Modified Sisal Fibers." *Journal of Applied Polymer Science* 94 (6): 2333–40. https://doi.org/10.1002/app.21203.

Mazani, N., S.M. Sapuan, M.L. Sanyang, A. Atiqah, and R.A. Ilyas. 2019. "Design and Fabrication of a Shoe Shelf from Kenaf Fiber Reinforced Unsaturated Polyester Composites." In *Lignocellulose for Future Bioeconomy*, edited by Hidayah Ariffin, S.M. Sapuan, and Mohd Ali Hassan, 1st ed., 315–32. Amsterdam, Netherlands: Elsevier. https://doi.org/10.1016/B978-0-12-816354-2.00017-7.

Megashah, L.N., H. Ariffin, M.R. Zakaria, and M.A. Hassan. 2018. "Properties of Cellulose Extract from Different Types of Oil Palm Biomass." *IOP Conference Series: Materials Science and Engineering* 368. https://doi.org/10.1088/1757-899X/368/1/012049.

Migneault, Sébastien, Ahmed Koubaa, Fouad Erchiqui, Abdelkader Chaala, Karl Englund, Cornelia Krause, and Michael Wolcott. 2008. "Effect of Fiber Length on Processing and Properties of Extruded Wood-Fiber/HDPE Composites." *Journal of Applied Polymer Science* 110 (2): 1085–92.

Mohanty, A.K., M. Misra, and L.T. Dreal. 2001. "Surface Modifications of Natural Fibres and Peformance of the Resulting Biocomposite." *Composite Interfaces* 8 (5): 313–43.

Mubarak, Asmaa Ali, R.A. Ilyas, Abu Hassan Nordin, Norzita Ngadi, and M.F.M. Alkbir. 2024. "Recent Developments in Sugarcane Bagasse Fibre-Based Adsorbent and Their Potential Industrial Applications: A Review." *International Journal of Biological Macromolecules* 277 (October): 134165. https://doi.org/10.1016/j.ijbiomac.2024.134165.

Mukherjee, Tapasi, and Nhol Kao. 2011. "PLA Based Biopolymer Reinforced with Natural Fibre: A Review." *Journal of Polymers and the Environment* 19 (3): 714–25. https://doi.org/10.1007/s10924-011-0320-6.

Nazrin, Asmawi, Salit Mohd Sapuan, Mohamed Yusoff Mohd Zuhri, Intan Syafinaz Mohamed Amin Tawakkal, and Rushdan Ahmad Ilyas. 2021. "Water Barrier and Mechanical Properties of Sugar Palm Crystalline Nanocellulose Reinforced Thermoplastic Sugar Palm Starch (TPS)/Poly(Lactic Acid) (PLA) Blend Bionanocomposites." *Nanotechnology Reviews* 10 (1): 431–42. https://doi.org/10.1515/ntrev-2021-0033.

Nechwatal, Axel, Klaus-Peter Mieck, and Thomas Reußmann. 2003. "Developments in the Characterization of Natural Fibre Properties and in the Use of Natural Fibres for Composites." *Composites Science and Technology* 63 (9): 1273–79.

Nishino, Takashi, Koichi Hirao, Masaru Kotera, Katsuhiko Nakamae, and Hiroshi Inagaki. 2003. "Kenaf Reinforced Biodegradable Composite." *Composites Science and Technology* 63 (9): 1281–86. https://doi.org/10.1016/S0266-3538(03)00099-X.

Norfarhana, A.S., R.A. Ilyas, and N. Ngadi. 2022. "A Review of Nanocellulose Adsorptive Membrane as Multifunctional Wastewater Treatment." *Carbohydrate Polymers* 291 (September): 119563. https://doi.org/10.1016/j.carbpol.2022.119563.

Nurazzi, N.M., M.R.M. Asyraf, A. Khalina, N. Abdullah, H.A. Aisyah, S. Ayu Rafiqah, F.A. Sabaruddin, et al. 2021. "A Review on Natural Fiber Reinforced Polymer Composite for Bullet Proof and Ballistic Applications." *Polymers* 13 (4): 646. https://doi.org/10.3390/polym13040646.

Nurazzi, N.M., A. Khalina, S.M. Sapuan, R.A. Ilyas, S. Ayu Rafiqah, and Z.M. Hanafee. 2020. "Thermal Properties of Treated Sugar Palm Yarn/Glass Fiber Reinforced Unsaturated Polyester Hybrid Composites." *Journal of Materials Research and Technology* 9 (2): 1606–18. https://doi.org/10.1016/j.jmrt.2019.11.086.

Nyambo, Calistor, Amar K. Mohanty, and Manjusri Misra. 2010. "Polylactide-Based Renewable Green Composites from Agricultural Residues and Their Hybrids." *Biomacromolecules* 11 (6): 1654–60. https://doi.org/10.1021/bm1003114.

Ochi, Shinji. 2008. "Mechanical Properties of Kenaf Fibers and Kenaf/PLA Composites." *Mechanics of Materials* 40 (4–5): 446–52. https://doi.org/10.1016/j.mechmat.2007.10.006.

Ogin, S.L., P. Brøndsted, and J. Zangenberg. 2016. "Composite Materials: Constituents, Architecture, and Generic Damage." In *Modeling Damage, Fatigue and Failure of Composite Materials*, edited by Ramesh Talreja and Janis Varna, 1st ed., 3–23. Elsevier Ltd. https://doi.org/10.1016/B978-1-78242-286-0.00001-7.

Oksman, Kristiina, Jackson A. Etang, Aji P. Mathew, and Mehdi Jonoobi. 2010. "Cellulose Nanowhiskers Separated from a Bio-Residue from Wood Bioethanol Production." *Biomass and Bioenergy* 35 (1): 146–52. https://doi.org/10.1016/j.biombioe.2010.08.021.

Oksman, Kristiina, M. Skrifvars, and J.F. Selin. 2003. "Natural Fibres as Reinforcement in Polylactic Acid (PLA) Composites." *Composites Science and Technology* 63 (9): 1317–24. https://doi.org/10.1016/S0266-3538(03)00103-9.

Okubo, Kazuya, Toru Fujii, and Erik T. Thostenson. 2009. "Multi-Scale Hybrid Biocomposite: Processing and Mechanical Characterization of Bamboo Fiber Reinforced PLA with Microfibrillated Cellulose." *Composites Part A: Applied Science and Manufacturing* 40 (4): 469–75. https://doi.org/10.1016/j.compositesa.2009.01.012.

Othman, Jameelah Alhad Salih, R.A. Ilyas, Abu Hassan Nordin, Norzita Ngadi, and M.F.M. Alkbir. 2024. "Recent Advancements in Bamboo Nanocellulose-Based Bioadsorbents and Their Potential in Wastewater Applications: A Review." *International Journal of Biological Macromolecules* 277 (October): 134451. https://doi.org/10.1016/j.ijbiomac.2024.134451.

Pappu, Asokan, Kim L. Pickering, and Vijay Kumar Thakur. 2019. "Manufacturing and Characterization of Sustainable Hybrid Composites Using Sisal and Hemp Fibres as Reinforcement of Poly (Lactic Acid) via Injection Moulding." *Industrial Crops and Products* 137 (November 2018): 260–69. https://doi.org/10.1016/j.indcrop.2019.05.040.

Plackett, David, Tom Løgstrup Andersen, Walther Batsberg Pedersen, and Lotte Nielsen. 2003. "Biodegradable Composites Based on L-Polylactide and Jute Fibres." *Composites Science and Technology* 63 (9): 1287–96. https://doi.org/10.1016/S0266-3538(03)00100-3.

Punia Bangar, Sneh, R.A. Ilyas, Nisha Chaudhary, Sanju Bala Dhull, Amreen Chowdhury, and Jose M. Lorenzo. 2023. "Plant-Based Natural Fibers for Food Packaging: A Green Approach to the Reinforcement of Biopolymers." *Journal of Polymers and the Environment* (June). https://doi.org/10.1007/s10924-023-02849-3.

Rahmadiawan, Dieter, Shih-Chen Shi, Hairul Abral, Mohammad Khalid Ilham, Eni Sugiarti, Ahmad Novi Muslimin, R.A. Ilyas, Remon Lapisa, and Nandy Setiadi Djaya Putra. 2024. "Comparative Analysis of the Influence of Different Preparation Methods on the Properties of TEMPO-Oxidized Bacterial Cellulose Powder Films." *Journal of Natural Fibers* 21 (1). https://doi.org/10.1080/15440478.2023.2301386.

Rajeshkumar, G., S. Arvindh Seshadri, G.L. Devnani, M.R. Sanjay, Suchart Siengchin, J. Prakash Maran, Naif Abdullah Al-Dhabi, et al. 2021. "Environment Friendly, Renewable and Sustainable Poly Lactic Acid (PLA) Based Natural Fiber Reinforced Composites – A Comprehensive Review." *Journal of Cleaner Production* 310 (April). https://doi.org/10.1016/j.jclepro.2021.127483.

Ramesh, P., B. Durga Prasad, and K.L. Narayana. 2020. "Effect of Fiber Hybridization and Montmorillonite Clay on Properties of Treated Kenaf/Aloe Vera Fiber Reinforced PLA Hybrid Nanobiocomposite." *Cellulose* 27 (12): 6977–93. https://doi.org/10.1007/s10570-020-03268-6.

Sabaruddin, Fatimah Athiyah, M.T. Paridah, S.M. Sapuan, R.A. Ilyas, Seng Hua Lee, Khalina Abdan, Norkhairunnisa Mazlan, Adlin Sabrina Muhammad Roseley, and H.P.S. Abdul Khalil. 2020. "The Effects of Unbleached and Bleached Nanocellulose on the Thermal and Flammability of Polypropylene-Reinforced Kenaf Core Hybrid Polymer Bionanocomposites." *Polymers* 13 (1): 116. https://doi.org/10.3390/polym13010116.

Sanchez-Garcia, Maria D., and Jose M. Lagaron. 2010. "On the Use of Plant Cellulose Nanowhiskers to Enhance the Barrier Properties of Polylactic Acid." *Cellulose* 17 (5): 987–1004. https://doi.org/10.1007/s10570-010-9430-x.

Sawpan, Moyeenuddin A., Kim L. Pickering, and Alan Fernyhough. 2011. "Improvement of Mechanical Performance of Industrial Hemp Fibre Reinforced Polylactide Biocomposites." *Composites Part A: Applied Science and Manufacturing*. https://doi.org/10.1016/j.compositesa.2010.12.004.

Sheltami, Rasha M., Ibrahim Abdullah, Ishak Ahmad, Alain Dufresne, and Hanieh Kargarzadeh. 2012. "Extraction of Cellulose Nanocrystals from Mengkuang Leaves (Pandanus Tectorius)." *Carbohydrate Polymers* 88 (2): 772–79. https://doi.org/10.1016/j.carbpol.2012.01.062.

Siakeng, Ramengmawii, Mohammad Jawaid, Hidayah Ariffin, S.M. Sapuan, Mohammad Asim, and Naheed Saba. 2019. "Natural Fiber Reinforced Polylactic Acid Composites: A Review." *Polymer Composites* 40 (2): 446–63. https://doi.org/10.1002/pc.24747.

Suriani, M.J., Fathin Sakinah Mohd Radzi, R.A. Ilyas, Michal Petrů, S.M. Sapuan, and C.M. Ruzaidi. 2021. "Flammability, Tensile, and Morphological Properties of Oil Palm Empty Fruit Bunches Fiber/Pet Yarn-Reinforced Epoxy Fire Retardant Hybrid Polymer Composites." *Polymers* 13 (8): 1282. https://doi.org/10.3390/polym13081282.

Syafri, Edi, Jamaluddin, Nasmi Herlina Sari, Melbi Mahardika, Putri Amanda, and Rushdan Ahmad Ilyas. 2022. "Isolation and Characterization of Cellulose Nanofibers from Agave Gigantea by Chemical-Mechanical Treatment." *International Journal of Biological Macromolecules* 200 (December 2021): 25–33. https://doi.org/10.1016/j.ijbiomac.2021.12.111.

Syafri, Edi, Anwar Kasim, Hairul Abral, and Alfi Asben. 2018. "Cellulose Nanofibers Isolation and Characterization from Ramie Using a Chemical-Ultrasonic Treatment." *Journal of Natural Fibers* 1–11. https://doi.org/10.1080/15440478.2018.1455073.

Syafri, Edi, Sudirman, Mashadi, Evi Yulianti, Deswita, Mochamad Asrofi, Hairul Abral, S.M. Sapuan, R.A. Ilyas, and Ahmad Fudholi. 2019. "Effect of Sonication Time on the Thermal Stability, Moisture Absorption, and Biodegradation of Water Hyacinth (Eichhornia Crassipes) Nanocellulose-Filled Bengkuang (Pachyrhizus Erosus) Starch Biocomposites." *Journal of Materials Research and Technology* 8 (6): 6223–31. https://doi.org/10.1016/j.jmrt.2019.10.016.

Teixeira, Eliangela de Morais, Thalita Jessika Bondancia, Kelcilene Bruna Ricardo Teodoro, Ana Carolina Corrêa, José Manoel Marconcini, and Luiz Henrique Caparelli Mattoso. 2011. "Sugarcane Bagasse Whiskers: Extraction and Characterizations." *Industrial Crops and Products* 33 (1): 63–66. https://doi.org/10.1016/j.indcrop.2010.08.009.

Thiruchitrambalam, M., A. Alavudeen, A. Athijayamani, N. Venkateshwaran, and A. Elaya Perumal. 2009. "Improving Mechanical Properties of Banana/Kenaf Polyester Hybrid Composites Using Sodium Laulryl Sulfate Treatment." *Materials Physics and Mechanics* 8 (2): 165–73.

Tibolla, Heloisa, Franciele Maria Pelissari, and Florencia Cecilia Menegalli. 2014. "Cellulose Nanofibers Produced from Banana Peel by Chemical and Enzymatic Treatment." *LWT – Food Science and Technology* 59 (2P2): 1311–18. https://doi.org/10.1016/j.lwt.2014.04.011.

Tokoro, Ryoko, Duc Minh Vu, Kazuya Okubo, Tatsuya Tanaka, Toru Fujii, and Takayasu Fujiura. 2008. "How to Improve Mechanical Properties of Polylactic Acid with Bamboo Fibers." *Journal of Materials Science* 43 (2): 775–87. https://doi.org/10.1007/s10853-007-1994-y.

Velde, Kathleen Van De, and Paul Kiekens. 2001. "Thermoplastic Pultrusion of Natural Fibre Reinforced Composites." *Composite Structures*. https://doi.org/10.1016/S0263-8223(01)00110-6.

Wambua, Paul, Jan Ivens, and Ignaas Verpoest. 2003. "Natural Fibres: Can They Replace Glass in Fibre Reinforced Plastics?" *Composites Science and Technology* 63 (9): 1259–64. https://doi.org/10.1016/S0266-3538(03)00096-4.

Yu, Tao, Yan Li, and Jie Ren. 2009. "Preparation and Properties of Short Natural Fiber Reinforced Poly(Lactic Acid) Composites." *Transactions of Nonferrous Metals Society of China (English Edition)*. https://doi.org/10.1016/S1003-6326(10)60126-4.

Yu, Tao, Jie Ren, Shumao Li, Hua Yuan, and Yan Li. 2010. "Effect of Fiber Surface-Treatments on the Properties of Poly(Lactic Acid)/Ramie Composites." *Composites Part A: Applied Science and Manufacturing* 41 (4): 499–505. https://doi.org/10.1016/j.compositesa.2009.12.006.

12 Applications of Natural Fiber–PLA Composites

12.1 INTRODUCTION

One of the most significant advantages of natural fiber-reinforced PLA composites is their potential to replace conventional petroleum-based products, which are known for their detrimental environmental impacts. The combination of PLA with natural fibers such as kenaf, hemp, and flax results in biocomposites that can be used in various industries from automotive and construction to packaging and consumer goods (Figure 12.1) (Asyraf et al. 2022). For instance, in the automotive sector, these composites are being increasingly utilized in nonstructural components, offering a sustainable alternative without compromising performance. Similarly, in the packaging industry, the use of PLA-based composites can reduce the reliance on nonbiodegradable plastics, contributing to a significant reduction in plastic waste.

In this chapter, we explore the diverse applications of natural fiber-reinforced PLA composites across various industries, emphasizing their mechanical and physical properties that make them suitable for different uses. The discussion will highlight the current challenges and limitations associated with these materials, as well as potential solutions and future research directions. By examining case studies and examples, we aim to provide a comprehensive understanding of how these biocomposites can be integrated into existing manufacturing processes and their potential to revolutionize industries by offering sustainable, high-performance alternatives to traditional materials.

12.2 BIOMEDICAL APPLICATIONS

12.2.1 Wound Management and Stent Applications

PLA and its copolymers have been used in a variety of wound care applications, including surgical sutures, wound healing after dental extractions, and avoiding postoperative adhesions. The rate of PLA degradation in sutures has recently been shown to depend on the magnitude of the applied stress. Due to viscoplastic flow causing creep rupture or fatigue failure, PLA exhibits premature failure at stress magnitudes that are substantially lower than the yield strength and the ultimate tensile strength of the content (Shih, Huang, and Chen 2010). Consequently, system failure can occur in some applications well before the material is expected to fail due to degradation in vivo.

During the crimp and expansion phases, a bioresorbable PLA stent, for example, is subjected to extremely high stresses and strains. The stent usually has significant viscoplastic residual strains after deployment. It must also preserve mechanical

 DOI: 10.1201/9781032678894-12

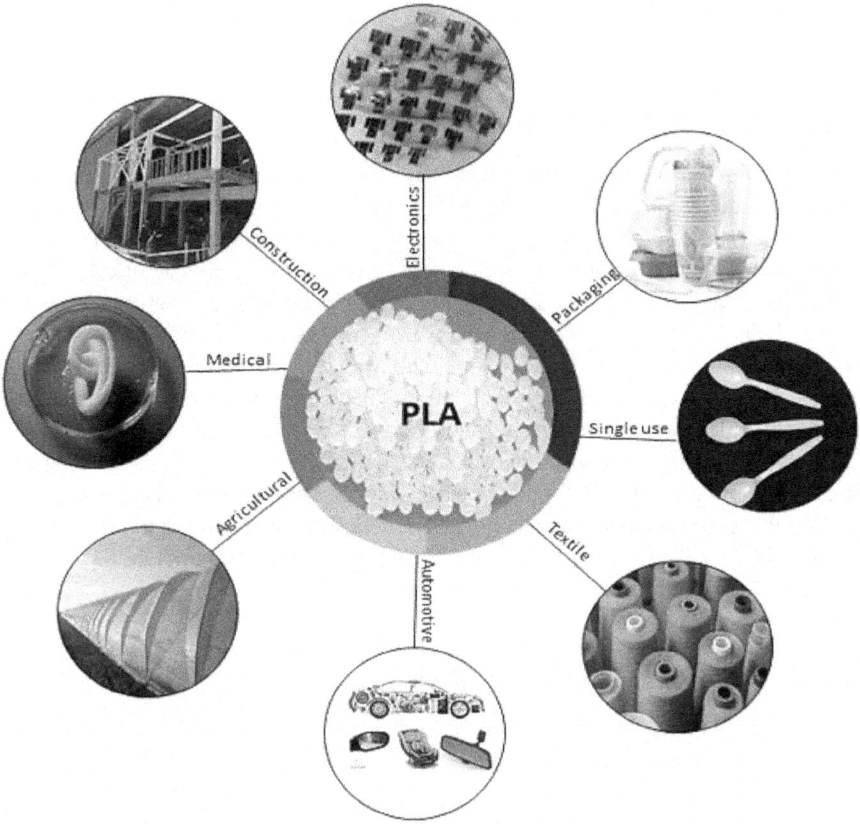

FIGURE 12.1 PLA applications in end-use industries and global market consumption.

Source: Asyraf et al. (2022).

integrity when subjected to a high number of load cycles with small stress and strain amplitudes (Moutee, Fortin, and Fafard 2007a). By altering the mechanical reaction of the devices to their in vivo conditions, the use of degradable, nonlinear, viscoplastic materials poses many new challenges to the production and use of bioresorbable stents. As a result, PLLA is the most widely used substrate for bioresorbable scaffolds/stents. Due to the lower stiffness and strength of PLLA relative to metals, the struts must normally be thicker than those used in traditional metal stents to achieve the desired radial strength (Duigou et al. 2008). Poor deliverability, platelet deposition, and vessel injury can result.

To resolve these issues, a thorough understanding of the mechanical response of the PLA implant/device is needed. This necessitates a thorough understanding of how the material's strength and stiffness shift over time as a result of time-dependent microstructural processes, how the material responds during deterioration, and how the material undergoes nonlinear viscoplastic deformations at finite deformations (Moutee, Fortin, and Fafard 2007b). Other uses, such as ligament and tendon repair, as well as urological surgery, require a longer retention of strength, and PLLA fibers are the preferred material.

12.2.2 PLA in Drug Delivery Systems

PLAs have been used to provide continuous drug release for a range of medicinal agents, including contraception, narcotic antagonists, local anesthetics, vaccines, peptides, and proteins, for extended periods of time. Erosion, diffusion, and swelling are three ways that polymeric drugs can be re-released. The breakage of ester bonds in PLA occurred at random due to hydrolytic ester cleavage, resulting in system erosion (Ilyas Rushdan et al. 2017). The hydrolytic products of this form of decomposition are then converted into nontoxic subproducts that are excreted by normal cellular activity and urine. Many medications, including antipsychotics, restenosis, hormones, oridonin, dermatotherapy, and protein, were encapsulated using PLA and its copolymers in the form of micro or nanoparticles. PLA particles were made using solvent evaporation and were found to be excellent candidates for a drug delivery system (Farah, Anderson, and Langer 2016). The mechanical stability and crystallinity degree of PLA in the formulation can be used to tune the challenge of managed drug release.

12.2.3 Orthopedic and Fixation Devices

The difficulties that come with using PLA in orthopedics are highly dependent on the position and form of the system that will be used. While most devices are made of steel or titanium to ensure that they can withstand in vivo loading for an extended period of time, polymers have an advantage over metal implants in that they pass stress to the damaged area over time, enabling the tissues to heal (Athanasiou, Niederauer, and Agrawal 1996). Another significant benefit is the avoidance of second surgeries to remove unnecessary hardware, which lowers medical costs and allows for incremental tissue function regeneration as the implant degrades naturally without the use of enzymes or catalysts (Mamun and Bledzki 2013).

PLA has previously been used to make biodegradable screws, fixation pins, plates, and suture anchors. These absorbable screws and pins have become increasingly popular in clinical practice, particularly in situations where high mechanical stiffness or strength is not needed. The knee, shoulder, foot and ankle, hand, wrist, elbow, pelvis, and zygomatic fractures are all important orthopedic areas. In other cases, high-performance PLA is needed, which has proven to be difficult to achieve. Controlling the L/D ratio in the polymer improved the mechanical properties of PLA, with the ratio of L/D 85/15 being polymerized and the prepared PLA being used to produce screws and fixation plates for fracture fixation (Athanasiou, Niederauer, and Agrawal 1996). The findings revealed that the plates could be used to repair fractures without the need for external help. These degradable devices had comparable success rates with those of metallic counterparts. However, in these applications, it is critical that the applied stress never exceeds the implant's ability, causing permanent deformation or premature failure due to viscoplastic flow or fracture.

12.2.4 Tissue Engineering and Regenerative Medicine

PLA consumption has increased exponentially in tissue engineering, one of the most exciting interdisciplinary and multidisciplinary research fields. In tissue engineering and regeneration, scaffold materials and fabrication technologies are critical (Lasprilla et al. 2012).

PLA matrix materials have sparked much interest as transplantation supports because they fade away from the transplantation site over time, leaving a perfect patch of natural neotissue behind. PLA has been studied for tissue engineering applications, such as bone scaffolds, due to its excellent biocompatibility. However, the existence of various tissues necessitates a material with a predetermined biodegradation profile. PLA's mechanical properties for tissue engineering have been stated to be improved using a variety of methods, including blending, composites formation, and co-polymerization (Sharma et al. 2021).

Three-dimensional porous PLA scaffolds for culturing various cell types have been developed for use in cell-based gene therapy for cardiovascular disorders, muscle tissue regeneration, bone and cartilage regeneration, and other treatments for cardiovascular, neurological, and orthopedic conditions (Nofar and Park 2014). In two other experiments, osteogenic stem cells were seeded on scaffolds made of this material and inserted in bone defects or subcutaneously to mimic both endochondral and intramembranous ossification processes in bone formation. Because of the high strength of PLLA mesh, 3D structures such as trays and cages can be developed. The PLA can take anywhere from 10 months to 4 years to degrade, depending on microstructural factors including chemical composition, porosity, and crystallinity, all of which can affect tensile strength. Furthermore, isolated cells can be stimulated to regenerate tissues and release drugs such as painkillers, anti-inflammatories, and antibiotics, which has recently prompted their investigation as cell transplantation scaffolds.

12.2.4.1 Tissue Engineering with Hydroxyapatite

PLA plays a crucial role in tissue engineering, particularly in bone regeneration and grafting applications. The current technologies aim to integrate engineered bone with native tissue, promoting both osteogenesis and angiogenesis. However, despite its advantages, PLA-based materials often face challenges due to their limited physiological functions (Wang et al. 2019). Issues such as immune responses, scaffold integration, and degradation must be addressed before implantation.

A promising approach combines PLA with hydroxyapatite (HA), which independently stimulates osteogenesis by activating osteoblasts and pre-osteoblastic cells (Russias et al. 2006; Bae et al. 2011). In composites, PLA enhances HA's mechanical properties while HA improves the flexural strength of PLA. Hatano et al. reported that a PLA/HA composite (80% HA) demonstrated an elastic modulus similar to human cortical bone (10 GPa), though exceeding this HA concentration compromised structural integrity (Hatano et al. 1999). The combination showed poor mechanical performance in vitro due to dissolution under physiological conditions. Despite these challenges, PLA offers excellent bioresorption, making it useful for bone grafting, particularly when tailored for regenerative therapies.

Other studies have shown that high-molecular-weight HA particles accelerate osteoblast development, while HA also alters PLA's physical properties. Russias et al. (2006) demonstrated that HA insertion increases PLA's surface roughness five-fold and improves hydrophilicity by reducing its water contact angle by 21%, thereby enhancing cell adhesion and osteoblastic activity. This method encourages bone growth by promoting cell proliferation and differentiation.

12.2.4.2 Tissue Engineering (Bone–Glass)

PLA demonstrates significant potential in bone tissue engineering, particularly when combined with calcium phosphate glass, which is highly biocompatible due to its low cytotoxicity. The chemical composition of calcium phosphate glass can be tailored to match the degradation rate of natural bone, enhancing its effectiveness in scaffolding for bone regeneration. The scaffold was created as a foam, where the glass was inserted into a PLA solution, resulting in a homogeneous composite that provided the necessary mechanical strength and interacted well with cells, promoting tissue integration (Navarro et al. 2004). Figure 12.2 shows SEM images of both the PLA-only and PLA/glass foam composites.

The authors improved the scaffolds by modulating porosity, mechanical strength, and bioactivity, and the PLA/calcium phosphate composites showed better integration with host tissues than the PLA controls, which demonstrated poor cell adherence. The porous interconnected veins in the composite facilitated osteoblast proliferation, angiogenesis, and nutrient sequestration. Adding bioglass to the composite increased porosity from 93% to 97%, improving cell interaction (Navarro et al. 2004).

Mechanical testing revealed that PLA/calcium phosphate composites exhibited superior compressive strength and modulus to that of the PLA-only controls. For instance, the glass-infused composite reached 120 MPa in compressive yield strength compared to 74.5 MPa for PLA foam. This 61% increase in strength demonstrates the successful integration of calcium phosphate glass. Additionally, the compressive modulus of the composite increased from 17.5 MPa in PLA controls to 20.2 MPa, further enhancing its mechanical stability (Navarro et al. 2004).

12.2.4.3 Tissue Engineering

PLA scaffolds also hold promise in regenerating epithelial cells, particularly in conditions like short bowel syndrome. In this context, PLA is often combined with polyglycolic acid (PGA) to create scaffolds for epithelial growth. When used in rat models with cardiovascular disease, PLA-coated PGA scaffolds successfully integrated with collagen and supported cyst formation, suggesting their potential for epithelial regeneration (Choi et al. 1998). Electrophysiological testing showed that mucosal cells grown on the scaffolds exhibited active ion transport, with the epithelial barrier strength comparable with that of natural tissue. Histological analysis, illustrated in Figure 12.3, revealed the presence of crypts and well-formed epithelial structures, indicating the potential of PGA/PLA scaffolds for future applications in epithelial regeneration. These findings highlight the experimental success of PGA/PLA composite scaffolds in supporting the growth and resilience of epithelial cells in the intestinal lining.

FIGURE 12.2 SEM micrographs of the fracture surface of PLA and PLA/glass foam; (bottom) glass particle covered with PLA.

Source: Navarro et al. (2004).

Moreover, PLA-based scaffolds are being explored for cardiovascular applications due to their tunable bioresorption properties. PLA grafts can provide mechanical support while gradually degrading to promote healing. For instance, synthetic PLA grafts demonstrated biocompatibility in arterial grafting and facilitated the growth of endothelial and stem cells without significantly impacting cell viability (Kabirian

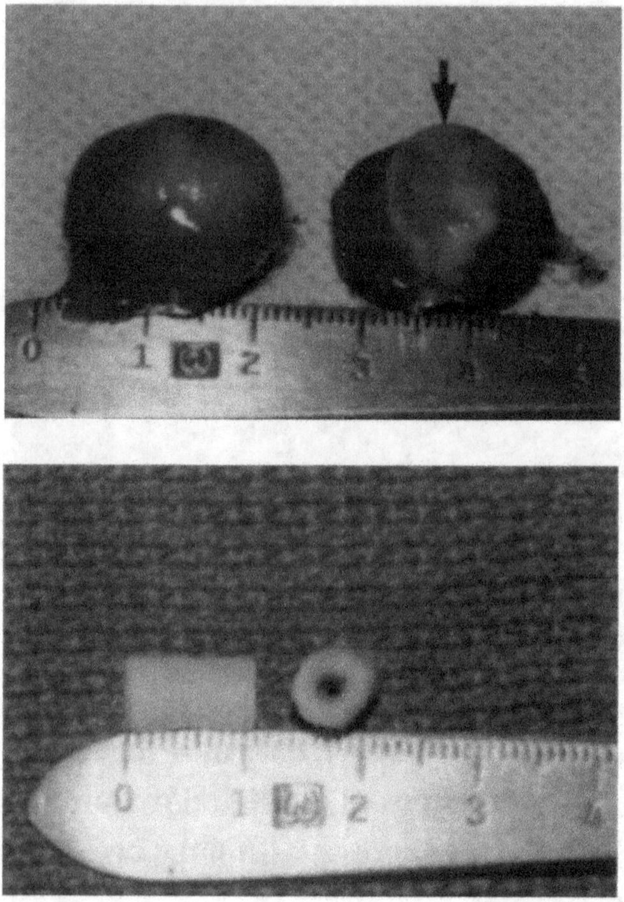

FIGURE 12.3 Sizes of the cysts for histological examination (right) relative to scaffolds (left).

Source: Choi et al. (1998).

et al. 2018). This makes PLA a promising material for improving outcomes in coronary artery disease and bypass surgeries, where traditional autografts pose risks such as infection and restenosis (Kabirian et al. 2018).

12.3 COMPONENTS IN ELECTRICAL TOWERS

PLA has been known due to its tremendous mechanical strength and stiffness to be applied in structural applications. In a recent review, Asyraf, Ishak, Sapuan, Yidris, et al. (2020b) established that PLA has a high potential to replace unsaturated polyester resin in pultruded cross arms. PLA had higher tensile strength and modulus than those of polyester, 49.6 MPa and 3.60 GPa, respectively (Zengwen et al. 2020). The

PLA polymer also showed good creep, quasi-static bending and electrical resistance properties, which are suitable for civil and energy engineering applications (Wang et al. 2018; Letcher and Waytashek 2014). The PLA seems to be suitable for cross arm building materials. This article also suggested that the PLA can be reinforced with flax fiber to produce cross arm in order to replace glass fiber reinforced unsaturated polyester composites.

12.4 AUTOMOTIVE

Plastics are now widely used in vehicles' interior and exterior components, such as bumpers, body panels, safety glass, and trims, as shown in Figure 12.4. In exterior applications, their lightweight nature and design flexibility allow the creation of complex shapes that would be difficult to manufacture with materials like metals. Plastics have revolutionized automotive design by providing durable, comfortable, and aesthetically pleasing interiors while enhancing occupant protection and reducing noise and vibration. Their strength, durability, corrosion resistance, and ability to endure high temperatures make them ideal for use in electrical systems, powertrains, fuel systems, chassis, and engine applications.

When selecting materials, vehicle manufacturers consider several factors, including environmental and safety regulations, production costs, mechanical properties, and the need for weight reduction. These criteria can vary depending on the type of vehicle and component, often requiring a balance of conflicting priorities to achieve an optimal design, and plastics continue to be essential for meeting the diverse and evolving demands of the automotive industry. With the automotive industry facing increasingly strict environmental regulations, the demand for ecofriendly alternatives

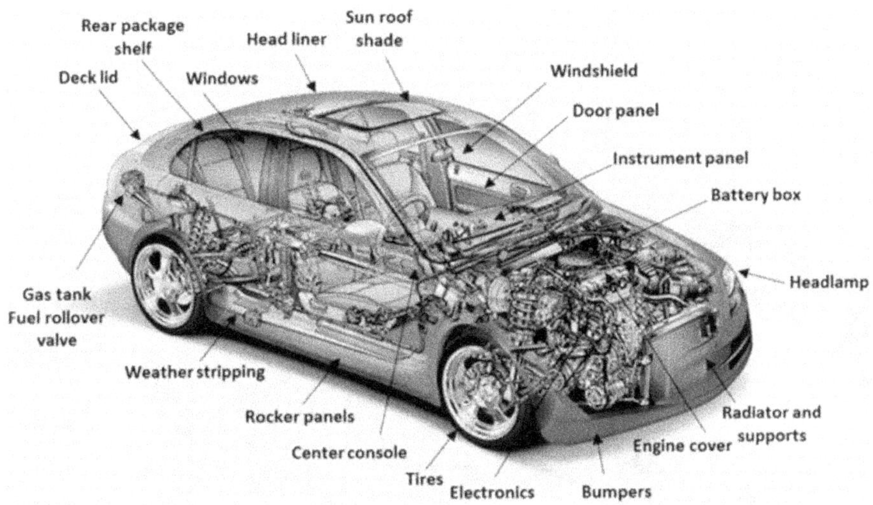

FIGURE 12.4 Potential replacement of plastic parts with PLA-based materials for automotive applications.

Source: Bouzouita et al. (2017).

is driving innovation in the use of next-generation bioplastics and biocomposites as sustainable vehicle components. PLA, known for its renewable nature, affordability, and impressive strength and rigidity, has emerged as a promising material for such applications. However, to compete with conventional petroleum-based plastics, PLA requires enhancements in key areas, including heat resistance, mechanical properties – particularly ductility and impact toughness and overall durability to meet the rigorous standards of the automotive sector (Bouzouita et al. 2017).

In response to the European Union's strict emissions guidelines, which set a target of 95 g/km of CO_2 by 2021 and an additional 15% reduction by 2025, biocomposites are emerging as a sustainable solution for the automotive industry. The use of biocomposites in vehicle manufacturing can help reduce the environmental impact of car production. Recognizing this, leading manufacturers like Mercedes-Benz have adopted natural fiber polymer composites in their vehicles. For instance, the E-class Mercedes-Benz utilized green composites in its door panels, resulting in significant weight reduction. In Malaysia, palm fiber composites are also gaining attention from local car manufacturers.

Modern vehicles consist of approximately 1,500 components, with 40% of these being polymeric materials, including around 600 plastic composite parts used in the body assembly. Figure 12.5 illustrates Ford's pioneering use of soy-based polymer body panels in their influential soybean car (Rahman et al., 2023). The plastic composite components in vehicles are strategically located in areas such as interior trims, external parts, fuel lines, under-hood systems, and structural components, demonstrating the versatility and importance of these materials in modern automotive design.

PLA is a well-known polymer in composites for its mechanical strength, stiffness and chemical resistance. Jung et al. (2011) used jute fiber as reinforcements in PLA polymer to enhance the heat resistance and impact strength of the composites.

FIGURE 12.5 Ford's soy-based composite car.

Source: Rahman et al. (2023).

They found that the alkali-treated of 10 wt.% jute fiber exhibited remarkable tensile and flexural strength, 55 MPa and 108 MPa, respectively. The alkali-treated jute/PLA composites displayed good interfacial adhesion between fiber and matrix. The researchers also implemented posttreatment annealing after the composites were fabricated. They discovered that the treated green composites had a remarkable increase in heat resistance. In the end, the PLA composites are potentially suitable for automotive parts such as indoor panels, engine cover, car spoiler, engine mounting rubber and antiroll bar (Ilyas et al. 2021; 2020; Shaharuzaman, Sapuan, and Mansor 2018; Mastura et al. 2017; Yusof et al. 2018). In order to implement PLA composites as automotive components, they should be well-designed using concurrent engineering techniques and finite element analysis in order to achieve optimum function ability and product property (Asyraf et al. 2019; u; Asyraf, Ishak, Sapuan, Yidris, et al. 2020a; Mansor et al. 2014; Azammi et al. 2018).

In a study conducted by Haryati et al. (2021) on the effect of chemically treated kenaf fiber on the mechanical and thermal properties of PLA composites via fused deposition modeling (FDM), the authors mixed 2.5 wt.% kenaf fiber with PLA using a twin-screw extruder to produce a 3D printing filament. They compared five parameters, including neat PLA (0 wt% fiber), untreated kenaf fiber composites, and three different silane treatments. Samples were printed according to ASTM standards for testing. The results showed that 1.0% silane after treatment with a 6% alkali solution enhanced interfacial bonding between the PLA matrix and the kenaf fibers. However, using a higher silane concentration (2.0%) caused fiber damage, leading to lower strength than with the 1.0% concentration. Untreated kenaf fiber composites exhibited the lowest strength due to poor interfacial bonding, which caused stress to distribute unevenly. Thus, optimizing the silane concentration is critical for natural fiber-reinforced composites, as it ensures proper bonding and enhances mechanical performance for advanced applications such as for automotive applications.

In the automotive sector, PLA hybrid biocomposites are increasingly being considered as a sustainable alternative to conventional synthetic composites for interior and exterior components. The automotive industry is under growing pressure to reduce the environmental impact of vehicles, not only through improved fuel efficiency but also by incorporating sustainable materials into vehicle design. PLA hybrid biocomposites offer the potential to reduce the overall weight of vehicles, which can contribute to improved fuel efficiency and reduced carbon emissions. Additionally, the use of natural fibers in these composites aligns with the industry's goals of increasing the use of renewable materials. Components such as door panels, dashboards, and interior trim can be manufactured using PLA hybrid biocomposites, offering both mechanical performance and aesthetic appeal. The ability to mold these materials into complex shapes further enhances their suitability for automotive applications. Furthermore, due to the excellent mechanical strength of natural fiber reinforced PLA composites, they are favored to be used in the automotive sector.

Researchers reported that natural fiber-reinforced PLA had comparable mechanical strength with that of normal laminated glass fiber-reinforced plastics, and the specific strength was three times higher than mild steel's (Sukmawan, Takagi, and Nakagaito 2016). The jute composite showed higher damping behavior, so jute composites can be practical options as automotive parts since they possessed low vibration and noise (Furtado et al. 2014). Due to its relatively cheap cost, good properties, environmental friendliness, and ease of manufacturing, natural fiber-reinforced PLA composite can be applied in a wide range of applications, including the aviation and automotive industries (Alkbir et al. 2016). In the automotive industry, coconut fiber is used to make furniture in cars, while cotton is used for noise cancellation. Sometimes, wood fiber is also used as furniture and in accessories in vehicles while flax, sisal, and hemp are applied in seatback linings and floor panels (Holbery and Houston 2006).

Notta-Cuvier et al. (2014) investigated the properties of PLA tailored for automotive applications with different specific additives, attempting to optimize PLA composites as a sustainable alternative to mineral-filled polypropylene. The researchers examined how PLA compounded with the plasticizer tributyl O-acetylcitrate (TBC), the impact modifier Biomax Strong®, and Cloisite® 25A influenced the tensile and impact behavior of the material. Automotive components require high rigidity, tensile strength, impact toughness, and significant elongation at break, similar to the properties of mineral-filled polypropylene. Additionally, materials must exhibit high nominal strain and load-bearing capabilities to effectively dissipate energy. Notta-Cuvier et al. highlighted the necessary tradeoff between high tensile rigidity and strength, and adequate ductility. Plasticizing PLA improves ductility but reduces tensile rigidity and strength. The quaternary composition of PLA + 10 wt% Biomax Strong® + 10 wt% TBC + Cloisite® 25A demonstrated promising ductility while retaining higher rigidity and strength compared to mineral-filled polypropylene. The content of Cloisite® 25A was crucial, with 1 wt% offering optimized strength, while 3 wt% improved ductility and rigidity. However, the Cloisite® content had minimal impact on tensile strength, suggesting a delicate balance between strengthening and enhancing ductility.

Öztürk (2020) studied automotive applications using renewable materials as an environmentally sustainable solution with a focus on weight reduction. Hemp, a non-woven natural fiber (NNF), and PLA fibers were used to create a continuous mat through the needle-punching method. This biobased mat was then hot-formed to enhance its mechanical properties. The fiber composition was a 50:50 ratio of NNF to PLA, with a final sheet density of 1300 g/m². After thermoforming, the biocomposite had a thickness of 2.2 mm, a flexural modulus of 2.1 GPa, tensile strength of 17 MPa, and 2.3% elongation at break. The mechanical performance was further tested under conditions of water immersion and at 40°C and 80°C. The Charpy impact strength was 21 kJ/m². The study's key finding is that replacing conventional reinforcement materials with renewable sources, such as hemp and PLA, offers a significant reduction in weight while maintaining acceptable mechanical strengths for automotive applications.

Motoc et al. (2021) explored the mechanical and thermal properties of additively manufactured PLA samples, driven by automotive industry demands, the influence of infill density on the material's performance. The researchers investigated how varying infill densities, ranging from 100% to 25%, affected the tensile strength, impact resistance and thermo-physical characteristics of PLA samples produced through 3D printing. The results showed a significant decrease in tensile strength and modulus by 49.9% and 42.0%, respectively, as infill density decreased from 100% to

25%. Impact strength also decreased by 56.0% at lower infill densities. The variation of Young's modulus with relative density was accurately modeled using a third-order polynomial function, consistent with the behavior of closed-cell thick edge rhombus cellular structures. Thermal properties were stable, with minimal changes in the coefficient of linear thermal expansion, showing a 6.34% decrease in specimens with the lowest infill density. The glass transition temperature was around 65 °C, and cold crystallization occurred near 80 °C, both unaffected by infill density, indicating stable performance across different process parameters.

12.5 PACKAGING

Numerous research studies have been conducted on replacing petroleum-based plastics with bio-based alternatives like cellulose, starch, chitosan, PHB, and PLA (Low et al. 2024; Bangar et al. 2024; Hawanis et al. 2024). These bioplastics offer significant environmental advantages, including biodegradability and reduced carbon footprints, making them promising candidates for sustainable plastic production (Jumaidin et al. 2021; Jumaidin, Adam, et al. 2019; Jumaidin, Saidi, et al. 2019). PLA is a suitable alternative as it promotes the usage of sustainable raw materials and decreases reliance on non-renewable fossil fuels (Jacob et al. 2020).

PLA packaging uses less plastic waste overall and contribute to environmental conservation efforts in the packaging industry. It degrades naturally, contributing to reduced environmental impact compared with conventional petrochemical-based plastics (Jamshidian, Tehrany, and Desobry 2013). PLA breaks down into smaller particles through biological reactions with microorganisms and bacteria, facilitating its degradation in various environments (Khosravi et al., 2020). This characteristic ensures that PLA packaging items do not end up as long-lasting waste in landfills or oceans, thus minimizing the negative impact on ecosystems and wildlife. By choosing PLA packaging solutions, businesses and consumers actively support the reduction of plastic waste and the transition toward more sustainable packaging practices (Yusoff et al. 2021). The US Food and Drug Administration (FDA) has designated PLA Generally Recognized as Safe (GRAS), indicating its suitability for direct food contact without posing health risks to consumers (Jamshidian, Tehrany, and Desobry 2013). Studies have shown that PLA does not migrate harmful substances into food products, providing a reliable and food-safe packaging option for various applications. The ease of processing PLA through methods like injection molding and extrusion offers practical advantages for manufacturers seeking sustainable packaging solutions. Given its versatility and compatibility with existing industrial techniques, PLA can be readily incorporated into various packaging applications without major modifications to production processes. This adaptability facilitates the broader adoption of PLA packaging across different sectors, enabling a smoother transition toward ecofriendly packaging alternatives and contributing to the collective effort in combating plastic waste and promoting environmental stewardship (Tawakkal et al. 2014).

Anderson and Shive noted that vivo conditions support the hydrolytic cleavage of PLA, with degradation typically taking between six months and two years to complete (Joseph et al. 2023). PLA begins to decompose via hydrolysis when it encounters moisture over a few months. This process occurs in two stages: Initially, the ester groups undergo nonenzymatic chain scission, lowering the molecular weight, a process that speeds up in the presence of acids or bases and is affected by temperature and moisture levels. In the second stage, the low molecular weight PLA exits the bulk polymer and is broken down by microorganisms, producing carbon dioxide, water, and humus (Feng et al. 2021).

Although crystalline PLA undergoes less hydrolysis than its amorphous configurations, semicrystalline PLA copolymers appear less prone to degradation than the contemporaneous semicrystalline counterpart as depicted in the aforementioned equations. Furthermore, the rate of degradation can be accelerated by the presence of acidic compounds that can also result in an increase in hydrolytic lipolysis, which is accentuated by the presence of basic molecules. By neutralizing carboxyl end groups, these basic molecules facilitate the base catalysis in the degradation process (DeStefano, Khan, and Tabada 2020).

PLA packaging is affected by moisture levels and temperature changes, which can affect the mechanical, barrier, and structural performance of the package over time. These parameters change the properties of PLA-based packaging, indicating the importance of environmental conditions for preserving the quality of PLA-based packaging. Decomposition of PLA packaging is directly influenced by factors like moisture absorption and temperature, which affect its protective potential for food items and the stability of the quality of the goods. With the increasing application of PLA packaging materials, it is very necessary to monitor and manage storage conditions to guarantee their functionality and durability (Swetha et al. 2023).

Despite these advantages, PLA faces the challenges of partial crystallization, lack of gas barrier and water vapor properties, and lower thermal resistance. Modifying PLA through copolymers, fillers, and plasticizers can help overcome these limitation (Vieira et al. 2011; Song et al. 2013) Adding nano-reinforcements to PLA composites improves mechanical, thermal, and barrier properties (Gong et al. 2014; Song et al. 2013), with ongoing efforts focused on optimizing filler dispersion and interaction with the PLA matrix for improved performance (Jacob et al. 2020).

There are many ways to enhance PLA properties for extending its applications in the packaging sector. Currently, melt blending has significant interest among researchers due to easy, cost-effective, and readily available processing technologies at the industrial scale (Maroufkhani et al. 2017), and PLA is recommended to be used with PHB because of its similar melting temperature and high crystallinity (Arrieta et al. 2017). PHB functions as a nucleating agent in PLA, which enhances the overall mechanical resistance and water barrier properties. PLA-PHB mixtures are good candidates for the controlled release of active compounds in the progress of active packaging systems. PLA-PHB blends demonstrated themselves as highly promising candidates for substituting petroleum-based polymers currently used for food packaging (Nazrin et al. 2020a).

Nazrin et al. (2021b) focused on the water barrier and mechanical properties of sugar palm crystalline nanocellulose-reinforced thermoplastic sugar palm starch (TPS)/PLA blend bionanocomposites, offering an ecofriendly alternative to conventional plastics. The highest solubility was observed in PLA20TPS80 (96.34%) and PLA40TPS60 (77.66%) due to higher plasticizer concentrations, allowing water molecules to penetrate more easily. Interestingly, increasing the PLA content did not always enhance water vapor permeability. Dynamic mechanical analysis showed a notable increase in storage modulus (E′) for PLA60TPS40 (53.2%), while PLA70TPS30 (10%) and PLA80TPS20 (0.6%) displayed minimal improvements. Impact strength significantly improved at PLA40TPS60 (33.13%), with only minor gains between 12% and 20% for higher PLA content. Overall, PLA60TPS40 exhibited promising functional properties, making it a suitable candidate for food packaging applications.

Nanocellulose exhibits remarkable mechanical properties that contribute to its effectiveness in reinforcing PLA. Nanocellulose is an exceptionally versatile material, renowned for its unique combination of rigidity, durability, and resilience. When incorporated as a filler, it greatly enhances the mechanical properties of composite materials, improving strength, flexibility, and structural integrity. This makes nanocellulose a highly valuable additive in a range of applications, from bioplastics and packaging to medical and industrial products, where both sustainability and performance are critical (Norfarhana, Ilyas, and Ngadi 2022; Norfarhana, Ilyas, et al. 2024; Mubarak et al. 2024; Othman et al. 2024; Ilyas et al. 2024; Norfarhana, Khoo, et al. 2024; Al-Fakih et al. 2024; Yusuf et al. 2024). Adding nanocellulose improves barrier features and flame retardancy over those of pure polymer materials (Bangar et al. 2022; Alias et al. 2021).

Cellulose nanocrystals possess tensile strength and elastic modulus values comparable with or even higher than those of Kevlar, increasing the strength and durability of PLA nanocomposites (Mishra, Sabu, and Tiwari 2018). Nanocellulose's large surface area-to-volume ratio and tunable surface properties make it ideal for diverse applications. Its specific surface area influences adsorption capacity, reactivity, and interactions with other materials (Dhali et al. 2021). The high surface area and porosity enhance filtration, separation, and adsorption processes (Arrieta et al. 2016). Adjusting the morphology and dispersion of nanocellulose in polymer matrices improves barrier properties against gas and water vapor, making it valuable in sustainable packaging solutions (Ferrer, Pal, and Hubbe 2017).

The renewability and sustainability of nanocellulose compared with other reinforcing materials make it an environmentally friendly option due to its abundant availability, nontoxicity, and biodegradability Nanocellulose offers an ecofriendly alternative without compromising mechanical strength and elasticity (Thomas et al. 2018). It can be chemically modified on its surface to enhance specific functionalities for targeted applications, using techniques such as carboxymethylation, amidation, esterification, etherylation, silylation, sulfonation, and phosphorylation (Thomas et al. 2018).

The primary benefits of using PLA/nanocellulose nanocomposites in food packaging include improved barrier properties, biodegradability, and processability. PLA-based composites with cellulose fibers show enhanced barrier properties against water vapor transportation, making them suitable for flexible food packaging applications (Popa et al. 2017). The biodegradable nature of PLA combined with the barrier-enhancing effect of cellulose fibers offers a sustainable packaging solution for the food industry. The synergistic effect of PLA reinforced with nanocellulose offers significant advantages in terms of enhanced barrier properties derived from the dispersion of nanoparticles, making them a promising option for food packaging applications.

12.6 CONSTRUCTION

The construction sector also offers broad potential usages of natural fiber composite. Researchers are evaluating building materials based on renewable resources such as natural fibers and their reinforcement in cement-based materials (Faruk et al. 2012). Structural beams and panels using biobased composite materials were developed, manufactured and tested, particularly on plant oil-based resins along with natural fiber composite, and composite building materials made from straw bales in the United States (Saravana Bavan and Mohan Kumar 2010).

Morales et al. (2017) focused on developing a biodegradable composite for structural applications by combining bamboo reinforcement with a suitable PLA biopolymer, utilizing an innovative mechanical extraction process. The resulting bamboo–PLA composite demonstrated impressive mechanical properties, with a specific Young's modulus more than double that of traditional e-glass/epoxy composites, while maintaining a low density of 0.91 g/cm^3. Although the composite's specific in-plane shear strength was lower than that of e-glass/epoxy, improvements in interfacial bonding through alternative resins could enhance performance. The composite exhibited good durability under environmental conditions, showing no significant degradation from high temperatures and humidity during testing. While its aviation applications may be limited, the composite shows promise for use in industries like housing and energy, particularly in components where cost-effectiveness and ecofriendliness are prioritized.

In another study, Oluwabunmi et al. (2020) investigated compostable, fully biobased foams using PLA and microcellulose fibrils (MCF) with the aim of addressing ecological and health concerns in the building industry. The study authors employed supercritical CO_2 (sc-CO_2) for physical foaming, ensuring that the resulting foams contained no volatile organic compounds. Results indicated that the presence of MCF significantly influenced the foams' properties, leading to initial improvements in mechanical performance and thermal insulation, followed by a decline as MCF concentration increased. The foams demonstrated enhanced compostability, achieving a maximum mineralization of 95% for PLA with 3 wt.% MCF, making them a sustainable alternative for building applications. Energy simulations showed that these bio-resourced foams increased heating and cooling needs by no more than 12% over conventional polyurethane insulation yet offered substantial environmental benefits due to their zero volatile organic compound manufacturing process. Ultimately, the authors highlighted that incorporating cellulose into PLA foams promotes biodegradation, ensuring compliance with international compostability standards at the end of their life cycle.

Bahar et al. (2023) conducted a study on the thermal and mechanical behavior of wood-PLA composites produced through additive manufacturing for use as building insulation. The research utilized a PLA blend containing 30% wood particles and explored various processing conditions, such as printing temperature and infill rate, to assess their impact on the composites' properties. Results indicated a minor reduction in tensile performance during the printing process, with the printing temperature having little effect on stiffness and other mechanical characteristics. The thermal properties were found to be significantly influenced by the infill rate, showing that while the 3D-printed wood-PLA material demonstrated favorable thermal attributes, its insulation performance remained 38% to 57% lower than conventional materials like glass wool and synthetic foams.

Bahar et al. (2023) highlighted that wood-PLA filaments exhibit considerable structural complexity due to inherent porosity, resulting in minimal mechanical performance loss upon printing, particularly in terms of stiffness. Although the thermal conductivity improved with less infill, achieving 50 mW/(mK) at 10% infill, the overall thermal performance was still unsatisfactory compared with traditional insulating materials. Consequently, while wood-PLA composites show potential as both structural and insulating materials, further reductions in infill rates are necessary for improved thermal efficiency, albeit with potential risks to mechanical stability.

12.7 FLAME RETARDANCE

Natural fiber-reinforced PLA composite has a huge potential usage to be used in the industry, but its flammability is a main concern; however, the flame-retardant test on natural fiber-reinforced PLA exhibited promising results. The altered flax fiber-reinforced PLA had a limiting oxygen index (LOI) of 26.1 with a UL-94 V-2 rating and the release of flammable gaseous during thermal decomposition of the PLA composite was suppressed after the addition of flame retardant, which enabled the modified composite to display high resistance to minor flammable sources of ignition (Zhang et al. 2018). The water absorption rate for all composites was observed to increase gently over the first 24 hours before levelling off, and the rate of water absorption was increased in all composites as the fiber content was increased. The absorption rate was found to decrease with successive alkali treatment of the fibers (Barkoula et al. 2008; Faiz Norrrahim et al. 2021).

Nazrin et al. focused on the flammability and physical stability of sugar palm crystalline nanocellulose (SPCNC)-reinforced TPS/PLA blend bionanocomposites, aimed at developing flame-retardant materials with enhanced biodegradability (Nazrin et al. 2021a; Asyraf et al. 2021). SPCNC was dispersed using sonication prior to starch gelatinization, and the resulting blend was cast into petri dishes before being melt-blended in five different PLA/TPS ratios using a Brabender mixer and then compression molded. Physical stability was assessed through soil degradation tests (over four months) and water uptake tests (over four weeks), using seawater, river water, and sewer water. The authors calculated the diffusion curves and coefficients of diffusion based on Fickian law.

Flammability was measured using LOI and flammability rate tests according to ASTM D635 and ASTM D2863 standards. Results showed that the PLA60TPS40 formulation (40% TPS) achieved a significant reduction (46%–69%) in maximum water uptake across all mediums. However, soil degradation only slightly increased (7.92%) for PLA70TPS30 (30% TPS), suggesting the effective reinforcement of SPCNC, which improved the dispersion of TPS within the PLA matrix; on the other hand, both PLA40TPS60 and PLA60TPS40 exhibited flammability rates and LOI. This study highlights the tradeoffs between biodegradability and flame resistance in developing ecofriendly bionanocomposites (Nazrin et al., 2021a).

12.8 3D AND 4D PRINTING PLA BIOCOMPOSITES

PLA is the main nature-based raw material for 3D printing; it is a wholly biodegradable thermoplastic polymer synthesized from renewable raw materials (Azlin et al. 2020; Nazrin et al. 2020b; Ganeshkumar et al. 2022). Ease of processing is a key advantage of PLA; it is one of the easiest materials to print, despite tending to slightly shrink after the 3D printing process (Chia and Wu 2014). The main advantage of PLA over acrylonitrile butadiene styrene (ABS) material is it does not need a heated platform during printing; hence, it can be printed at low temperatures between 190 and 230 °C (Mazzanti, Malagutti, and Mollica 2019). PLA also does not require complex post-processing since it can be treated with acetone or sanded when necessary, and the supports are usually very easily removable. Many manufacturers fabricated PLA filament to be used with 3D and 4D printing such as Prusament, Amolen, Polymaker PolyMax, Proto Pasta, MatterHackers, Fillamentum, Colorfabb, Sunlu, and Paramount 3D. One of the most prominent PLA filament manufacturers is an Austrian company called WeforYou, and the German company Evonik has developed PLA for the medical sector. The American company NaturaWorks is another large producer of biopolymers, and Corbion, based in the Netherlands, has centralized the development of high-performance resins with PLA.

Since PLA is suitable for interaction with foods, it is used as a replacement for petroleum-based plastics for packaging applications, especially in the food industry (Jamshidian et al. 2010). PLA can be utilized in 3D printing using FDM, which manufactures parts through the extrusion of thermoplastic filaments, and PLA is one of the commonly used materials for this technology. Composites are exceptionally advantageous in manufacturing lightweight with strong mechanical properties parts. The fibers contribute mechanical strength to parts without compromising weight, which is the primary factor of recognizing composites as fiber reinforced materials (Ilyas and Sapuan 2020a, 2020b).

There are two classes of reinforcements, short fibers (Figure 12.6) and continuous fibers (Figure 12.7). With short fibers, chopped fibers consisting of segments less than a millimeter in length are mixed with PLA, ABS, or nylon into 3D printing plastics to raise the stiffness and to a lesser extent the strength of components. Continuous addition of the fibers to the thermoplastics produced stronger parts. Commonly, the most utilized fiber in 3D printing is carbon fiber, followed by glass, Kevlar, and natural fibers. There are also various hybrid materials that combine plastics with powders to give new color, finish, or extra material properties. Figure 12.8 shows the large-scale printing process with poplar/PLA composites. These materials are typically fabricated from 70% PLA biopolymer and another 30% hybrid natural material. For instance, wood-based filaments, such as poplar, bamboo, jute, cork, wood dust, and kenaf are combined with PLA to provide a more organic texture to the hybrid filament, which in line with UN Sustainable Development Goals (I) to encourage inclusive and sustainable industrialization and foster innovation and (II) to safeguard sustainable consumption and production patterns.

Nasir et al. (2022) investigated the thermal and rheological properties of biocomposite filaments made from sugar palm fiber (SPF) and PLA for FDM 3D printing. They focused on the effects of chemical treatments, specifically alkaline and silane,

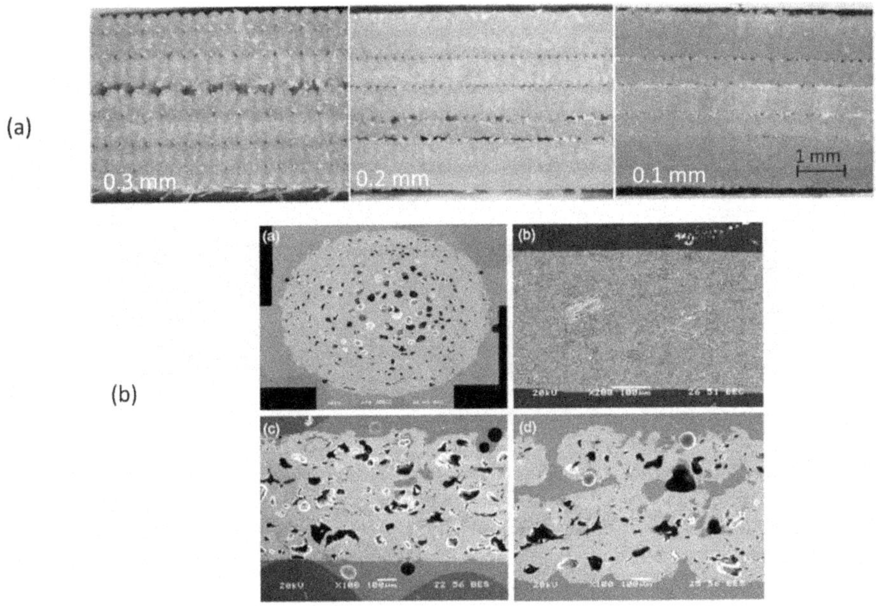

FIGURE 12.6 Short natural fiber-reinforced PLA: (a) effect of layer height on wood, 0.3, 0.2, and 0.1 mm. Higher porosity content was observed with higher layer height; b) scanning electron micrograph of printed wood/PLA biocomposites with raw filament cross-section.

Source: Adapted from Le Duigou et al. (2016) and Ayrilmis et al. (2019) with copyright permission from Elsevier.

to enhance the adhesion between SPF and PLA while removing impurities, thus improving the composite's suitability for FDM printing. The thermal properties of the treated composites were analyzed using thermogravimetric analysis (TGA) and differential scanning calorimetry (DSC), while the melt flow index and rheological tests were performed to assess the material's flow behavior during the extrusion process. TGA results showed improved thermal stability for SPF/PLA composites treated with NaOH and silane, with a final residue of 0.4% at 789.5 °C, indicating strong thermal resistance. The DSC analysis revealed that the chemical treatments did not significantly alter the melting temperature, with all samples displaying a melting point around 155 °C; however, untreated SPF/PLA composites exhibited the highest degradation temperature of 383.2 °C. Furthermore, the NaOH and silane-treated SPF/PLA composite had the highest melt flow rate at 17.6 g/min, suggesting better printability for FDM applications. Although rheological tests showed no significant impact of chemical treatment on the material's shear stress and viscosity, the study concludes that NaOH and silane treatments enhance the thermal degradation and flow properties of SPF/PLA composites, making them more suitable for FDM 3D printing. This development of biodegradable filaments has promising applications across various industries, including biomedicine, automotive, aviation,

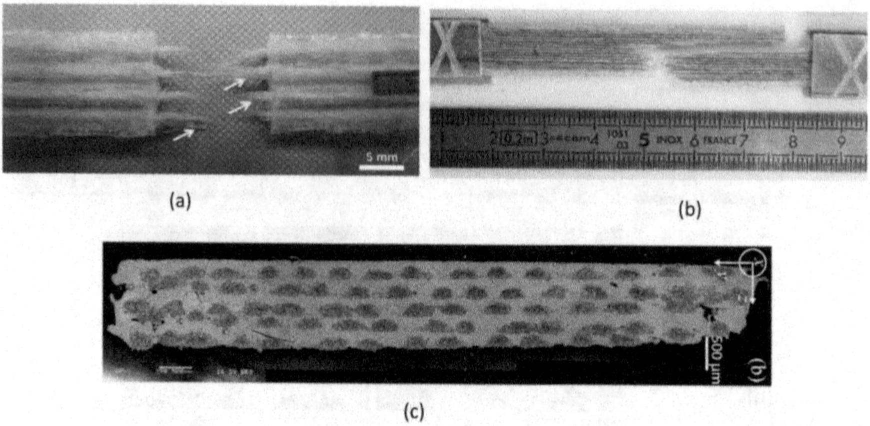

FIGURE 12.7 Continuous natural fiber: (a) fractured continuous unidirectional jute fibers/PLA sample printed with in-nozzle reinforced with various fiber debonding; (b) fractured continuous unidirectional flax fiber/PLA sample printed with pre-impregnated filaments with a transverse crack followed by propagation along the tensile axis; (c) SEM micrograph of a cross-section of flax/PLA composite.

Source: Adapted with permission from Le Duigou et al. (2019) and Matsuzaki et al. (2016).

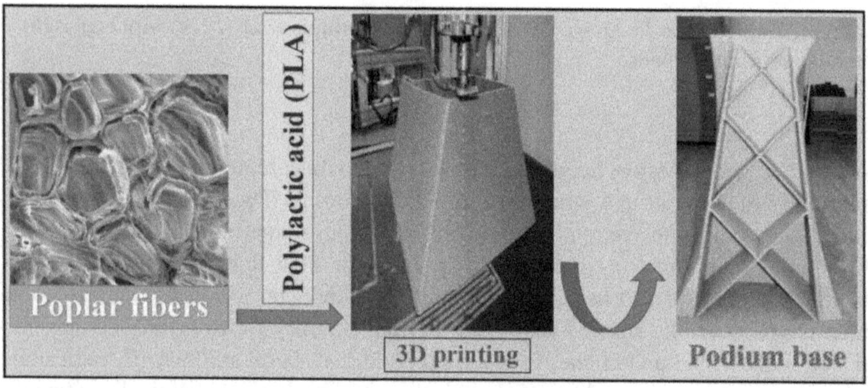

FIGURE 12.8 Large-scale printing with poplar/PLA composites.

Source: Adapted from Zhao et al. (2019).

and household appliances. Further research is recommended to explore the relationship between thermal, rheological, and mechanical properties to optimize filament performance.

Four-dimensional (4D) printing is a relatively recent trend to develop 3D-printed structures that can change their shape or properties over time (Champeau et al. 2020a; Spiegel et al. 2020; Zolfagharian, Kaynak, and Kouzani 2020). The difference is that

4D-printed objects can transform themselves over time, while 3D-printed objects maintain fixed shape like any plastic or metal parts. The fourth dimension of 4D is the transformation over time, where 4D printing technology offers an output of smart structures by using new manufacturing techniques of 3D printing, advanced materials, and customized design. 4D printed objects need a stimulus to begin the deformation phase; the trigger can be water, heat, light, or magnetic field (Quanjin et al. 2020; Ma et al. 2020; Subash and Kandasubramanian 2020; Zafar and Zhao 2020). The concept of 4D printing is a smart structure that consists of rigid materials connected with expandable elements, or it can also be a whole structure made from expandable materials depending on what materials properties are needed and what are the applications (Chu et al. 2020; Zolfagharian et al. 2020a).

The expandable elements can change their shape when exposed to certain stimuli, and this causes the hard parts to move or rotate, resulting in the whole structure transformation. The expandable element of smart materials can be hydrogel and elements with shape memory. A hydrogel is a polymeric material that is capable of absorbing a large amount of water (Yang et al. 2020; Nautiyal, Sahu, and Gupta 2020). It can be programmed to expand or shrink when there are changes in the external environmental conditions (Zheng et al. 2020; Champeau et al. 2020b). Hydrogels are biocompatible and easily modified. Elements with shape memory can be considered smart materials due to their capability to return to their original shape from a deformed shape when stimuli are applied. Figure 12.9 shows the moisture-induced deployable structure based on curved-line folding inspired by Aldrovanda (Le Duigou et al. 2017a).

Hygromorph biocomposite (HBC) actuators used the transport properties of plant fibers to generate an out-of-plane displacement when a moisture gradient was present. Le Duigou et al. 2017b) developed a theoretical actuating response (curvature) formulation of maleic anhydride polypropylene (MAPP)/plant fibers (flax, jute, kenaf, and coir) based on bimetallic actuators (Figure 12.9). The result showed that the actuation was directly related to the fibers biochemical composition and microstructure, in which flax and jute fibers were observed to be the best candidates to be used in HBCs.

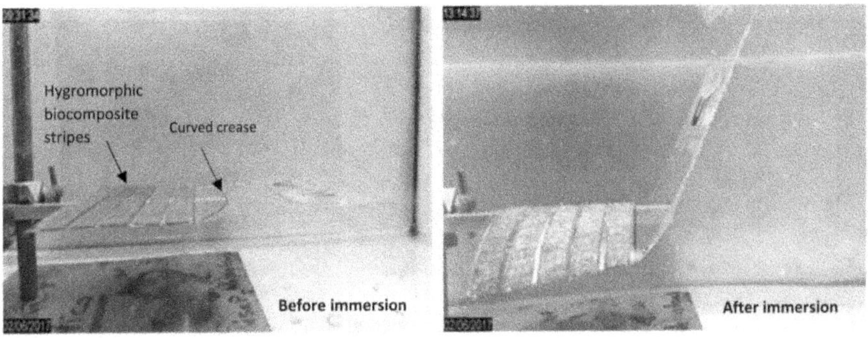

FIGURE 12.9 Moisture-induced deployable structure based on curved-line folding inspired by Aldrovanda.

Source: Adapted from Le Duigou et al. (2017b) with copyright permission from Elsevier.

Poppinga et al. (2020a) researched plant movements as concept generators for the development of biomimetic compliant mechanisms using Lilium 'Casa Blanca' or lily flower model. The structure of the hygroscopic made up of wood-based hygromorph composite was translated into a 4D-printed mechanism or printed under warm and dry conditions at a temperature of 21 °C and relative humidity of 18%. The composite then underwent deformation when submerged in water at the temperature of 19 °C. The submerged composite resulted in edge growth-driven actuation as identified from the petals of the lily flower (Figure 12.10). Their novel biomimetic compliant mechanisms highlighted the feasibility of modern printing techniques for designing and developing versatile tailored motion responses for technical applications.

Alief et al. (2019) modeled the shape memory properties of 4D-printed PLA for application of disk spacer in minimally invasive spinal fusion. They observed that specific PLA structure possessed thermal shape memory behavior that can be thermomechanically trained into temporary shape and returned to its permanent shape when heated. Based on the simulation result, nonuniformed hollow spaces pattern displayed favorable results that could be a reference for future researchers seeking to design suitable patterns for 4D PLA models (Figure 12.11). The possible advanced applications of 4D printing are medical devices for stents placed into blood vessels, drug capsules that release medicine, home appliances for control and that adjust according to humidity and heat, footwear and clothes, implants for humans and animals made from biocompatible materials, soft robots that can be activated without reliance on an electric device, smart valves, and sensors for infrastructure lines (Martins et al. 2020; Zolfagharian et al. 2020b; Miao et al. 2020; Fer and Becker 2020; Durga Prasad Reddy and Sharma 2020; L. Wang, Leng, and Du 2020).

12.9 CHALLENGES AND SOLUTIONS IN APPLICATION

PLA has gained widespread attention as a biodegradable and renewable polymer, offering a promising alternative to traditional petroleum-based plastics. The increasing global demand for sustainable materials has driven the growth of bioplastics,

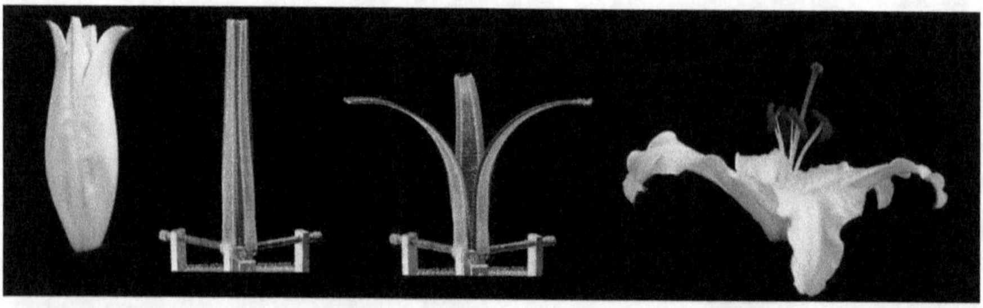

FIGURE 12.10 Edge growth causing a strain gradient in Lilium 'Casa Blanca' drives the flower-opening mechanism, a principle that is translated into a 4D printed mechanism with a wood-based hygromorph composite edge.

Source: Adapted from Poppinga et al. (2020b) with copyright permission from Oxford University Press.

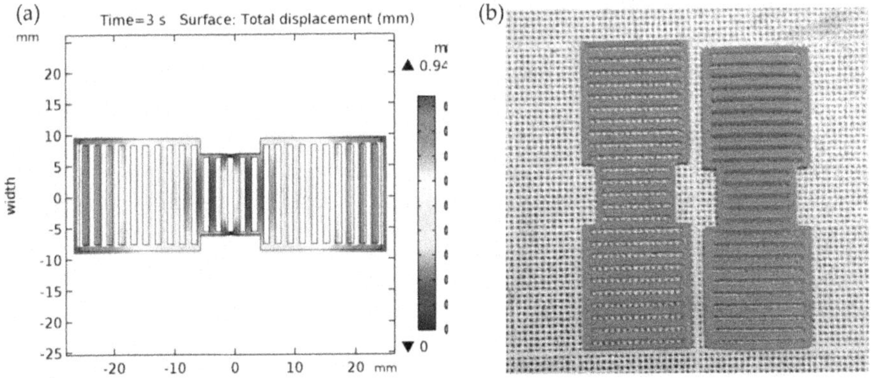

FIGURE 12.11 (a) PLA model deformation after 3 seconds. Uniformed vertical hollow spaces of 1 mm width; (b) side-by-side comparison before and after adding stimulus. Left: a PLA model before adding stimulus and right: after adding the stimulus.

Source: Adapted from Alief et al. (2019) with copyright permission from AIP Conference Proceedings.

with PLA positioned as a leader in this space due to its biodegradability and versatile applications. Despite these advantages, several challenges still hinder its broader adoption, particularly in achieving durability and improving its mechanical properties for high-performance applications. This section addresses these challenges and explores potential solutions that have been investigated to overcome them.

12.9.1 Market Growth and Production Challenges

The global demand for bioplastics, particularly PLA, is projected to grow by 36% over the next five years, as shown in recent market analyses. To meet this growing demand, new production facilities are being established worldwide. For example, companies like Total-Corbion and COFCO are setting up large-scale PLA plants, with others like BBCA Galactic and Hisun following suit. These plants are designed to produce tens of thousands of tons of PLA annually, reflecting the increasing need for sustainable materials in various industries. However, the production process for PLA still faces challenges, such as high energy consumption and resource dependency, which need further optimization to ensure economic viability.

12.9.2 Mechanical and Thermal Limitations

One of the main technical challenges associated with PLA is its relatively poor mechanical strength, heat stability, and water barrier properties compared to conventional thermoplastics. These limitations restrict its use in high-performance applications that require durable materials. To address these issues, several modification strategies have been explored. For example, the use of chain extenders such as epoxides, anhydrides, and isocyanates can improve PLA's crystallinity and mechanical properties. However, the presence of unreacted nonbiodegradable agents in the final product can be problematic, particularly for applications requiring full biodegradability. Additionally, researchers have explored the blending of PLA with other polymers, such as PGA, to enhance its properties. PGA offers superior heat distortion temperature, gas barrier properties, and mechanical strength, making it an ideal supplement to PLA. While blending PLA with PGA has shown promise, particularly in biomedical applications, the high cost of PGA has limited its widespread adoption.

12.9.3 Enhancing Durability through Additives and Processing

To achieve durable applications, PLA requires enhancements in strength, toughness, and thermal stability. Traditional methods to improve toughness, such as incorporating impact modifiers or nucleating agents, often come at the cost of reduced stiffness or strength. Therefore, an optimized strategy is needed to balance these properties. Compatibilizers like Janus materials or hydrophilic–lipophilic surfactant blends have shown potential in reinforcing the mechanical properties of PLA-based blends. These materials can create strong bonds between different phases of a polymer blend, improving both interfacial adhesion and phase structure. Processing methods also play a crucial role in enhancing PLA's durability. For instance, low-temperature and high-pressure processing of polymer blends can create materials with superior mechanical properties. However, further research is required to optimize these techniques for commercial-scale production.

12.9.4 Sustainability and Economic Viability

While PLA has a favorable environmental profile due to its renewable origin, the cost of production remains a barrier to its large-scale adoption. To address this, researchers are investigating low-cost substrates and more efficient microorganisms for lactic acid production, which is the precursor to PLA. Additionally, novel copolymers and nanomaterials such as titanium dioxide (TiO_2), magnesium oxide (MgO), and zinc oxide (ZnO)enhance the strength and stability of PLA, making it more competitive with traditional plastics in various industries. Moreover, regulatory measures and increased public awareness of environmental issues have contributed to the growing acceptance of bioplastics. Governments in several countries have introduced policies to promote the use of biodegradable plastics, while simultaneously restricting the use of conventional plastics. These efforts are essential for fostering a circular economy and encouraging the widespread adoption of PLA in applications ranging from packaging to automotive and medical products.

12.9.5 Applications and Future Prospects

The development of durable PLA-based composites opens opportunities for high-performance applications in industries such as automotive, biomedical, electrical appliances, textiles, and housing. PLA's compatibility with various fibers, both natural and synthetic, further enhances its applicability in these sectors. Advanced production techniques like electrospinning, 3D printing, and coating are being integrated into PLA manufacturing, paving the way for innovative applications. Future research will continue to focus on improving the physical, mechanical, and thermal properties of PLA to expand its use in durable applications. By addressing the current challenges and optimizing production processes, PLA has the potential to become a key material in the transition toward sustainable and ecofriendly alternatives in the global market.

12.10 CONCLUSIONS

PLA is a versatile biodegradable polymer with the potential to replace conventional petroleum-based plastics, driven by growing concerns about environmental sustainability and plastic waste. PLA's biodegradability, biocompatibility, and mechanical properties make it a promising alternative, but challenges remain, particularly in enhancing its durability and thermal stability for high-performance applications.

One of the key limitations of PLA is its relatively poor mechanical strength and heat resistance compared to traditional plastics. This limits its use in demanding sectors like automotive and aviation. Researchers have explored solutions such as the incorporation of additives, including chain extenders and nucleating agents, to improve PLA's crystallinity, toughness, and overall mechanical properties. However, the use of nonbiodegradable additives poses a risk to the environmental benefits of PLA, requiring careful management of these modifications.

Improving processing techniques is another area of focus. Low-temperature, high-pressure processing and advanced methods such as electrospinning and 3D printing have shown potential to enhance the mechanical properties of PLA, expanding its application range. Compatibilizers like Janus materials and surfactant blends can improve interfacial adhesion in polymer blends, further strengthening PLA's mechanical properties for durable applications. The growing global demand for bioplastics, particularly PLA, is driving significant investments in production capacity. With a projected 36% growth in the bioplastics market over the next five years, companies across the United States, Europe, and China are scaling up PLA production. This increased capacity, coupled with ongoing research, will enable the development of cost-effective, durable PLA-based products for various industries.

Looking to the future, PLA's success depends on balancing performance, sustainability, and cost. Ongoing research into advanced materials such as nanocomposites and innovative copolymers will help address the challenges of enhancing PLA's properties while maintaining its biodegradability. Additionally, efforts to reduce production costs through more efficient processes, including the use of low-cost feedstocks and microbial fermentation techniques, are critical to making PLA commercially viable on a large scale. In conclusion, PLA is well positioned to play a leading role in the transition to sustainable plastics. While challenges remain in enhancing its mechanical properties and reducing production costs, advancements in research and technology are likely to overcome these barriers. The increasing emphasis on reducing plastic waste and adopting biodegradable alternatives makes PLA a key material in the global effort to create ecofriendly products for industries ranging from packaging to healthcare.

REFERENCES

Al-Fakih, Ghassan O.A., R.A. Ilyas, A. Atiqah, M.S.N. Atikah, R. Saidur, Alain Dufresne, Mohd Shahneel Saharudin, Hairul Abral, and S.M. Sapuan. 2024. "Advanced Functional Materials Based on Nanocellulose/Mxene: A Review." *International Journal of Biological Macromolecules* 278 (October): 135207. https://doi.org/10.1016/j.ijbiomac.2024.135207.

Alias, Aisyah Humaira, Mohd Nurazzi Norizan, Fatimah Athiyah Sabaruddin, Muhammad Rizal Muhammad Asyraf, Mohd Nor Faiz Norrahim, Ahmad Rushdan Ilyas, Anton M. Kuzmin, et al. 2021. "Hybridization of MMT/Lignocellulosic Fiber Reinforced Polymer Nanocomposites for Structural Applications: A Review." *Coatings* 11 (11): 1355. https://doi.org/10.3390/coatings11111355.

Alief, Nindya Aprilia, Sugeng Supriadi, and Yudan Whulanza. 2019. "Modelling the Shape Memory Properties of 4D Printed Polylactic Acid (PLA) for Application of Disk Spacer in Minimally Invasive Spinal Fusion." *AIP Conference Proceedings* 2092: 020005. https://doi.org/10.1063/1.5096673.

Alkbir, M.F.M., S.M. Sapuan, A.A. Nuraini, and M.R. Ishak. 2016. "Fibre Properties and Crashworthiness Parameters of Natural Fibre-Reinforced Composite Structure: A Literature Review." *Composite Structures*. https://doi.org/10.1016/j.compstruct.2016.01.098.

Arrieta, Marina Patricia, E. Fortunati, N. Burgos, M.A. Peltzer, J. López, and L. Peponi. 2016. *Nanocellulose-Based Polymeric Blends for Food Packaging Applications. Multifunctional Polymeric Nanocomposites Based on Cellulosic Reinforcements*. Elsevier Inc. https://doi.org/10.1016/B978-0-323-44248-0.00007-9.

Arrieta, Marina Patricia, María Dolores Samper, Miguel Aldas, and Juan López. 2017. "On the Use of PLA-PHB Blends for Sustainable Food Packaging Applications." *Materials* 10 (9): 1008.

Asyraf, Muhammad Rizal Muhammad, M.R. Ishak, M.N.F. Norrahim, N.M. Nurazzi, S.S. Shazleen, R.A. Ilyas, M. Rafidah, and M.R. Razman. 2021. "Recent Advances of Thermal Properties of Sugar Palm Lignocellulosic Fibre Reinforced Polymer Composites." *International Journal of Biological Macromolecules* 193 (December): 1587–99. https://doi.org/10.1016/j.ijbiomac.2021.10.221.

Asyraf, Muhammad Rizal Muhammad, M.R. Ishak, S.M. Sapuan, and N. Yidris. 2019. "Conceptual Design of Creep Testing Rig for Full-Scale Cross Arm Using TRIZ-Morphological Chart-Analytic Network Process Technique." *Journal of Materials Research and Technology* 8 (6): 5647–58. https://doi.org/10.1016/j.jmrt.2019.09.033.

Asyraf, Muhammad Rizal Muhammad, M.R. Ishak, S.M. Sapuan, and N. Yidris. 2020. "Conceptual Design of Multi-Operation Outdoor Flexural Creep Test Rig Using Hybrid Concurrent Engineering Approach." *Journal of Materials Research and Technology* 9 (2): 2357–68. https://doi.org/10.1016/j.jmrt.2019.12.067.

Asyraf, Muhammad Rizal Muhammad, M.R. Ishak, S.M. Sapuan, N. Yidris, R.A. Ilyas, M. Rafidah, and M.R. Razman. 2020a. "Evaluation of Design and Simulation of Creep Test Rig for Full-Scale Cross Arm Structure." *Advances in Civil Engineering*. https://doi.org/10.1155/2019/6980918.

Asyraf, Muhammad Rizal Muhammad, M.R. Ishak, S.M. Sapuan, N. Yidris, R.A. Ilyas, M. Rafidah, and M.R. Razman. 2020b. "Potential Application of Green Composites for Cross Arm Component in Transmission Tower: A Brief Review." Edited by Yulin Deng. *International Journal of Polymer Science* 2020 (December): 1–15. https://doi.org/10.1155/2020/8878300.

Asyraf, Muhammad Rizal Muhammad, Agusril Syamsir, Nazirul Mubin Zahari, Abu Bakar Mohd Supian, Mohamad Ridzwan Ishak, Salit Mohd Sapuan, Shubham Sharma, et al. 2022. "Product Development of Natural Fibre-Composites for Various Applications: Design for Sustainability." *Polymers* 14 (5): 920. https://doi.org/10.3390/polym14050920.

Athanasiou, Kyriacos A., Gabriele G. Niederauer, and C. Mauli Agrawal. 1996. "Sterilization, Toxicity, Biocompatibility and Clinical Applications of Polylactic Acid/Polyglycolic Acid Copolymers." *Biomaterials* 17 (2): 93–102. https://doi.org/10.1016/0142-9612(96)85754-1.

Ayrilmis, Nadir, Mirko Kariz, Jin Heon Kwon, and Manja Kitek Kuzman. 2019. "Effect of Printing Layer Thickness on Water Absorption and Mechanical Properties of 3D-Printed Wood/PLA Composite Materials." *The International Journal of Advanced Manufacturing Technology* 102 (5–8): 2195–2200. https://doi.org/10.1007/s00170-019-03299-9.

Azammi, A.M. Noor, S.M. Sapuan, M.R. Ishak, and Mohamed T.H. Sultan. 2018. "Conceptual Design of Automobile Engine Rubber Mounting Composite Using TRIZ-Morphological Chart-Analytic Network Process Technique." *Defence Technology* (May). https://doi.org/10.1016/j.dt.2018.05.009.

Azlin, M.N.M., S.M. Sapuan, E.S. Zainudin, M.Y.M. Zuhri, and R.A. Ilyas. 2020. "Natural Polylactic Acid-Based Fiber Composites: A Review." In *Advanced Processing, Properties, and Applications of Starch and Other Bio-Based Polymers*, edited by Faris M. Al-Oqla and S.M. Sapuan, 1st ed., 21–34. Oxford: Elsevier. https://doi.org/10.1016/B978-0-12-819661-8.00003-2.

Bae, Ji-Young, Jong-Eun Won, Jong-Sub Park, Hae-Hyoung Lee, and Hae-Won Kim. 2011. "Improvement of Surface Bioactivity of Poly(Lactic Acid) Biopolymer by Sandblasting with Hydroxyapatite Bioceramic." *Materials Letters* 65 (19–20): 2951–55. https://doi.org/10.1016/j.matlet.2011.06.023.

Bahar, Anis, Ameur El Amine Hamami, Ferhat Benmahiddine, Sofiane Belhabib, Rafik Belarbi, and Sofiane Guessasma. 2023. "The Thermal and Mechanical Behaviour of Wood-PLA Composites Processed by Additive Manufacturing for Building Insulation." *Polymers* 15 (14): 3056. https://doi.org/10.3390/polym15143056.

Bangar, Sneh Punia, M.M. Harussani, R.A. Ilyas, Adeleke Omodunbi Ashogbon, Arashdeep Singh, Monica Trif, and Seid Mahdi Jafari. 2022. "Surface Modifications of Cellulose Nanocrystals: Processes, Properties, and Applications." *Food Hydrocolloids* 130 (September): 107689. https://doi.org/10.1016/j.foodhyd.2022.107689.

Bangar, Sneh Punia, William Scott Whiteside, Amreen Chowdhury, R.A. Ilyas, and Anil Kumar Siroha. 2024. "Recent Advancements in Functionality, Properties, and Applications of Starch Modification with Stearic Acid: A Review." *International Journal of Biological Macromolecules* 280 (November): 135782. https://doi.org/10.1016/j.ijbiomac.2024.135782.

Barkoula, N.M., B. Alcock, N.O. Cabrera, and T. Peijs. 2008. "Flame-Retardancy Properties of Intumescent Ammonium Poly(Phosphate) and Mineral Filler Magnesium Hydroxide in Combination with Graphene." *Polymers and Polymer Composites* 16 (2): 101–13. https://doi.org/10.1002/pc.

Bouzouita, Amani, Delphine Notta-Cuvier, Jean-Marie Raquez, Franck Lauro, and Philippe Dubois. 2017. "Poly(Lactic Acid)-Based Materials for Automotive Applications." *Industrial Applications of Poly (Lactic Acid)* 177–219. https://doi.org/10.1007/12_2017_10.

Champeau, Mathilde, Daniel Alves Heinze, Thiago Nunes Viana, Edcarlos Rodrigues de Souza, Anne Cristine Chinellato, and Silvia Titotto. 2020a. "4D Printing of Hydrogels: A Review." *Advanced Functional Materials* 30 (31): 1910606. https://doi.org/10.1002/adfm.201910606.

Champeau, Mathilde, Daniel Alves Heinze, Thiago Nunes Viana, Edcarlos Rodrigues de Souza, Anne Cristine Chinellato, and Silvia Titotto. 2020b. "4D Printing of Hydrogels: A Review." *Advanced Functional Materials* 30 (31): 1910606. https://doi.org/10.1002/adfm.201910606.

Chia, Helena N., and Benjamin M. Wu. 2014. "High-Resolution Direct 3D Printed PLGA Scaffolds: Print and Shrink." *Biofabrication* 7 (1): 015002. https://doi.org/10.1088/1758-5090/7/1/015002.

Choi, Rosa S., Martin Riegler, Charbalos Pothoulakis, Byung Soo Kim, David Mooney, Martin Vacanti, and Joseph P. Vacanti. 1998. "Studies of Brush Border Enzymes, Basement Membrane Components, and Electrophysiology of Tissue-Engineered Neo-intestine." *Journal of Pediatric Surgery* 33 (7): 991–97. https://doi.org/10.1016/S0022-3468(98)90520-6.

Chu, Honghui, Wenguang Yang, Lujing Sun, Shuxiang Cai, Rendi Yang, Wenfeng Liang, Haibo Yu, and Lianqing Liu. 2020. "4D Printing: A Review on Recent Progresses." *Micromachines* 11 (9): 796. https://doi.org/10.3390/mi11090796.

DeStefano, Vincent, Salaar Khan, and Alonzo Tabada. 2020. "Applications of PLA in Modern Medicine." *Engineered Regeneration* 1 (April): 76–87. https://doi.org/10.1016/j.engreg.2020.08.002.

Dhali, Kingshuk, Mehran Ghasemlou, Fugen Daver, Peter Cass, and Benu Adhikari. 2021. "A Review of Nanocellulose as a New Material Towards Environmental Sustainability." *Science of the Total Environment* 775: 145871. https://doi.org/10.1016/j.scitotenv.2021.145871.

Duigou, Antoine Le, A. Barbé, E. Guillou, and M. Castro. 2019. "3D Printing of Continuous Flax Fibre Reinforced Biocomposites for Structural Applications." *Materials and Design*. https://doi.org/10.1016/j.matdes.2019.107884.

Duigou, Antoine Le, M. Castro, R. Bevan, and N. Martin. 2016. "3D Printing of Wood Fibre Biocomposites: From Mechanical to Actuation Functionality." *Materials & Design* 96 (April): 106–14. https://doi.org/10.1016/j.matdes.2016.02.018.

Duigou, Antoine Le, Isabelle Pillin, Alain Bourmaud, Peter Davies, and Christophe Baley. 2008. "Effect of Recycling on Mechanical Behaviour of Biocompostable Flax/Poly(l-Lactide) Composites." *Composites Part A: Applied Science and Manufacturing* 39 (9): 1471–78. https://doi.org/10.1016/j.compositesa.2008.05.008.

Duigou, Antoine Le, Samuel Requile, Johnny Beaugrand, Fabrizio Scarpa, and Mickael Castro. 2017a. "Natural Fibres Actuators for Smart Bio-Inspired Hygromorph Biocomposites." *Smart Materials and Structures* 26 (12): 125009. https://doi.org/10.1088/1361-665X/aa9410.

Duigou, Antoine Le, Samuel Requile, Johnny Beaugrand, Fabrizio Scarpa, and Mickael Castro. 2017b. "Natural Fibres Actuators for Smart Bio-Inspired Hygromorph Biocomposites." *Smart Materials and Structures* 26 (12): 125009. https://doi.org/10.1088/1361-665X/aa9410.

Durga Prasad Reddy, R., and Varun Sharma. 2020. "Additive Manufacturing in Drug Delivery Applications: A Review." *International Journal of Pharmaceutics* 589 (November): 119820. https://doi.org/10.1016/j.ijpharm.2020.119820.

Faiz Norrrahim, Mohd Nor, Rushdan Ahmad Ilyas, Norizan Mohd Nurazzi, Mohd Saiful Asmal Rani, Mahamud Siti Nur Atikah, and Siti Shazra Shazleen. 2021. "Chemical Pretreatment of Lignocellulosic Biomass for the Production of Bioproducts: An Overview." *Applied Science and Engineering Progress* (July): 1–18. https://doi.org/10.14416/j.asep.2021.07.004.

Farah, Shady, Daniel G. Anderson, and Robert Langer. 2016. "Physical and Mechanical Properties of PLA, and Their Functions in Widespread Applications – A Comprehensive Review." *Advanced Drug Delivery Reviews* 107 (December): 367–92. https://doi.org/10.1016/j.addr.2016.06.012.

Faruk, Omar, Andrzej K. Bledzki, Hans-Peter Fink, and Mohini Sain. 2012. "Biocomposites Reinforced with Natural Fibers: 2000–2010." *Progress in Polymer Science* 37 (11): 1552–96. https://doi.org/10.1016/j.progpolymsci.2012.04.003.

Feng, Pei, Jiye Jia, Mingyang Liu, Shuping Peng, Zhenyu Zhao, and Cijun Shuai. 2021. "Degradation Mechanisms and Acceleration Strategies of Poly (Lactic Acid) Scaffold for Bone Regeneration." *Materials and Design* 210: 110066. https://doi.org/10.1016/j.matdes.2021.110066.

Fer, Gaëlle Le, and Matthew L. Becker. 2020. "4D Printing of Resorbable Complex Shape-Memory Poly(Propylene Fumarate) Star Scaffolds." *ACS Applied Materials & Interfaces* 12 (20): 22444–52. https://doi.org/10.1021/acsami.0c01444.

Ferrer, Ana, Lokendra Pal, and Martin Hubbe. 2017. "Nanocellulose in Packaging: Advances in Barrier Layer Technologies." *Industrial Crops and Products* 95: 574–82. https://doi.org/10.1016/j.indcrop.2016.11.012.

Furtado, Samuel C.R., A.L. Araújo, Arlindo Silva, Cristiano Alves, and A.M.R. Ribeiro. 2014. "Natural Fibre-Reinforced Composite Parts for Automotive Applications." *International Journal of Automotive Composites* 1 (1): 18. https://doi.org/10.1504/ijautoc.2014.064112.

Ganeshkumar, S., S. Dharani Kumar, U. Magarajan, S. Rajkumar, B. Arulmurugan, Shubham Sharma, Changhe Li, R.A. Ilyas, and Mohamed Fathy Badran. 2022. "Investigation of Tensile Properties of Different Infill Pattern Structures of 3D-Printed PLA Polymers: Analysis and Validation Using Finite Element Analysis in ANSYS." *Materials* 15 (15): 5142. https://doi.org/10.3390/ma15155142.

Gong, X., L. Pan, C.Y. Tang, L. Chen, Z. Hao, W.C. Law, X. Wang, C.P. Tsui, and C. Wu. 2014. "Preparation, Optical and Thermal Properties of CdSe–ZnS/poly (Lactic Acid)(PLA) Nanocomposites." *Composites Part B: Engineering* 66: 494–99.

Haryati, Aida, Nadlene Razali, Michal Petrů, Mastura Taha, Noryani Muhammad, and Rushdan Ahmad Ilyas. 2021. "Effect of Chemically Treated Kenaf Fibre on Mechanical and Thermal Properties of PLA Composites Prepared Through Fused Deposition Modeling (FDM)." *Polymers* 13 (19): 3299. https://doi.org/10.3390/polym13193299.

Hatano, K., H. Inoue, T. Kojo, T. Matsunaga, T. Tsujisawa, C. Uchiyama, and Y. Uchida. 1999. "Effect of Surface Roughness on Proliferation and Alkaline Phosphatase Expression of Rat Calvarial Cells Cultured on Polystyrene." *Bone* 25 (4): 439–45. https://doi.org/10.1016/S8756-3282(99)00192-1.

Hawanis, Husain Siti Nor, Rushdan Ahmad Ilyas, Rafidah Jalil, Rushdan Ibrahim, Rohah Abdul Majid, and Nur Hafizah Ab Hamid. 2024. "Extraction and Characterization of Morphological, Physical, Physiochemical, Thermal, and Chemical Composition of Five Bamboo Malaysian Species." *Cellulose* 31 (13): 7941–52. https://doi.org/10.1007/s10570-024-06090-6.

Holbery, James, and Dan Houston. 2006. "Natural-Fiber-Reinforced Polymer Composites in Automotive Applications." *JOM* 58 (11): 80–86. https://doi.org/10.1007/s11837-006-0234-2.

Ilyas, R.A., M.R.M. Asyraf, L. Rajeshkumar, Habib Awais, Amna Siddique, Khubab Shaker, Yasir Nawab, and Mat Uzir Wahit. 2024. "A Review of Bio-Based Nanocellulose Epoxy Composites." *Journal of Environmental Chemical Engineering* 12 (5): 113835. https://doi.org/10.1016/j.jece.2024.113835.

Ilyas, R.A., and S.M. Sapuan. 2020a. "Biopolymers and Biocomposites: Chemistry and Technology." *Current Analytical Chemistry* 16 (5): 500–3.

Ilyas, R.A., and S.M. Sapuan. 2020b. "The Preparation Methods and Processing of Natural Fibre Bio-Polymer Composites." *Current Organic Synthesis* 16 (8): 1068–70. https://doi.org/10.2174/157017941608200120105616.

Ilyas, R.A., S.M. Sapuan, M.N.F. Norrrahim, T.A.T. Yasim-Anuar, A. Kadier, M.S. Kalil, M.S.N. Atikah, et al. 2021. "Nanocellulose/Starch Biopolymer Nanocomposites: Processing, Manufacturing, and Applications." In *Advanced Processing, Properties, and Application of Strach and Other Bio-Based Polymer*, edited by F.M. Al-Oqla, 1st ed. Amsterdam, Netherlands: Elsevier Inc. https://doi.org/10.1016/B978-0-12-189661-8.00006-8.

Ilyas, R.A., M.S. Sapuan, Mohd Nurazzi Norizan, Mohd Nor Faiz Norrrahim, R. Ibrahim, M.S.N. Atikah, M.R.M. Huzaifah, et al. 2020. "Macro to Nanoscale Natural Fiber Composites for Automotive Components: Research, Development, and Application." In *Biocomposite and Synthetic Composites for Automotive Applications*, edited by M.S. Sapuan and R.A. Ilyas. Amsterdam, Netherlands: Woodhead Publishing Series.

Ilyas Rushdan, Ahmad, Mohd Sapuan Salit, Muhammed Lamin Sanyang, and Mohamad Ridzwan Ishak. 2017. "Nanocrystalline Cellulose As Reinforcement For Polymeric Matrix Nanocomposites and Its Potential Applications: A Review." *Current Analytical Chemistry* 13 (October). https://doi.org/10.2174/1573411013666171003155624.

Jacob, Jissy, Usman Lawal, Sabu Thomas, and Ravi Babu Valapa. 2020. *Biobased Polymer Composite from Poly(Lactic Acid): Processing, Fabrication, and Characterization for Food Packaging: Processing and Development of Polysaccharide-Based Biopolymers for Packaging Applications.* Elsevier Inc. https://doi.org/10.1016/B978-0-12-818795-1.00004-6.

Jamshidian, Majid, Elmira Arab Tehrany, and Stéphane Desobry. 2013. "Antioxidants Release from Solvent-Cast PLA Film: Investigation of PLA Antioxidant-Active Packaging." *Food and Bioprocess Technology* 6 (6): 1450–63. https://doi.org/10.1007/s11947-012-0830-9.

Jamshidian, Majid, Elmira Arab Tehrany, Muhammad Imran, Muriel Jacquot, and Stéphane Desobry. 2010. "Poly-Lactic Acid: Production, Applications, Nanocomposites, and Release Studies." *Comprehensive Reviews in Food Science and Food Safety* 9 (5): 552–71. https://doi.org/10.1111/j.1541-4337.2010.00126.x.

Joseph, Tomy Muringayil, Anoop Kallingal, Akshay Maniyeri Suresh, Debarshi Kar Mahapatra, Mohamed S. Hasanin, Józef Haponiuk, and Sabu Thomas. 2023. "3D Printing of Polylactic Acid: Recent Advances and Opportunities." *International Journal of Advanced Manufacturing Technology* 125 (3–4): 1015–35. https://doi.org/10.1007/s00170-022-10795-y.

Jumaidin, Ridhwan, Nur Wardah Adam, Rushdan Ahmad Ilyas, Mohamed Saiful Firdaus Hussin, Mastura Mohammad Taha, Muhd Ridzuan Mansor, Umar Al Amani Azlan, and Mohd Shukri Yob. 2019. "Water Transport and Physical Properties of Sugarcane Bagasse Fibre Reinforced Thermoplastic Potato Starch Biocomposite." *Journal of Advanced Research in Fluid Mechanics and Thermal Sciences* 61 (2): 273–81.

Jumaidin, Ridhwan, Nuraliah Ahmad Diah, R.A. Ilyas, Roziela Hanim Alamjuri, and Fahmi Asyadi Md Yusof. 2021. "Processing and Characterisation of Banana Leaf Fibre Reinforced Thermoplastic Cassava Starch Composites." *Polymers* 13 (9): 1420. https://doi.org/10.3390/polym13091420.

Jumaidin, Ridhwan, Zulhelmi Asyul Sutan Saidi, Rushdan Ahmad Ilyas, Mohd Nazri Ahmad, Mohammad Khalid Wahid, Mohd Yuhazri Yaakob, Nurul Ain Maidin, Mohd Hidayat Ab Rahman, and Mohd Hairizal Osman. 2019. "Characteristics of Cogon Grass Fibre Reinforced Thermoplastic Cassava Starch Biocomposite: Water Absorption and Physical Properties." *Journal of Advanced Research in Fluid Mechanics and Thermal Sciences* 62 (1): 43–52.

Jung, Jae-Won, Seong-Ho Kim, Si-Hwan Kim, Jong-Kyoo Park, and Woo-Il Lee. 2011. "Research on the Development of the Properties of PLA Composites for Automotive Interior Parts." *Journal of the Korean Society for Composite Materials* 24 (3): 1–5. https://doi.org/10.7234/kscm.2011.24.3.001.

Kabirian, Fatemeh, Bartek Ditkowski, Ali Zamanian, Ruth Heying, and Masoud Mozafari. 2018. "An Innovative Approach Towards 3D-Printed Scaffolds for the Next Generation of Tissue-Engineered Vascular Grafts." *Materials Today: Proceedings* 5 (7): 15586–94. https://doi.org/10.1016/j.matpr.2018.04.167.

Khosravi, A., A. Fereidoon, M.M. Khorasani, G. Naderi, M.R. Ganjali, P. Zarrintaj, M.R. Saeb, and T.J. Gutiérrez. 2020. "Soft and Hard Sections from cellulose-Reinforced Poly (Lactic Acid)-Based Food Packaging Films: A Critical Review." *Food Packaging and Shelf Life* 23: 100429.

Lasprilla, Astrid J.R., Guillermo A.R. Martinez, Betânia H. Lunelli, André L. Jardini, and Rubens Maciel Filho. 2012. "Poly-Lactic Acid Synthesis for Application in Biomedical Devices – A Review." *Biotechnology Advances* 30 (1): 321–28. https://doi.org/10.1016/j.biotechadv.2011.06.019.

Letcher, Todd, and Megan Waytashek. 2014. "Material Property Testing of 3D-Printed Specimen in Pla on an Entry-Level 3D Printer." In *ASME International Mechanical Engineering Congress and Exposition, Proceedings (IMECE)* (Vol. 2A). ASME. https://doi.org/10.1115/IMECE2014-39379.

Low, L.Q., R.A. Ilyas, R. Jalil, H.S.N. Hawanis, R. Ibrahim, H.A.A. Azriena, M.Y.M. Zuhri, et al. 2024. "Physical and Mechanical Properties Enhancement of Beaten Oil Palm Trunk Pulp and Paper by Optimizing Starch Addition: Towards Sustainable Packaging Solutions." *Industrial Crops and Products* 221 (December): 119232. https://doi.org/10.1016/j.indcrop.2024.119232.

Ma, SuQian, YunPeng Zhang, Meng Wang, YunHong Liang, Lei Ren, and LuQuan Ren. 2020. "Recent Progress in 4D Printing of Stimuli-Responsive Polymeric Materials." *Science China Technological Sciences* 63 (4): 532–44. https://doi.org/10.1007/s11431-019-1443-1.

Mamun, Abdullah A., and Andrzej K. Bledzki. 2013. "Micro Fibre Reinforced PLA and PP Composites: Enzyme Modification, Mechanical and Thermal Properties." *Composites Science and Technology* 78: 10–17. https://doi.org/10.1016/j.compscitech.2013.01.013.

Mansor, M.R., S.M. Sapuan, E.S. Zainudin, A.A. Nuraini, and A. Hambali. 2014. "Conceptual Design of Kenaf Fiber Polymer Composite Automotive Parking Brake Lever Using Integrated TRIZ-Morphological Chart-Analytic Hierarchy Process Method." *Materials and Design* 54: 473–82. https://doi.org/10.1016/j.matdes.2013.08.064.

Maroufkhani, Mahshid, Ali Asghar Katbab, Wangcheng Liu, and Jinwen Zhang. 2017. "Polylactide (PLA) and Acrylonitrile Butadiene Rubber (NBR) Blends: The Effect of ACN Content on Morphology, Compatibility and Mechanical Properties." *Polymer* 115: 37–44. https://doi.org/10.1016/j.polymer.2017.03.025.

Martins, Sávio Santana, Ana Catarina Jorge Evangelista, Ahmed W.A. Hammad, Vivian W.Y. Tam, and Assed Haddad. 2020. "Evaluation of 4D BIM Tools Applicability in Construction Planning Efficiency." *International Journal of Construction Management* (November): 1–14. https://doi.org/10.1080/15623599.2020.1837718.

Mastura, M.T., S.M. Sapuan, M.R. Mansor, and A.A. Nuraini. 2017. "Conceptual Design of a Natural Fibre-Reinforced Composite Automotive Anti-Roll Bar Using a Hybrid Approach." *International Journal of Advanced Manufacturing Technology* 91 (5–8): 2031–48. https://doi.org/10.1007/s00170-016-9882-8.

Matsuzaki, Ryosuke, Masahito Ueda, Masaki Namiki, Tae-Kun Jeong, Hirosuke Asahara, Keisuke Horiguchi, Taishi Nakamura, Akira Todoroki, and Yoshiyasu Hirano. 2016. "Three-Dimensional Printing of Continuous-Fiber Composites by in-Nozzle Impregnation." *Scientific Reports* 6 (1): 23058. https://doi.org/10.1038/srep23058.

Mazzanti, Valentina, Lorenzo Malagutti, and Francesco Mollica. 2019. "FDM 3D Printing of Polymers Containing Natural Fillers: A Review of Their Mechanical Properties." *Polymers* 11 (7): 1094. https://doi.org/10.3390/polym11071094.

Miao, Shida, Haitao Cui, Timothy Esworthy, Bhushan Mahadik, Se-jun Lee, Xuan Zhou, Sung Yun Hann, John P. Fisher, and Lijie Grace Zhang. 2020. "4D Self-Morphing Culture Substrate for Modulating Cell Differentiation." *Advanced Science* 7 (6): 1902403. https://doi.org/10.1002/advs.201902403.

Mishra, Raghvendra Kumar, Arjun Sabu, and Santosh K. Tiwari. 2018. "Materials Chemistry and the Futurist Eco-Friendly Applications of Nanocellulose: Status and Prospect." *Journal of Saudi Chemical Society* 22 (8): 949–78. https://doi.org/10.1016/j.jscs.2018.02.005.

Morales, Angel Pozo, Alfredo Güemes, Antonio Fernandez-Lopez, Veronica Carcelen Valero, and Sonia de La Rosa Llano. 2017. "Bamboo-Polylactic Acid (PLA) Composite Material for Structural Applications." *Materials* 10 (11): 1–22. https://doi.org/10.3390/ma10111286.

Motoc, Dana Luca, Horatiu Teodorescu Draghicescu, Daniela Florea, Ion Preda, and Nicolae Ispas. 2021. "Insights into Mechanical and Thermal Properties of Additively Manufactured PLA Samples Triggered by Automotive Industry Demands." *Materiale Plastice* 57 (4): 193–201. https://doi.org/10.37358/MP.20.4.5418.

Moutee, Mohssine, Yves Fortin, and Mario Fafard. 2007a. "A Global Rheological Model of Wood Cantilever as Applied to Wood Drying." *Wood Science and Technology* 41 (3): 209–34. https://doi.org/10.1007/s00226-006-0106-5.

Moutee, Mohssine, Yves Fortin, and Mario Fafard. 2007b. "A Global Rheological Model of Wood Cantilever as Applied to Wood Drying." *Wood Science and Technology* 41 (3): 209–34. https://doi.org/10.1007/s00226-006-0106-5.

Mubarak, Asmaa Ali, R.A. Ilyas, Abu Hassan Nordin, Norzita Ngadi, and M.F.M. Alkbir. 2024. "Recent Developments in Sugarcane Bagasse Fibre-Based Adsorbent and Their Potential Industrial Applications: A Review." *International Journal of Biological Macromolecules* 277 (October): 134165. https://doi.org/10.1016/j.ijbiomac.2024.134165.

Nasir, Mohd Hakim Mohd, Mastura Mohammad Taha, Nadlene Razali, Rushdan Ahmad Ilyas, Victor Feizal Knight, and Mohd Nor Faiz Norrrahim. 2022. "Effect of Chemical Treatment of Sugar Palm Fibre on Rheological and Thermal Properties of the PLA Composites Filament for FDM 3D Printing." *Materials* 15 (22): 1–16. https://doi.org/10.3390/ma15228082.

Nautiyal, Ujjwal, Nandini Sahu, and Diksha Gupta. 2020. "Hydrogel: Preparation, Characterization and Applications." *Asian Pacific Journal of Nursing and Health Sciences* 3 (1). https://doi.org/10.46811/apjnh/3.1.1.

Navarro, M., M.P. Ginebra, J.A. Planell, S. Zeppetelli, and L. Ambrosio. 2004. "Development and Cell Response of a New Biodegradable Composite Scaffold for Guided Bone Regeneration." *Journal of Materials Science: Materials in Medicine* 15 (4): 419–22. https://doi.org/10.1023/B:JMSM.0000021113.88702.9d.

Nazrin, Asmawi, Salit Mohd Sapuan, M.Y.M. Zuhri, R.A. Ilyas, R. Syafiq, and S.F.K. Sherwani. 2020a. "Nanocellulose Reinforced Thermoplastic Starch (TPS), Polylactic Acid (PLA), and Polybutylene Succinate (PBS) for Food Packaging Applications." *Frontiers in Chemistry* 8 (213): 1–12. https://doi.org/10.3389/fchem.2020.00213.

Nazrin, Asmawi, Salit Mohd Sapuan, M.Y.M. Zuhri, R.A. Ilyas, R. Syafiq, and S.F.K. Sherwani. 2020b. "Nanocellulose Reinforced Thermoplastic Starch (TPS), Polylactic Acid (PLA), and Polybutylene Succinate (PBS) for Food Packaging Applications." *Frontiers in Chemistry* 8 (213): 1–12. https://doi.org/10.3389/fchem.2020.00213.

Nazrin, Asmawi, Salit Mohd Sapuan, Mohamed Yusoff Mohd Zuhri, Intan Syafinaz Mohamed Amin Tawakkal, and Rushdan Ahmad Ilyas. 2021a. "Flammability and Physical Stability of Sugar Palm Crystalline Nanocellulose Reinforced Thermoplastic Sugar Palm Starch/Poly(Lactic Acid) Blend Bionanocomposites." *Nanotechnology Reviews* 11 (1): 86–95. https://doi.org/10.1515/ntrev-2022-0007.

Nazrin, Asmawi, Salit Mohd Sapuan, Mohamed Yusoff Mohd Zuhri, Intan Syafinaz Mohamed Amin Tawakkal, and Rushdan Ahmad Ilyas. 2021b. "Water Barrier and Mechanical Properties of Sugar Palm Crystalline Nanocellulose Reinforced Thermoplastic Sugar Palm Starch (TPS)/Poly(Lactic Acid) (PLA) Blend Bionanocomposites." *Nanotechnology Reviews* 10 (1): 431–42. https://doi.org/10.1515/ntrev-2021-0033.

Nofar, Mohammadreza, and Chul B. Park. 2014. "Poly (Lactic Acid) Foaming." *Progress in Polymer Science* 39 (10): 1721–41. https://doi.org/10.1016/j.progpolymsci.2014.04.001.

Norfarhana, A.S., R.A. Ilyas, and Norzita Ngadi. 2022. "A Review of Nanocellulose Adsorptive Membrane as Multifunctional Wastewater Treatment." *Carbohydrate Polymers* 291 (September): 119563. https://doi.org/10.1016/j.carbpol.2022.119563.

Norfarhana, A.S., R.A. Ilyas, Norzita Ngadi, and Mohd Hafiz Dzarfan Othman. 2024. "Innovative Ionic Liquid Pretreatment Followed by Wet Disk Milling Treatment Provides Enhanced Properties of Sugar Palm Nano-Fibrillated Cellulose." *Heliyon* 10 (6): e27715. https://doi.org/10.1016/j.heliyon.2024.e27715.

Norfarhana, A.S., P.S. Khoo, R.A. Ilyas, N.H. Ab Hamid, H.A. Aisyah, Mohd Nor Faiz Norrahim, V.F. Knight, et al. 2024. "Exploring of Cellulose Nanocrystals from Lignocellulosic Sources as a Powerful Adsorbent for Wastewater Remediation." *Journal of Polymers and the Environment* (April). https://doi.org/10.1007/s10924-024-03227-3.

Notta-Cuvier, D., J. Odent, R. Delille, M. Murariu, F. Lauro, J.M. Raquez, B. Bennani, and P. Dubois. 2014. "Tailoring Polylactide (PLA) Properties for Automotive Applications: Effect of Addition of Designed Additives on Main Mechanical Properties." *Polymer Testing* 36 (June): 1–9. https://doi.org/10.1016/j.polymertesting.2014.03.007.

Oluwabunmi, Kayode, Nandika Anne D'Souza, Weihuan Zhao, Tae-Youl Choi, and Thomas Theyson. 2020. "Compostable, Fully Biobased Foams Using PLA and Micro Cellulose for Zero Energy Buildings." *Scientific Reports* 10 (1): 17771. https://doi.org/10.1038/s41598-020-74478-y.

Othman, Jameelah Alhad Salih, R.A. Ilyas, Abu Hassan Nordin, Norzita Ngadi, and M.F.M. Alkbir. 2024. "Recent Advancements in Bamboo Nanocellulose-Based Bioadsorbents and Their Potential in Wastewater Applications: A Review." *International Journal of Biological Macromolecules* 277 (October): 134451. https://doi.org/10.1016/j.ijbiomac.2024.134451.

Öztürk, Sinan. 2020. "The Investigation of Polylactic Acid Based Natural Fiber Reinforced Biocomposites for Automotive Applications." *Kırklareli Üniversitesi Mühendislik ve Fen Bilimleri Dergisi* (June): 21–31. https://doi.org/10.34186/klujes.626590.

Popa, Elisabeta Elena, Maria Rapa, Ovidiu Popa, Gabriel Mustatea, Vlad Ioan Popa, Amalia Carmen Mitelut, and Mona Elena Popa. 2017. " Polylactic Acid/Cellulose Fibres Based Composites for Food Packaging Applications." *Materiale Plastice* 4: 673–77.

Poppinga, Simon, David Correa, Bernd Bruchmann, Achim Menges, and Thomas Speck. 2020a. "Plant Movements as Concept Generators for the Development of Biomimetic Compliant Mechanisms." *Integrative and Comparative Biology* 60 (4): 886–95. https://doi.org/10.1093/icb/icaa028.

Poppinga, Simon, David Correa, Bernd Bruchmann, Achim Menges, and Thomas Speck. 2020b. "Plant Movements as Concept Generators for the Development of Biomimetic Compliant Mechanisms." *Integrative and Comparative Biology* 60 (4): 886–95. https://doi.org/10.1093/icb/icaa028.

Quanjin, Ma, M.R.M. Rejab, M.S. Idris, Nallapaneni Manoj Kumar, M.H. Abdullah, and Guduru Ramakrishna Reddy. 2020. "Recent 3D and 4D Intelligent Printing Technologies: A Comparative Review and Future Perspective." *Procedia Computer Science* 167: 1210–19. https://doi.org/10.1016/j.procs.2020.03.434.

Rahman, M. Z., Rahman, M., Mahbub, T., Ashiquzzaman, M., Sagadevan, S., & Hoque, M. E. (2023). Advanced biopolymers for automobile and aviation engineering applications. *Journal of Polymer Research, 30*(3): 106. https://doi.org/10.1007/s10965-023-03440-z.

Russias, J., E. Saiz, R.K. Nalla, K. Gryn, R.O. Ritchie, and A.P. Tomsia. 2006. "Fabrication and Mechanical Properties of PLA/HA Composites: A Study of in Vitro Degradation." *Materials Science and Engineering: C* 26 (8): 1289–95. https://doi.org/10.1016/j.msec.2005.08.004.

Saravana Bavan, D., and G.C. Mohan Kumar. 2010. "Potential Use of Natural Fiber Composite Materials in India." *Journal of Reinforced Plastics and Composites* 29 (24): 3600–13. https://doi.org/10.1177/0731684410381151.

Shaharuzaman, Mohd Adrinata, Salit Mohd Sapuan, and Muhd Ridzuan Mansor. 2018. "Prioritizing the Product Design Specification of Side-Door Impact Beam Using Analytic Hierarchy Process Method." *Mechanical Engineering Research Day* 34–35.

Sharma, Shubham, P. Sudhakara, Jujhar Singh, R.A. Ilyas, M.R.M. Asyraf, and M.R. Razman. 2021. "Critical Review of Biodegradable and Bioactive Polymer Composites for Bone Tissue Engineering and Drug Delivery Applications." *Polymers* 13 (16): 2623. https://doi.org/10.3390/polym13162623.

Shih, Yeng Fong, Chien Chung Huang, and Po Wei Chen. 2010. "Biodegradable Green Composites Reinforced by the Fiber Recycling from Disposable Chopsticks." *Materials Science and Engineering A* 527 (6): 1516–21. https://doi.org/10.1016/j.msea.2009.10.024.

Song, X., X. Zhang, H. Wang, F. Liu, S. Yu, and S. Liu. 2013. "Methanolysis of Poly (Lactic Acid)(PLA) Catalyzed by Ionic Liquids." *Polymer Degradation and Stability* 98 (12): 2760–64.

Spiegel, Christoph A., Marc Hippler, Alexander Münchinger, Martin Bastmeyer, Christopher Barner-Kowollik, Martin Wegener, and Eva Blasco. 2020. "4D Printing at the Microscale." *Advanced Functional Materials* 30 (26): 1907615. https://doi.org/10.1002/adfm.201907615.

Subash, Alsha, and Balasubramanian Kandasubramanian. 2020. "4D Printing of Shape Memory Polymers." *European Polymer Journal* 134 (July): 109771. https://doi.org/10.1016/j.eurpolymj.2020.109771.

Sukmawan, Romi, Hitoshi Takagi, and Antonio Norio Nakagaito. 2016. "Strength Evaluation of Cross-Ply Green Composite Laminates Reinforced by Bamboo Fiber." *Composites Part B: Engineering*. https://doi.org/10.1016/j.compositesb.2015.08.072.

Swetha, T. Angelin, Abhispa Bora, K. Mohanrasu, P. Balaji, Rathinam Raja, Kumar Ponnuchamy, Govarthanan Muthusamy, and A. Arun. 2023. "A Comprehensive Review on Polylactic Acid (PLA) – Synthesis, Processing and Application in Food Packaging." *International Journal of Biological Macromolecules* 234 (November 2022): 123715. https://doi.org/10.1016/j.ijbiomac.2023.123715.

Tawakkal, Intan S.M.A., Marlene J. Cran, Joseph Miltz, and Stephen W. Bigger. 2014. "A Review of Poly(Lactic Acid)-Based Materials for Antimicrobial Packaging." *Journal of Food Science* 79 (8). https://doi.org/10.1111/1750-3841.12534.

Thomas, Bejoy, Midhun C. Raj, B.K. Athira, H.M. Rubiyah, Jithin Joy, Audrey Moores, Glenna L. Drisko, and Clément Sanchez. 2018. "Nanocellulose, a Versatile Green Platform: From Biosources to Materials and Their Applications." *Chemical Reviews* 118 (24): 11575–625. https://doi.org/10.1021/acs.chemrev.7b00627.

Vieira, A.C., J.C. Vieira, R.M. Guedes, and A.T. Marques. 2010. "March. Degradation and Viscoelastic Properties of PLA-PCL, PGA-PCL, PDO and PGA fibres." In *Materials Science Forum*, vol. 636, 825–32. Trans Tech Publications Ltd.

Wang, Guilong, Long Wang, Lun Howe Mark, Vahid Shaayegan, Guizhen Wang, Huiping Li, Guoqun Zhao, and Chul B. Park. 2018. "Ultralow-Threshold and Lightweight Biodegradable Porous PLA/MWCNT with Segregated Conductive Networks for High-Performance Thermal Insulation and Electromagnetic Interference Shielding Applications." *ACS Applied Materials and Interfaces* 10 (1): 1195–203. https://doi.org/10.1021/acsami.7b14111.

Wang, Jing, Menglu Wang, Fuying Chen, Yihang Wei, Xuening Chen, Yong Zhou, Xiao Yang, Xiangdong Zhu, Chongqi Tu, and Xingdong Zhang. 2019. "Nano-Hydroxyapatite Coating Promotes Porous Calcium Phosphate Ceramic-Induced Osteogenesis Via BMP/Smad Signaling Pathway." *International Journal of Nanomedicine* 14 (October): 7987–8000. https://doi.org/10.2147/IJN.S216182.

Wang, Linlin, Jinsong Leng, and Shanyi Du. 2020. "4D Printing of Shape Memory Polymers and Their Composites: Research Status and Application Progress." *Harbin Gongye Daxue Xuebao/Journal of Harbin Institute of Technology*. https://doi.org/10.11918/202003059.

Yang, Jiawei, Ruobing Bai, Baohong Chen, and Zhigang Suo. 2020. "Hydrogel Adhesion: A Supramolecular Synergy of Chemistry, Topology, and Mechanics." *Advanced Functional Materials* 30 (2): 1901693. https://doi.org/10.1002/adfm.201901693.

Yusof, N.S.B., S.M. Sapuan, M.T.H. Sultan, and M. Jawaid. 2018. "Concept Generation of Sugar Palm/Glass Fiber Reinforced Thermoplastic Polyurethane Hybrid Composite Automotive Crash Box." *Journal of Advanced Research in Materials Science* 49 (1): 10–17.

Yusoff, Noorul Hidayah, Kaushik Pal, Thinakaran Narayanan, and Fernando Gomes de Souza. 2021. "Recent Trends on Bioplastics Synthesis and Characterizations: Polylactic Acid (PLA) Incorporated with Tapioca Starch for Packaging Applications." *Journal of Molecular Structure* 1232. https://doi.org/10.1016/j.molstruc.2021.129954.

Yusuf, J., S.M. Sapuan, Umer Rashid, R.A. Ilyas, and M.R. Hassan. 2024. "Thermal, Mechanical, and Morphological Properties of Oil Palm Cellulose Nanofibril Reinforced Green Epoxy Nanocomposites." *International Journal of Biological Macromolecules* 278 (October): 134421. https://doi.org/10.1016/j.ijbiomac.2024.134421.

Zafar, Muhammad Qasim, and Haiyan Zhao. 2020. "4D Printing: Future Insight in Additive Manufacturing." *Metals and Materials International*. https://doi.org/10.1007/s12540-019-00441-w.

Zengwen, Cao, Zhifeng Lu, Hongwei Pan, Junjia Bian, Lijing Han, Huiliang Zhang, Lisong Dong, and Yuming Yang. 2020. "Structuring Poly (Lactic Acid) Film with Excellent Tensile Toughness Through Extrusion Blow Molding." *Polymer* 187. https://doi.org/10.1016/j.polymer.2019.122091.

Zhang, Lu, Zhi Li, Ye Tang Pan, Adriana Pérez Yáñez, Shuang Hu, Xiu Qin Zhang, Rui Wang, and De Yi Wang. 2018. "Polydopamine Induced Natural Fiber Surface Functionalization: A Way Towards Flame Retardancy of Flax/Poly(Lactic Acid) Biocomposites." *Composites Part B: Engineering*. https://doi.org/10.1016/j.compositesb.2018.07.037.

Zhao, Xianhui, Halil Tekinalp, Xianzhi Meng, Darby Ker, Bowie Benson, Yunqiao Pu, Arthur J. Ragauskas, et al. 2019. "Poplar as Biofiber Reinforcement in Composites for Large-Scale 3D Printing." *ACS Applied Bio Materials* 2 (10): 4557–70. https://doi.org/10.1021/acsabm.9b00675.

Zheng, Yijun, Zhijun Chen, Qiyang Jiang, Jun Feng, Si Wu, and Aránzazu del Campo. 2020. "Near-Infrared-Light Regulated Angiogenesis in a 4D Hydrogel." *Nanoscale* 12 (25): 13654–61. https://doi.org/10.1039/D0NR02552F.

Zolfagharian, Ali, Akif Kaynak, Mahdi Bodaghi, Abbas Z. Kouzani, Saleh Gharaie, and Saeid Nahavandi. 2020a. "Control-Based 4D Printing: Adaptive 4D-Printed Systems." *Applied Sciences* 10 (9): 3020. https://doi.org/10.3390/app10093020.

Zolfagharian, Ali, Akif Kaynak, and Abbas Kouzani. 2020. "Closed-Loop 4D-Printed Soft Robots." *Materials & Design* 188 (March): 108411. https://doi.org/10.1016/j.matdes.2019.108411.

Zolfagharian, Ali, M.A. Parvez Mahmud, Saleh Gharaie, Mahdi Bodaghi, Abbas Z. Kouzani, and Akif Kaynak. 2020b. "3D/4D-Printed Bending-Type Soft Pneumatic Actuators: Fabrication, Modelling, and Control." *Virtual and Physical Prototyping* 15 (4): 373–402. https://doi.org/10.1080/17452759.2020.1795209.

Index

C

N

R

S

T

U

V

W

X

Y

Z

For Product Safety Concerns and Information please contact our EU representative GPSR@taylorandfrancis.com
Taylor & Francis Verlag GmbH, Kaufingerstraße 24, 80331 München, Germany

www.ingramcontent.com/pod-product-compliance
Lightning Source LLC
LaVergne TN
LVHW011115300926
849160LV00008B/388

* 9 7 8 1 0 3 2 6 7 8 8 8 7 *